ADVANCES IN CATALYSIS

VOLUME 22

Advisory Board

G. K. BORESKOV
Novosibirsk, U.S.S.R.

M. BOUDART
Stanford, California

M. CALVIN
Berkeley, California

P. H. EMMETT
Baltimore, Maryland

J. HORIUTI
Sapporo, Japan

W. JOST
Göttingen, Germany

G. NATTA
Milan, Italy

E. K. RIDEAL
London, England

P. W. SELWOOD
Santa Barbara, California

H. S. TAYLOR
Princeton, New Jersey

ADVANCES IN CATALYSIS

VOLUME 22

Edited By

D. D. ELEY
*The University
Nottingham, England*

HERMAN PINES
*Northwestern University
Evanston, Illinois*

PAUL B. WEISZ
*Mobil Research and
Development Corporation
Princeton, New Jersey*

1972

ACADEMIC PRESS • NEW YORK AND LONDON

COPYRIGHT © 1972, BY ACADEMIC PRESS, INC.
ALL RIGHTS RESERVED
NO PART OF THIS BOOK MAY BE REPRODUCED IN ANY FORM,
BY PHOTOSTAT, MICROFILM, RETRIEVAL SYSTEM, OR ANY
OTHER MEANS, WITHOUT WRITTEN PERMISSION FROM
THE PUBLISHERS.

ACADEMIC PRESS, INC.
111 Fifth Avenue, New York, New York 10003

United Kingdom Edition published by
ACADEMIC PRESS, INC. (LONDON) LTD.
24/28 Oval Road, London NW1 7DD

LIBRARY OF CONGRESS CATALOG CARD NUMBER: 49-7755

PRINTED IN THE UNITED STATES OF AMERICA

Contents

CONTRIBUTORS.. vii
PREFACE... ix

Hydrogenation and Isomerization over Zinc Oxide
R. J. KOKES AND A. L. DENT

I.	Introduction..	1
II.	The Active Sites...	4
III.	Mechanism of Ethylene Hydrogenation...............................	16
IV.	Reactions with Propylene...	29
V.	Reactions of Butene...	41
VI.	Reactions of Acetylenes...	46
VII.	Concluding Remarks...	47
	References..	48

Chemisorption Complexes and Their Role in Catalytic Reactions on Transition Metals
Z. KNOR

I.	Introduction..	51
II.	General Formation of the Problem...................................	52
III.	Chemisorption Complexes...	57
IV.	Characterization of the Metal Surfaces.............................	65
V.	Concluding Remarks...	71
	References..	71

Influence of Metal Particle Size in Nickel-on-Aerosil Catalysts on Surface Site Distribution, Catalytic Activity, and Selectivity
R. VAN HARDEVELD AND F. HARTOG

I.	Introduction..	75
II.	Statistics of Surface Atoms and Surface Sites....................	77
III.	Infrared Studies on the Adsorption of N_2, CO, and CO_2.....	86
IV.	Deuteration and Exchange of Benzene.............................	100
V.	Conclusions...	110
VI.	Preparation and Characterization of Catalysts.....................	110
	References..	110

Adsorption and Catalysis on Evaporated Alloy Films
R. L. MOSS AND L. WHALLEY

I.	Introduction..	115
II.	Preparation of Alloy Films..	117

III.	Characterization of Alloy Films	134
IV.	Adsorption and Catalysis on Alloy Films	147
V.	Conclusions	184
	References	185

Heat-Flow Microcalorimetry and Its Application to Heterogeneous Catalysis

P. C. GRAVELLE

I.	Introduction	191
II.	Principle of Heat-Flow Calorimetry	194
III.	Some Heat-Flow Microcalorimeters That Can Be Used in Heterogeneous Catalysis Research	196
IV.	Theory of Heat-Flow Calorimetry	206
V.	Analysis of the Calorimetric Data	214
VI.	Measurement of the Differential Heats of Gas–Solid Interactions	226
VII.	Applications of Heat-Flow Microcalorimetry in Heterogeneous Catalysis	237
VIII.	Conclusions	259
	References	261

Electron Spin Resonance in Catalysis

JACK H. LUNSFORD

I.	Introduction	265
II.	Theoretical Basis	267
III.	Experimental Considerations	282
IV.	Applications	295
	Appendix A: Quantum Mechanics for Spin Systems	326
	Appendix B: Determination of Energy Levels from the Spin Hamiltonian	328
	Appendix C: The g Tensor	332
	Appendix D: The Hyperfine Tensor	336
	References	340

AUTHOR INDEX 345
SUBJECT INDEX 357
CONTENTS OF PREVIOUS VOLUMES 365

Contributors

Numbers in parentheses indicate the pages on which authors' contributions begin.

A. L. DENT, *Department of Chemical Engineering, Carnegie-Mellon University, Pittsburgh, Pennsylvania* (1)

P. C. GRAVELLE, *Department of Chimie-Physique, Institut de Recherches sur la Catalyse, CNRS, Villeurbanne, France* (191)

F. HARTOG, *Catalysis Department, Central Laboratory DSM, Geleen, The Netherlands* (75)

R. J. KOKES, *Department of Chemistry, The Johns Hopkins University, Baltimore, Maryland* (1)

Z. KNOR, *Institute of Physical Chemistry, Czechoslovak Academy of Sciences, Prague, Czechoslovakia* (51)

JACK H. LUNSFORD, *Department of Chemistry, Texas A&M University, College Station, Texas* (265)

R. L. MOSS, *Warren Spring Laboratory, Stevenage, England* (115)

R. VAN HARDEVELD, *Catalysis Department, Central Laboratory DSM, Geleen, The Netherlands* (75)

L. WHALLEY, *Warren Spring Laboratory, Stevenage, England* (115)

Preface

The complexity and inhomogenicity of catalytic sites of metals and metal oxides make it difficult to interpret the mechanism of catalytic reactions on solid surfaces. Investigations that may lead to a better characterization of adsorbed species on catalytic sites could add much to our understanding of heterogeneous catalysis.

The present volume contains six articles reviewing the approaches and progress made in describing catalytic sites and chemisorbed species.

Kokes and Dent by combining IR spectroscopy with hydrogen isotope techniques, and by applying kinetic and stereochemical considerations were able to determine the intermediate surface species in the hydrogenation and isomerization of simple olefins over zinc oxide.

Van Hardeveld and Hartog describe the effect of metal particle size on the properties of a metal on carrier catalyst. They have related the adsorptive and catalytic properties of metal crystals to crystal size and to the structure of the crystal surface.

Knor in reviewing "Chemisorption Complexes and their Role in Catalytic Reactions on Transition Metals" concludes that even in the simplest systems many different chemisorbed particles originate on the surface during the catalytic reactions. These particles can interact with each other and also with gaseous reaction components to influence the course of a particular reaction. The author suggests that a modification of the surface properties of the metals, as for example by forming alloys, might be a promising direction of further fundamental research.

The subject of "Adsorption and Catalysis on Evaporated Alloy Films" is reviewed and Moss and Whalley conclude that phase separation caused a variety of complications which makes it difficult to define the nature of catalytic activity.

The chapter "Electron Spin Resonance in Catalysis" by Lunsford was prompted by the extensive activity in this field since the publication of an article on a similar subject in Volume 12 of this serial publication. This chapter is limited to paramagnetic species that are reasonably well defined by means of their spectra. It contains applications of ESR technique to the study of adsorbed atoms and molecules, and also to the evaluation of surface effects. The application of ESR to the determination of the state of transition metal ions in catalytic reactions is also discussed.

P. C. Gravelle reviews "Heat-Flow Microcalorimetry" and shows its applications to the study of adsorption and heterogeneous catalysis.

HERMAN PINES

Hydrogenation and Isomerization over Zinc Oxide

R. J. KOKES

Department of Chemistry
The Johns Hopkins University
Baltimore, Maryland

and

A. L. DENT

Department of Chemical Engineering
Carnegie-Mellon University
Pittsburgh, Pennsylvania

I. Introduction	1
II. The Active Sites	4
A. Hydrogen Adsorption	4
B. Poisoning Experiments	9
C. Addition of H_2–D_2 Mixtures to Ethylene	12
D. A Model for the Active Sites	13
III. Mechanism of Ethylene Hydrogenation	16
A. Goals	16
B. Kinetics	16
C. Ethylene on Zinc Oxide	19
D. Intermediates in Ethylene Hydrogenation	23
IV. Reactions with Propylene	29
A. Propylene on Zinc Oxide. The π-Allyl	29
B. Reactions of π-Allyls	37
V. Reactions of Butene	41
A. Stereochemical Considerations	41
B. Butene on Zinc Oxide	42
C. Isomerization of *cis*-Butene	45
VI. Reactions of Acetylenes	46
Acetylenes on Zinc Oxide	46
VII. Concluding Remarks	47
References	48

I. Introduction

Hydrogenation and isomerization of olefins are classic examples of heterogeneous catalytic reactions. Both metals (*1, 2*) and oxides (*3, 4*) are catalysts for these reactions, but the metals, especially group VIII metals, have been studied in the most detail. Thus, although this review deals with such reactions over an oxide, it is useful to consider first the more extensive results obtained for metals.

Hydrogenation of ethylene over metals presumably occurs via the following steps (1, 2):

$$\underset{*}{H} + \underset{*\ *}{CH_2-CH_2} \underset{k_{-1}}{\overset{k_1}{\rightleftharpoons}} \underset{*}{CH_2-CH_3} + 2^*, \tag{1}$$

$$\underset{*}{CH_2-CH_3} + \underset{*}{H} \xrightarrow{k_2} C_2H_6(g) + 2^*, \tag{2}$$

where * represents a surface atom. For more complex olefins a variation on (1), alkyl reversal, offers a pathway for the double-bond isomerization that accompanies hydrogenation of higher olefins, viz:

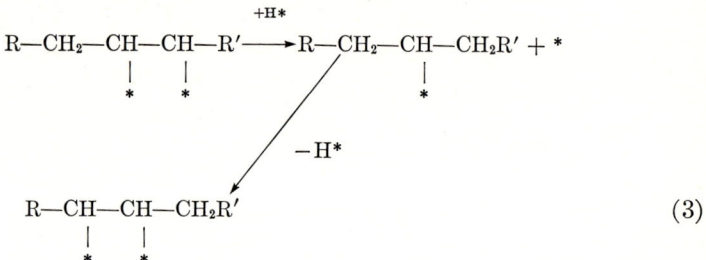

$$\tag{3}$$

Observations consistent with this scheme include (1):

(a) Active catalysts for ethylene hydrogenation also catalyze H_2–D_2 equilibration. This suggests hydrogen adsorbs dissociatively in agreement with (1) and (2).

(b) Reaction of light ethylene with deuterium leads to exchange with ethylene and an isotopic smear in the resulting ethanes, i.e., a distribution of ethanes of the form $C_2H_{6-x}D_x$, $0 \leq x \leq 6$. This suggests that (1) is indeed reversible.

(c) Efficient isomerization of olefins requires hydrogen as a cocatalyst as suggested by (3) (5).

Detailed examination of the results of isomerization, hydrogenation, and related exchange reactions reveals defects in the above scheme (6, 7), but despite its shortcomings, it remains the best simple approximation to the truth.

Over zinc oxide (*8–14*) and chromia (*3*), which are also hydrogenation catalysts, we find the following features:

(a) Active catalysts for ethylene hydrogenation also catalyze H_2–D_2 equilibration. This suggests that hydrogen adsorbs dissociatively in agreement with (1) and (2) (*3, 9*).

(b) Reaction of light ethylene with deuterium leads to $C_2H_4D_2$ and *no* exchange with ethylene. Thus step 1, if it occurs, is *not* reversible (*3, 8, 10*).

(c) Efficient isomerization of olefins does *not* require hydrogen as a cocatalyst. Thus Eq. (3) is *not* applicable (*3, 12, 14*).

These features do not preclude (1) and (2) (or their near equivalent) as a pathway for hydrogenation provided (1) is irreversible, but they do preclude alkyl reversal as an isomerization pathway. Over chromia, Burwell *et al.* (*3*) suggest that allyl species may furnish an isomerization pathway, viz:

$$\begin{array}{c} R{-}CH_2{-}CH{=}CH{-}R' \rightarrow R{-}\underset{*}{CH}{-}CH{=}CH{-}R' + \underset{*}{H} \\ \downarrow \\ R{-}CH{=}CH{-}CH_2{-}R' \leftarrow R{-}CH{=}CH{-}\underset{*}{CH}R' + H^* \end{array} \quad (4)$$

In homogeneous (*15, 16*) and heterogeneous (*17*) base catalyzed reactions, such species, with anionic character, are firmly established as intermediates. It seems clear that a similar species must be regarded as a prime candidate for an intermediate in isomerization over zinc oxide. Regardless of what the isomerization intermediate is, however, it is clearly different from that involved in isomerization over metals.

At least for ethylene hydrogenation, catalysis appears to be simpler over oxides than over metals. Even if we were to assume that Eqs. (1) and (2) told the whole story, this would be true. In these terms over oxides the hydrocarbon surface species in the addition of deuterium to ethylene would be limited to C_2H_4 and C_2H_4D, whereas over metals a multiplicity of species of the form $C_2H_{4-x}D_x$ and $C_2H_{5-x}D_x$ would be expected. Adsorption (*18*) and IR studies (*19*) reveal that even with ethylene alone, metals are complex. When a metal surface is exposed to ethylene, self-hydrogenation and dimerization occur. These are surface reactions, not catalysis; in other words, the extent of these reactions is determined by the amount of surface available as a reactant. The over-all result is that a metal surface exposed to an olefin forms a variety of carbonaceous species of variable stoichiometry. The presence of this variety of relatively inert species confounds attempts to use physical techniques such as IR to characterize reactive surface species. As a consequence, evidence for the surface species and intermediates over metals is based primarily on mechanistic inference. Once again, with zinc oxide the behavior is simpler. Self-hydrogenation and dimerization do not occur on adsorption of ethylene (10). This comparative simplicity of ethylene adsorption and its reaction with deuterium suggest that IR studies may be especially meaningful with oxides. Such studies have been carried out only recently, but results to date (10–14) suggest that this optimistic outlook is at least partially warranted.

It has been more than 30 years since Woodman and Taylor (20) reported that zinc oxide was an effective catalyst for the hydrogenation of ethylene. Since then a number of reports (21–26) have appeared dealing with this and related aspects of the chemistry of zinc oxide. The only thorough studies of the kinetics of ethylene hydrogenation, however, were carried out by Teichner's group (22–25). These studies dealt primarily with reactions at elevated temperatures, i.e., roughly 100–400°C. Under these conditions the reaction is complex; orders and activation energies change with temperature, and formation of ethylene residues with consequent poisoning is evident. The temperature range spanned by these studies includes the range in which adsorption of hydrogen is complex (27–31) and includes partial reduction of the catalyst (32); hence, the attendant complexities are not unexpected. In this review we confine our consideration to reactions at or below 100°C, a region where the comparative simplicity cited in the foregoing paragraphs holds.

II. The Active Sites

A. Hydrogen Adsorption

Oxides that are hydrogenation catalysts also chemisorb hydrogen. These oxides must first be "activated," usually by high temperature degassing or hydrogen pretreatment, before they are active for either hydrogenation or hydrogen adsorption (20–31). Thus, the sites for hydrogen adsorption appear to be the active sites for hydrogenation, and hydrogen adsorption provides a probe with which to study the active sites. One can reasonably hope that if we understand hydrogen adsorption on one of these oxides, we may extrapolate reasonably to other oxide systems. As pointed out by Parravano and Boudart (27): "It is important to note that the principal features of hydrogen chemisorption, which are summarized above, apply equally well to absorbents other than zinc oxide, for instance, to chromium oxide. A satisfactory explanation, therefore, must not depend on the specific properties of zinc oxide." Accordingly, we shall devote our discussion

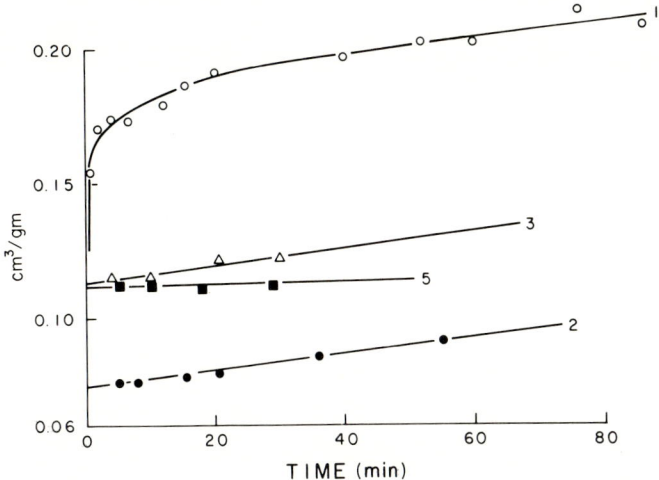

Fig. 1. Adsorption of hydrogen versus time ($P = 128$ mm): ○, run 1, virgin catalyst; ●, run 2, after run 1 + 2 min degas; △, run 3, after run 2 + 15 min degas; ■, run 5, after run 4 (16 hr adsorption) + 30 min degas.

first to the nature of hydrogen adsorption on zinc oxide at room temperature (10).

Adsorption of hydrogen at room temperature and 128 mm on freshly activated zinc oxide is shown by the uppermost curve in Fig. 1. Rapid initial adsorption is followed by a slow adsorption that continues for days. This behavior is typical for oxides (27). Following the last point shown for run 1, the sample was evacuated for 2 min, hydrogen was readmitted, and the amount of adsorption was measured as a function of time (curve 2). After the initial rapid readsorption of 0.075 cm³/gm, the slow process resumes at a rate equal to that when run 1 was terminated. The slow process was followed for 1 hr; then the catalyst was evacuated for 15 min and the readsorption was again measured (curve 3). With the longer evacuation the initial rapid readsorption is greater, i.e., 0.113 cm³/gm, but once again the slow process resumes at a rate comparable to that at the end of run 2. The catalyst was then degassed and the readsorption was followed for 20 hr; at this point the slow adsorption, still apparent for extended time periods, was trivial over a period of 1 hr. The catalyst was evacuated for 0.5 hr and readsorption was studied (curve 5). Since slow adsorption was not observed, the amount of rapid adsorption is the total adsorption. The amount of rapid readsorption in run 5 (0.112 cm³/gm) is

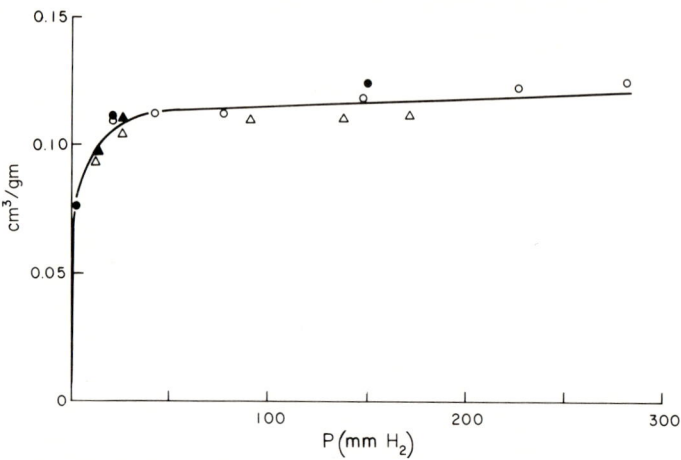

Fig. 2. Type I hydrogen isotherms: ○, hydrogen activation; △, oxygen activation (solid symbols denote desorption points).

essentially the same as the zero-time intercept for run 3 (0.113 cm³/gm), but quite a bit more than the zero-time intercept for run 2 (0.074 cm³/gm).

These and other data (10) show that hydrogen chemisorption is operationally of two types: Type I chemisorption which is removed by evacuation for 15 min at room temperature, and type II chemisorption which is *not* removed by evacuation at room temperature even after several hours. The type I chemisorption appears to be independent of the amount of type II chemisorption (compare runs 3 and 5). Figure 2 shows an isotherm for type I adsorption, as defined. This is a typical curve for chemisorption and suggests that type I chemisorption occurs on sites corresponding to roughly 5% of the BET V_m value. (The designation type I and type II chemisorption was chosen in preference to "fast" and "slow" because not all of the type II chemisorption is slow. For example, the amount of adsorption in curve 1 of Fig. 1 is 0.154 cm³/gm after 2 min. We would estimate *at least* one-third of this adsorption is type II. Thus, some type II irreversible chemisorption is quite rapid.)

Figure 3 shows the spectrum of zinc oxide in the presence and absence of hydrogen. The background scan shows three strong bands at 3665, 3616, and 3445 cm⁻¹ as well as a weak band at 3635 cm⁻¹; all of these bands are due to residual surface hydroxyls. (We shall reserve the term hydroxyl for bands in the background; the term OH shall be used for bands formed by chemisorption.) The spectrum in hydrogen shows two strong bands at 3489 and 1709 cm⁻¹; these bands, first reported by Eischens *et al.* (*32*), are assigned to an OH and ZnH species. By studies of the kind depicted in Figs. 1 and 2 it has been shown that only type 1 chemisorption gives rise to these bands in the IR (*10*).

Hydrogenation of ethylene at room temperature over zinc oxide occurs at rates of the order 0.05 cm^3/min gm. Both type I and type II chemisorption have initial rates comparable to this; hence, on this basis alone, either or both could be effective in hydrogenation of ethylene. The issue can be resolved by tracer experiments (*10*). First, the freshly degassed catalyst is exposed to deuterium overnight; type I deuterium is removed by brief evacuation and reaction is carried out with $H_2:C_2H_4$. The participation in hydrogenation of type I hydrogen and type II deuterium can then be assayed by the amount of deuterium appearing in the ethane.

Results of such an experiment are summarized in Table I. The second column gives the composition in the system at the start of the hydrogenation; amounts of type I and type II chemisorption were estimated from adsorption data. The third column gives the composition in the system at the end of the hydrogenation run; the amount of type II D_2 adsorption remaining after reaction was estimated by isotopic dilution of hydrogen exhaustively exchanged with the catalyst after the run. The salient feature of these results is that even though the turnover of hydrogen on the surface was nearly fourfold greater than the type II deuterium adsorption, less

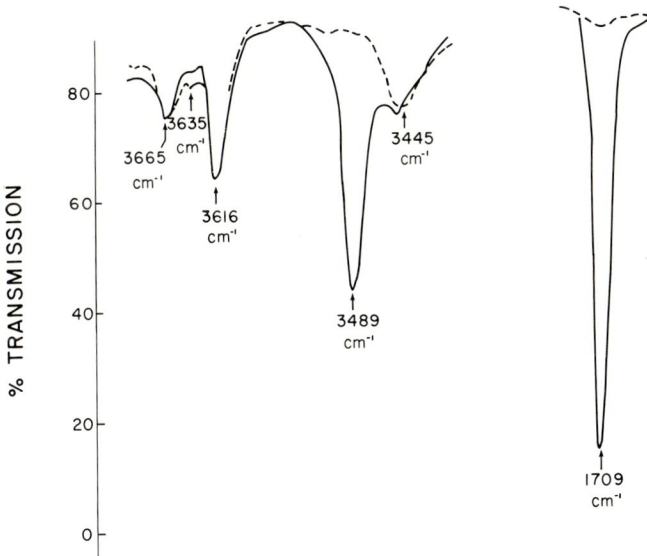

FIG. 3. Spectrum of hydrogen on zinc oxide ($P = 100$ mm; dashed line is background.)

TABLE I

Participation of Type II Hydrogen in Hydrogenation

Components	Standard temperature and pressure (cm²)			
	$t = 0$ (hr)	$t = 0.5$ (hr)	Calc.[a]	Calc.[a]
$C_2H_4(g)$	1.86	1.13	1.13	1.13
$C_2H_6(g)$	0.00	0.72	0.73	0.55
$C_2H_4D_2(g)$[b]	0.00	0.01	0.00	0.18
$H_2(g)$	3.55	2.77	2.82	2.82
$D_2(g)$[b]	0.00	0.05	0.00	0.00
$D_2(II)$	0.20	0.13	0.20	0.02
$H_2(II)$	0.00	0.07	0.00	0.18
$H_2(I)$	0.10	0.10	0.10	0.10

[a] See text.
[b] Any deuterium content is reported as $C_2H_4D_2$ and D_2 equivalent.

than 2% of the product stemmed from this deuterium. Column four lists the expected product distribution if type II chemisorption is inert; column five lists the expected product distribution if type II chemisorption is the kinetic equivalent of type I. Clearly, the view that type II chemisorption is inert is closest to the truth. It is evident from these data that some exchange of type II deuterium with gaseous hydrogen does occur: Note the deuterium appearing in the gas phase. Perhaps this occurs via reactions of gaseous hydrogen with type II deuterium, i.e., via a Rideal–Eley mechanism. Regardless of the exchange mechanism, however, it appears that the deuterium content of the ethane (1.9%) is roughly comparable to the deuterium content of the unreacted hydrogen (1.5%); hence, a major portion of the deuterated ethane could have formed from gaseous deuterium produced by exchange of hydrogen with type II deuterium rather than by direct reaction.

In sum these results show that type I hydrogen adsorption occurs reversibly on a limited number of sites with the formation of Zn–H and OH bonds. It is this type I hydrogen that reacts with ethylene to form ethane. Type II hydrogen plays a passive role at room temperature. It does not participate as a reactant; it does not contribute to the IR bands; it causes no measurable change in the conductivity of the catalyst.[1] Accordingly, type I hydrogen provides the best gage of what is happening to the active sites.

[1] Conductivity changes show up as background shifts in the IR spectra.

B. Poisoning Experiments

Many of the early workers on hydrogenation and related reactions on zinc oxide reported that oxygen adsorption poisoned the catalyst (*21, 22, 26, 27*) but that the activity could be restored by high temperature evacuation or hydrogen treatment. Studies correlating adsorption to the solid state properties of zinc oxide (*31, 33*), a metal-excess, n-type semiconductor, had shown that oxygen adsorbed and reacted with conduction electrons and/or donors and degassing or hydrogen treatment (at elevated temperatures) restored the conductivity. Accordingly, it is not surprising that the active sites were assumed to stem from the structural features of zinc oxide that led to its semiconductivity. These conclusions applied even to studies of hydrogenation of ethylene at moderate temperatures. For example, Taylor and Wethington (*21*), faced with a decrease in hydrogenation rate at 0°C by two orders of magnitude on oxygen pretreatment, concluded: "It is clear that catalytic activity, like electrical conductivity is closely related to oxygen deficiency of the oxide." Later experiments by Aigueperse and Teichner (*22, 23*) at higher temperatures showed that doping with lithia or gallia (which decreases and increases, respectively, the conductivity) had little effect on the activity, a result that led to the conclusion (*23*): "The electronic structure of catalysts is therefore without influence on the catalytic activity." However, they also observed poisoning with oxygen, an observation that led them to remark (*23*): "The formation of a nonstoichiometric oxide is a necessary condition for catalytic activity in the hydrogenation of ethylene." Thus, it has become part of our tradition to assume that the active sites are a result of nonstoichiometry.

The transmission of zinc oxide in the infrared can provide a quantitative measure of the number of charge carriers. Thomas (*34*) has applied the theoretical equations to experimental data for single crystals and found they hold quite well. We have found the theoretical predictions hold qualitatively even for pressed powders (*10*) insofar as pretreatment known to increase the conductivity decreases the transmission and vice versa. These effects are quite pronounced. For example, a sample of zinc oxide degassed at high temperatures will show an increase in transmission by a factor of 2–4 when exposed to oxygen and degassed at room temperature even though the amount of chemisorbed oxygen is thought to be only about 0.005 cm^3/gm (*33, 35*). Similarly, treatment with hydrogen at high temperatures makes the sample essentially opaque. Thus, in IR studies we can monitor not only the adsorbed species but also the changes in oxidation state, i.e., the nonstoichiometry of the catalyst.

Studies of the IR spectra of adsorbed hydrogen as a function of catalyst oxidation state showed it had little effect in the intensity of the IR bands.

In fact, the spectrum shown in Fig. 3 is that of a sample roasted in dry oxygen at 450°C for several hours, cooled in oxygen to room temperature, and briefly evacuated. From the high transmission we know this sample has the low conductivity characteristic of near-stoichiometric zinc oxide; yet, the spectrum associated with type I chemisorption is quite evident. Thus, it appears that oxygen has no deleterious effect on the sites for type I adsorption. Further evidence on this point is revealed by the isotherm in Fig. 2. One set of points is for a catalyst activated by high temperature treatment in hydrogen; the other set of points is for a catalyst treated at high temperatures in oxygen. The conclusion is unavoidable: *The amount of type I chemisorption, which occurs on active sites, is not a function of the nonstoichiometry of zinc oxide.*

Studies of ethylene adsorption and activity for hydrogenation activity as a function of pretreatment are consistent with the foregoing results. When a catalyst is activated by roasting in dry oxygen at 450°C for several hours, followed by cooling in oxygen to room temperature and brief evacuation, neither the amount of ethylene adsorption nor the activity are significantly less than these for a catalyst activated by high temperature treatment in vacuum or in hydrogen. Severe poisoning effects due to oxygen exposure appear only when the catalyst history is such that it has residual chemisorbed hydrogen. These poisoning effects are likely to be due to reaction of the chemisorbed hydrogen with oxygen to form water or its pre-

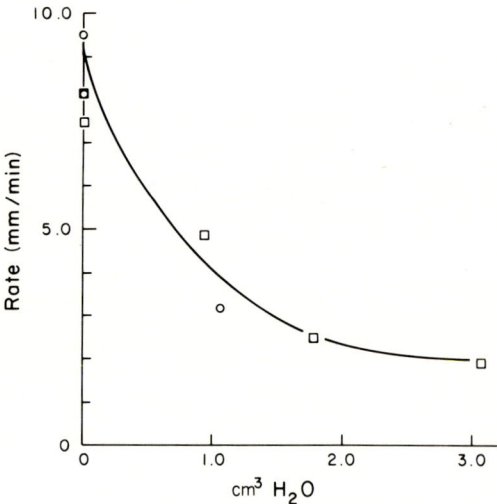

Fig. 4. Activity versus water adsorption: □, hydrogen adsorption sequence; ○, ethylene adsorption sequence.

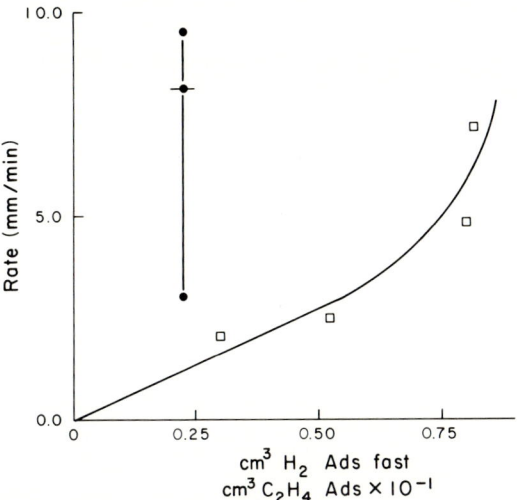

FIG. 5. Correlation of rate with adsorption: □, hydrogen adsorption sequence; ●, ethylene adsorption sequence.

cursor, which is a poison.[2] Thus, nonstoichiometry and the resultant semiconductivity appear to play no significant role in adsorption and hydrogenation catalysis on zinc oxide at room temperature.

Figure 4 shows the effect of water adsorption on the catalytic activity. (In this study uniform distribution of the water in the catalyst bed was achieved by annealing at 300°C followed by quenching.) It is clear that water is a strong poison. If we assume a cross section for water of 10.6 Å^2 (36), monolayer coverage corresponds to about 14 cm^3; on this basis, the amount of water adsorption needed to halve the activity corresponds to about 5% coverage, which is roughly the number of sites that adsorb type I hydrogen.

Infrared studies show that when water is adsorbed on the surface, the background intensity in the hydroxyl region increases; new bands may appear but hydrogen-bonding effects make such conclusions uncertain. If such a catalyst is then exposed to hydrogen (or deuterium), no bands due to adsorbed hydrogen (or deuterium) are observed. Thus, adsorption of water apparently occurs on the active sites and blocks out type I chemisorption.

[2] Although oxygen "poisoning" of hydrogenation at room temperature appears to arise primarily from interaction with residual hydrogen, we need not conclude that the same must be true for the high temperature hydrogenation studies of Aigueperse and Teichner (23).

Figure 5 summarizes the results of an experiment in which the changes in rate due to water adsorption are correlated with the attendant changes in ethylene and type I hydrogen chemisorption. These data clearly show that the decrease in rate due to water poisoning is due to reduction in type I hydrogen chemisorption rather than changes in the amount of ethylene adsorption. These conclusions are borne out by IR studies; a catalyst poisoned by water shows little change in the spectrum of adsorbed ethylene.

C. Addition of H_2–D_2 Mixtures to Ethylene

Twigg (37) studied the reaction of ethylene with a 50:50 H_2–D_2 mixture over nickel. He found that the deuterium distribution in the product ethane was the same for the preequilibrated mixture (about 50% HD) as for the unequilibrated mixture (about 0% HD). This result was not due to rapid preequilibration of the isotopic hydrogen in the gas phase; in the presence of ethylene the equilibration rate is much slower than hydrogenation. These observations, however, could stem from the rapid alkyl reversal depicted in Eq. (1) which, in effect, premixes atomically the reactant hydrogen. The observations of Meyer and Burwell (38), however, cannot be rationalized in this manner. These authors showed that addition of deuterium to 2-butyne over palladium yielded about 98% cis-2-butene-2,3-d_2. They also showed that although H_2–D_2 equilibration over palladium was poisoned by the presence of the alkyne, addition of a 50:50 H_2–D_2 mixture (unequilibrated) to 2-butyne resulted in cis-2-butene with a random distribution of deuterium. Since the clean dideutero addition to the butyne precludes the occurrence of the analog of alkyl reversal, we must conclude the H_2–D_2 reactant undergoes isotopic self-mixing prior to reaction even though very little HD appears in the gas phase. We can interpret these results as follows. All empty metal sites dissociate hydrogen. Even in the presence of reacting hydrocarbon, there are enough such sites adjacent to permit isotopic mixing prior to reaction. If reaction with adsorbed hydrocarbon is much more rapid than desorption, this would lead to the observed inhibition of gas phase H_2–D_2 exchange even though the hydrogen that adds to the hydrocarbon is isotopically equilibrated.

Over zinc oxide it is clear that only a limited number of sites are capable of type I hydrogen adsorption. This adsorption on a Zn—O pair site is rapid with a half-time of less than 1 min; hence, it is fast enough so that H_2–D_2 equilibration (half-time 8 min) can readily occur via type I adsorption. If the active sites were clustered, one might expect the reaction of ethylene with H_2–D_2 mixtures to yield results similar to those obtained for the corresponding reaction with butyne-2 over palladium: That is, despite the clean dideutero addition of deuterium to ethylene, the eth-

TABLE II

Deuterium Distribution in Ethane^a

% Reaction	Reactant	C_2H_6 %	C_2H_5D %	$C_2H_4D_2$ %
10	Pure D_2	0.0	0.0	99.9
6	H_2–D_2	45.1	16.7	38.2
6	H_2–D_2	45.1	16.9	38.1
6	H_2–HD–D_2^{b}	30.0	43.0	26.9

^{a} Conner and Kokes (9).
^{b} Analysis showed that equilibration was not complete. The initial composition was (H_2–HD–D_2) 34:36:30. After the run, the composition was 33:40:27.

ane product from the H_2–D_2 mixture would contain a random distribution of deuterium. If on the other hand this limited number of sites were widely separated, one would expect the product ethane to contain only $C_2H_4D_2$ and C_2H_6 provided the H_2:D_2 exchange was completely poisoned by the presence of ethylene. Actually, the rate of H_2–D_2 equilibration is not completely poisoned but is reduced by the presence of ethylene to one-third its value in the absence of ethylene. Accordingly, to see which of the above cases prevails, we must look at the products of hydrogenation by isotopic mixtures at low conversions so that the effect of exchange is minimized. Table II lists the products obtained for experiments in which equilibrated and nonequilibrated H_2–D_2 mixtures were reacted with ethylene. It is quite clear that the mixtures are not randomized prior to addition; the distribution of deuterium in the ethanes reflects the gas phase composition inasmuch as hydrogen maintains its molecular identity. Thus, these results suggest that oxides differ from metals in that the active sites are widely separated.

A Rideal–Eley mechanism in which gas phase hydrogen bombards and reacts directly with adsorbed ethylene is clearly an alternative to the foregoing interpretation, but we believe this alternative is unlikely for zinc oxide. On a catalyst poisoned with water, the amount of ethylene adsorption and its IR spectrum are virtually unchanged (13); yet, such a catalyst is inactive for hydrogenation despite the fact that gaseous hydrogen is still bombarding the apparently unchanged adsorbed ethylene. The fact that this inactive catalyst shows no type I hydrogen adsorption clearly suggests the type I hydrogen is the *sine qua non* for the unpoisoned reaction.

D. A Model for the Active Sites

The active site for type I hydrogen adsorption appears to consist of isolated, noninteracting Zn—O pair sites which are not affected by the oxida-

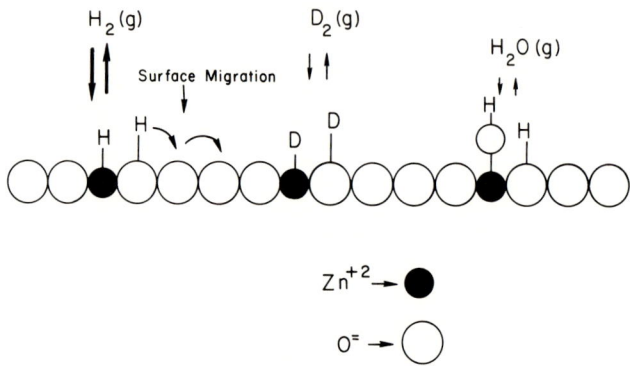

FIG. 6. Adsorption of H_2, D_2, and H_2O on the active sites (schematic).

tion state of the catalyst. As a working hypothesis, it has been suggested that the sites consist of a zinc ion in the trigonal holes in a close-packed layer of oxide ions (10). For zinc oxide these trigonal holes are about 0.58 Å in radius, which is nearly big enough to accommodate zinc ions with an effective radius of 0.70 Å. If the zinc ion in the trigonal site and the surrounding oxide ions are coplanar, the outermost plane tangent to the oxide spheres (radius 1.33 Å) will be 0.63 Å above the surface of the zinc sphere; hence, the zinc ion would appear to be buried in the oxide layer. Such a structure could reasonably arise from surface reconstruction of the $000\bar{1}$ planes of zinc oxide; evidence for similar reconstruction in GaAs and GaSb, which has the related zinc blende structure, has been reported by MacRae (39). The advantage of this view is that the sites stem from rearrangements to stabilize the $000\bar{1}$ planes and do not require deviations from stiochiometry. We shall use this picture to systematize the discussion that follows with the clear recognition that it is speculative and provides a more detailed picture of the active sites than is actually needed.

The essential features of the above picture are shown schematically in Fig. 6. Hydrogen or deuterium adsorbs and desorbs rapidly on these sites (half-time less than 1 min); hence, the slower H_2–D_2 equilibration (half-time about 8 min) appears to be determined by the site-to-site migration required for exchange. Poisoning experiments show that water also prefers these sites; in agreement with the limited IR data, Fig. 6 shows the adsorbed water yields surface hydroxyls (10).

Adsorption of ethylene as an olefinic species would not be likely to occur on the zinc half of the active site. A rigid ethylene molecule could not approach the sequestered zinc ions because of steric restrictions; hence, ethylene would be confined primarily to the oxide part of this layer. In line with this the spectrum of deuterium (at 200 mm) in the presence of ethylene (20 mm) shows distortion of the ZnD band but no change in total intensity, whereas the OD band drops to about 60% of the intensity it had in the absence of ethylene (10). Once reaction of ethylene with hydrogen occurs, the intermediate, presumed to be an ethyl radical, forms a σ bond. Such a highly directed bond could form with the sequestered zinc. Alkyl reversal over metals, which occurs readily, is usually pictured as a trade of a σ carbon—hydrogen bond for a σ carbon—metal bond and a metal—hydrogen bond. The strength of the σ carbon—metal bond can be gaged by the heat of adsorption of ethylene, typically about 50 kcal or more for metals (1). Only for comparatively high values of the heat of adsorption of ethylene can we expect alkyl reversal to be energetically favorable. Over zinc oxide the heat of adsorption of ethylene is moderately low, 14 kcal (13). As a consequence, alkyl reversal is not favored energetically, and deuterium addition to ethylene yields only dideuteroethane. These results are compatible with our picture of the active sites. Ethylene cannot adsorb to form two σ-metal—carbon bonds, with their relatively large bond strength, unless metal sites are adjacent; hence, with isolated active sites only a weakly bound olefinic species can form and its low heat of adsorption precludes alkyl reversal.

Let us now look at the chemistry of the reaction of water and hydrogen with the active sites. When water reacts with the active site, it seems quite clear that this should be viewed as heterolytic fission of an OH bond with the proton adding to the oxide ion and the hydroxide ion adding to the zinc ion. This is shown schematically below:

$$H_2O + O^{2-}\!\!-\!\!Zn^{2+}\!\!-\!\!O^{2-} \rightarrow O^{2-}\!\!-\!\!\underset{|}{Zn^+}\!\!-\!\!\underset{|}{O^-} \atop \overset{OH\ \ \ H}{} \quad (5)$$

Thus, in this reaction the active site functions as an acid–base pair and the adsorption of water is an acid–base reaction. The driving force for this reaction is the resulting reduction of the charge separation. In a similar fashion we can view hydrogen adsorption as heterolytic fission at the highly polar active site, viz:

$$H_2 + O^{2-}\!\!-\!\!Zn^{2+}\!\!-\!\!O^{2-} \rightarrow O^{2-}\!\!-\!\!\underset{|}{Zn^+}\!\!-\!\!\underset{|}{O^-} \atop \overset{H\ \ \ H}{} \quad (6)$$

In Eq. (6) the hydrogen attached to the oxygen has protonic character and the hydrogen attached to the zinc has hydridic character. Such bonds, with considerable ionic character, would be expected to yield intense bands in the IR, whereas a largely covalent bond would yield relatively weak IR bands. Thus, the relatively intense ZnH band observed (Fig. 3) is consistent with the polar character suggested in this formulation.

The above is formally equivalent to the picture of a coordinatively unsaturated surface (CUS) put forward by Burwell *et al.* (*3*) in their discussion of chromia. The acid–base formalism does have the advantage of drawing attention to the analogy of acid and base catalyzed reactions. If a hydrocarbon undergoes reaction at these sites via loss of a proton to the oxide site, the reaction should be analogous to a base catalyzed reaction; if it undergoes reaction via the loss of a hydride to the zinc site or addition of a proton from the oxide site, the reaction should be analogous to an acid catalyzed reaction. This view, which we find useful, is implicit in the discussion that follows.

III. Mechanism of Ethylene Hydrogenation

A. Goals

Hydrogenation of ethylene presumably involves the reaction of two surface species, adsorbed hydrogen and ethylene, to form an intermediate surface species, the surface ethyl, which reacts further with hydrogen to form the product ethane. In traditional mechanistic studies the objective is to indicate these species as shown in Eqs. (1) and (2) and to deal with the implied consequences in terms of kinetics and tracer studies. Ideally, it would be most valuable to specify also the structure and reactivity of surface species and intermediates. When this goal is achieved, we can truly assert that the mechanism is well understood. The observability of reacting hydrogen by IR spectroscopy, the comparative simplicity of the addition of deuterium to ethylene, the reproducibility of the activity, the simplicity of olefin adsorption, and the excellent transmission properties combine to make hydrogenation over zinc oxide a prime candidate for approaching these goals viz a combined infrared-kinetic study. A portion of these goals, specification of the structure of adsorbed reacting hydrogen, has been reached largely due to the work of Eischens *et al.* (*32*). The goals that remain are specification of the kinetics, the structure of adsorbed ethylene, and the structure of the intermediate.

B. Kinetics

Consider the elementary steps enumerated in Section II. Hydrogen adsorption and site-to-site migration occur by the following sequence (*10*):

$$H_2(g) + ZnO \rightleftarrows H\text{—}ZnO\text{—}H, \tag{7}$$

$$H\text{—}ZnO\text{—}H + O \rightleftarrows H\text{—}ZnO + OH, \tag{8}$$

$$H\text{—}ZnO + O \rightleftarrows ZnO + OH, \tag{9}$$

where ZnO and O represent the active sites and the intervening oxide sites, respectively. Ethylene adsorption occurs on all bare oxide sites, viz:

$$O + C_2H_4(g) \rightleftarrows O\text{—}C_2H_4, \tag{10}$$

$$ZnO + C_2H_4(g) \rightleftarrows ZnO\text{—}C_2H_4, \tag{11}$$

$$H\text{—}ZnO + C_2H_4(g) \rightleftarrows H\text{—}ZnO\text{—}C_2H_4. \tag{12}$$

The actual reaction steps are:

$$H\text{—}ZnO\text{—}C_2H_4 \xrightarrow{k_{13}} C_2H_5\text{—}ZnO\ (I), \tag{13}$$

$$C_2H_5\text{—}ZnO + OH \xrightarrow{k_{14}} C_2H_6 + ZnO + O, \tag{14}$$

Since deuterium addition to ethylene yields $C_2H_4D_2$, both of the last steps are irreversible and (13) can be taken as the rate-determining step. Accordingly, if all prior steps are at equilibrium, we can write for the rate R of ethane formation:

$$R = k_{13}\,(H\text{—}ZnO\text{—}C_2H_4)$$

$$= k_{13}K_{12}\,(H\text{—}ZnO)[C_2H_4(g)]$$

$$= k_{13}\frac{K_{12}}{K_{11}}\frac{(HZnO)(ZnO\text{—}C_2H_4)}{(ZnO)}$$

but

$$\frac{(HZnO)}{(ZnO)} = \left(\frac{K_7K_8}{K_9}\right)^{1/2}(H_2)^{1/2},$$

and hence,

$$R = k_{13}\frac{K_{12}}{K_{11}}\left(\frac{K_7K_8}{K_9}\right)^{1/2}(H_2)^{1/2}(ZnO\text{—}C_2H_4) \tag{15}$$

Thus, the rate should be half-order in hydrogen pressure and proportional to the amount of ethylene adsorbed on the oxide half of the active site under reaction conditions.

Figure 7 shows a plot of initial hydrogenation rate versus the square root of hydrogen pressure for several different ethylene pressures (10). Since there is some dependence on ethylene pressure and the activity varies from one sequence to another, we have multiplied the rates for each sequence by a scale factor. These sequences, which represent a 10-fold range of ethylene pressures and a 25-fold range of hydrogen pressures, combine to give a well-defined straight line nearly through the origin. Thus, the order with respect to hydrogen is consistent with the proposed sequence.

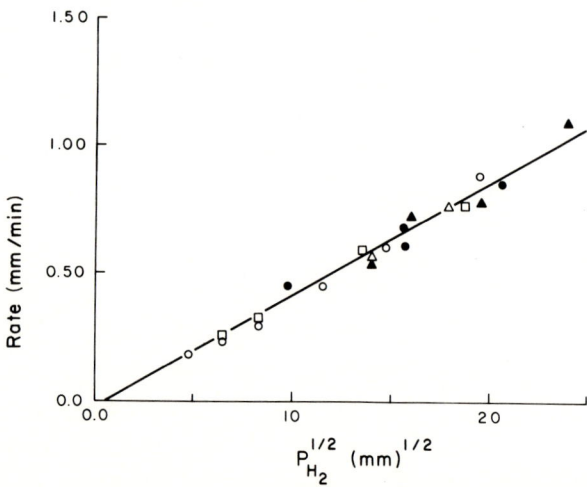

Fig. 7. Order with respect to hydrogen: ○, sequence 9A, $P_{C_2H_4} = 12.7$ mm, scale factor × 1; □, sequence 9B, $P_{C_2H_4} = 27.4$ mm, scale factor × 1.12; △, sequence 9C, $P_{C_2H_4} = 71$ mm, scale factor × 0.90; ▲, sequence 9D, $P_{C_2H_4} = 129$ mm, scale factor × 0.87; ● sequence 6A, $P_{C_2H_4} = 24$ mm, scale factor × 0.85.

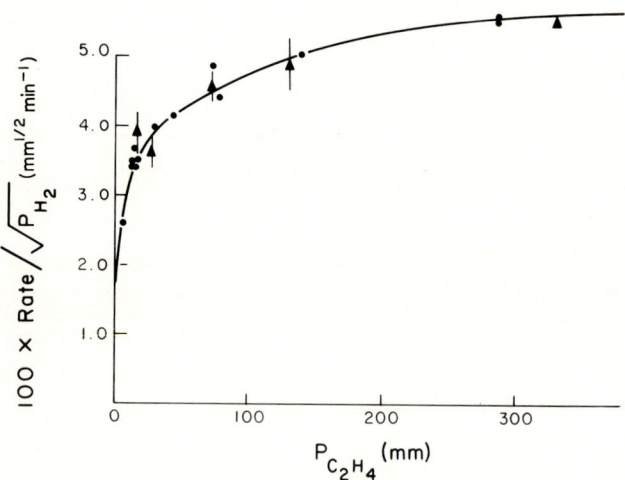

Fig. 8. Order with respect to ethylene: ●, sequence 9: ▲, average values for different sequences shown in Fig. 1.

In another sequence of runs both ethylene and hydrogen pressure were varied. Dependence on hydrogen was factored out by dividing the initial rate by the square root of the hydrogen pressure. A plot of the factored rate versus ethylene pressure is shown in Fig. 8 (*10*). The triangular points with error bars show the average of the factored rates for the sequences depicted in Fig. 7. Since both hydrogen and ethylene pressures were varied, this plot not only establishes the order in ethylene but furnishes a further test of the proposed order with respect to hydrogen. The shape of the curve is similar to that found for saturation chemisorption; in fact, it shows saturation at pressures similar to those for equilibrium ethylene chemisorption (*10*). Thus, the order in ethylene is also consistent with the proposed reaction sequence.

C. Ethylene on Zinc Oxide

Ethylene adsorption at room temperature is rapid and reversible. Even after prolonged exposure to the catalyst, the ethylene is recoverable as such by brief evacuation (*10*). The isotherms are nonlinear and show some evidence of saturation at 0.5–0.6 cm^3/gm, a value roughly five times that of the type I hydrogen. Since the adsorption is quite weak, it would seem that this adsorption is, in part, physical adsorption. To investigate this possibility, adsorption of ethylene (boiling point $-104°C$) was compared to that of ethane (boiling point $-89°C$) (*13*). By traditional criteria physical adsorption of ethane should be greater than that of ethylene, and the comparison of the relative adsorption should let us assay what fraction of the ethylene adsorption is physical.

Ethylene adsorption at 0°C and 1 mm is about two orders of magnitude greater than that for ethane. The initial heat of adsorption of ethylene is about 14.0 kcal/mole, a value about four times the heat of liquefaction (3.2 kcal/mole); by way of contrast the heat of adsorption is about 5 kcal/mole for ethane, only slightly greater than the heat of liquefaction (3.5 kcal/mole). Thus, by the usual criteria, ethylene adsorption is chemical at low averages. Plots for ethylene of the heat of adsorption vs coverage show the value is roughly constant at low coverage, falls precipitously at about 0.25 cm^3/gm and reaches values consistent with physical adsorption at coverages above about 0.30 cm^3/gm (*13*). If we assume the physically adsorbed component of the ethylene to be similar to that found for ethane, we would expect the physically adsorbed ethylene to be roughly proportional to the pressure, whereas judging from the isotherm, the chemically adsorbed ethylene (ca 0.3 cm^3/gm) is nearly independent of pressure above about 25 mm.

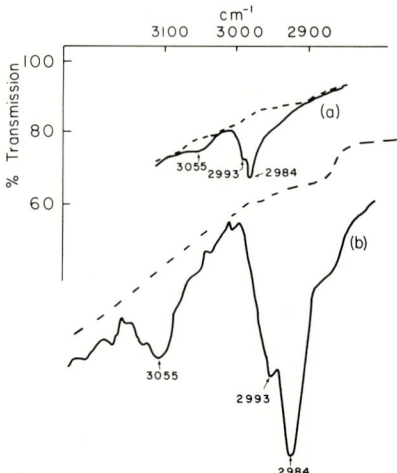

FIG. 9. Infrared spectrum of adsorbed ethylene at 100 mm: (a) normal scale spectrum; (b) expanded scale spectrum (see text).

Figures 9 and 10 show the spectrum of adsorbed ethylene at a pressure of 100 mm in the C—H stretching and deformation region, respectively (*13*). (No bands were observed in the ZnH or OH region; hence, ethylene adsorbs without appreciable dissociation.) In each figure the upper curve gives the spectrum without amplification; the lower figure was run with a fivefold ordinate expansion and a twofold expansion in the wavelength scale. In the stretching region (Fig. 9) the strongest band is at 2984 cm^{-1}; progressively weaker bands are found at 2993, 3055, and 3125 cm^{-1}. (The last band is not shown in Fig. 9). In the deformation region (Fig. 10) there are two strong bands at 1521 and 1327 cm^{-1} due to the zinc oxide itself and a weak band at 1300 cm^{-1} caused by perturbation of the surface by adsorbed molecules. The band at 1600 cm^{-1} is assigned to the $>$C$=$C$<$ stretch of chemisorbed ethylene; bands at 1451 and 1438 cm^{-1} are assigned to CH deformations. Support for this assignment is supplied by the spectrum of adsorbed perdeuteroethylene, which shows only one band in this region at 1495 cm^{-1}. This band, clearly a double bond stretch, shows a shift comparable to the shift in the C$=$C frequency that occurs on deuteration of gaseous ethylene, i.e., from 1623 to 1515 cm^{-1} (*40*). Shifts on deuteration for carbon—hydrogen deformation bands are much larger, of the order of 400 cm^{-1} and would place these bands below 1150 cm^{-1}, the transmission cutoff for zinc oxide; hence, as expected, no other bands are seen in this region. In the C—D stretching region, however, bands corresponding to those in Fig. 9 are evident.

The dependence of intensity of the observed bands on pressure is significant. Except for a relatively small intercept the intensities of bands at about 3130 and 3060 cm^{-1} are roughly proportional to pressure between 22 and 240 mm, whereas the intensities of bands at 1451, 1438, 1600, and 2984 cm^{-1} are insensitive to pressure in this region. (The band at 2993 cm^{-1} seems to behave similarly to the bands above 3000 cm^{-1}, but its overlap with the band at 2984 cm^{-1} makes analysis difficult.) Thus, the bands above 3000 cm^{-1} (and perhaps the band at 2993 cm^{-1}) are primarily due to physically adsorbed ethylene and only in part due to chemisorbed ethylene. By way of contrast the remaining bands stem primarily from chemisorbed ethylene.

The band at 1600 cm^{-1} due to a double-bond stretch shows that chemisorbed ethylene is olefinic; C—H stretching bands above 3000 cm^{-1} support this view. Interaction of an *olefin* with a surface with appreciable heat suggests π-bonding is involved. Powell and Sheppard (*41*) have noted that the spectrum of olefins in π-bonded transition metal complexes appears to involve fundamentals similar to those of the free olefin. Two striking differences occur. First, infrared forbidden bands for the free olefin become allowed for the lower symmetry complex; second, the fundamentals of ethylene corresponding to ν_2 and ν_3 shift much more than the other fundamentals. In Table III we compare the fundamentals observed for liquid ethylene (*42*) and a π-complex (*43*) to those observed for chemisorbed ethylene. Two points are clear from Table III. First, bands forbidden in the IR for gaseous ethylene are observed for chemisorbed ethyl-

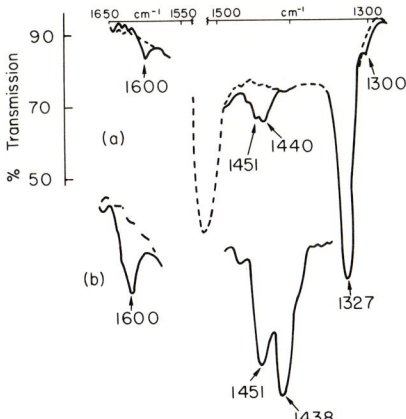

Fig. 10. Infrared spectrum of adsorbed ethylene at 100 mm: (a) normal scale spectrum; (b) expanded scale spectrum (see text).

TABLE III

Observed Bands for Chemisorbed Ethylene and Related Systems

Type[a] vibration	$C_2H_4(l)$[b] (cm^{-1})	π-Complex[c] (cm^{-1})	Adsorbed ethylene[d] (cm^{-1})
9	3105 (I)	3094	3140[e] (3122)
5	3075 (R)	3079	3055 (3078)
1	3008 (R)	3013	2993 (3000)
11	2980 (I)	2988	2984
2	1621 (R)	1518	1600
12	1435 (I)	1426	1438
3	1342 (R)	1419	1451

[a] Numbering is the same as in Herzberg (*40*).

[b] For gaseous ethylene bands labeled I and R are infrared and Raman active, respectively. See Stoicheff (*42*) for liquid ethylene.

[c] For the salt $KPtCl_3(C_2H_4)$ (*43*).

[d] Some of these bands shift as a function of pressure. The numbers listed are for the lowest pressures of ethylene (22 mm); the numbers in parentheses are the positions at the highest ethylene pressures (240 mm). In part these bands are due to physically adsorbed ethylene.

[e] This band is broad and weak. The position is uncertain to ± 10 cm^{-2}.

ene; second, the ν_3 and ν_2 bands are both shifted, although the former is shifted more and the latter is shifted less than in transition metal complexes. Nevertheless, this similarity of the spectrum of adsorbed ethylene to the spectrum of π-complexes strongly suggests chemisorbed ethylene is "π-bonded" to the surface.

The shift in the C=C frequency, ν_2, for adsorbed ethylene relative to that in the gas phase is 23 cm^{-1}. This is much greater than the 2 cm^{-1} shift that is observed on liquefaction (*42*) but is less than that found for complexes of silver salts (*44*) (about 40 cm^{-1}) or platinum complexes (*43*) (105 cm^{-1}). Often there is a correlation of the enthalpy of formation of complexes of ethylene to this frequency shift (*44, 45*). If we use the curve showing this correlation for heat of adsorption of ethylene on various molecular sieves (*45*), we find that a shift of 23 cm^{-1} should correspond to a heat of adsorption of 13.8 kcal. This value is in excellent agreement with the value of 14 kcal obtained for isosteric heats at low coverage. Thus, this comparison reinforces the conclusion that ethylene adsorbed on zinc oxide is best characterized as an olefin π-bonded to the surface, i.e., a surface π-complex.

D. Intermediates in Ethylene Hydrogenation

Addition of deuterium to ethylene is ideally suited to a search for intermediates by IR techniques. The only hydrocarbon surface species expected to be present during reaction are C_2H_4 and C_2H_4D, provided ethane adsorption is trivial. It is relatively easy to arrange experimental conditions so that only the adsorbed hydrocarbon species contribute to the spectrum; if such a spectrum shows a paraffinic C—D band under reaction conditions, it should arise from the intermediate. Initial experiments (11), carried out with an IR cell in a circulating system, showed that when deuterium was admixed to ethylene in contact with the catalyst, a single new CD band did appear together with at least one CH band; both sets of bands were in the region characteristic of paraffins. If the run was carried out to completion (in excess D_2), the new bands stayed constant during reaction but disappeared when the reaction was complete. If the reaction was interrupted by condensation of the hydrocarbons but circulation of deuterium was continued, these new bands disappeared at a rate estimated to be comparable to that of the steady state reaction. Thus, these new bands behave as if they arose from the intermediate.

Characterization of the reaction intermediate is facilitated by studies in a flow system in which the sample cell and a reference cell are mounted in series in a double beam spectrometer (13). Not only can we observe the intermediate bands under rigorous steady state conditions, but we can monitor the conversion by sampling the effluent. In addition, the reference cell assures the spectrum we see is that of surface species. Primitive analysis of the kinetics reveals the intermediate is favored by relatively high ethylene pressures; hence, use of a reference cell to cancel contributions of the gas phase is an important factor.

Figure 11(b) shows the C—H stretching region of the spectrum of ad-

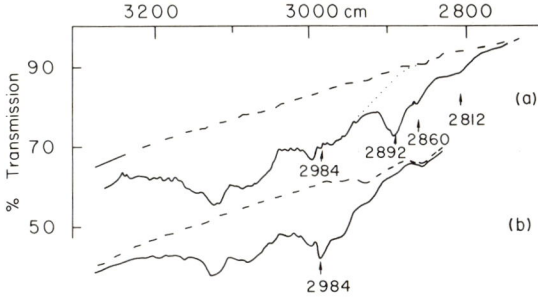

Fig. 11. Spectrum of ethylene in the presence and absence of hydrogen: (a) $P_{H_2} = 420$ mm Hg, $P_{C_2H_4} = 340$ mm Hg; (b) C_2H_4 at 240 mm Hg (displaced ordinate).

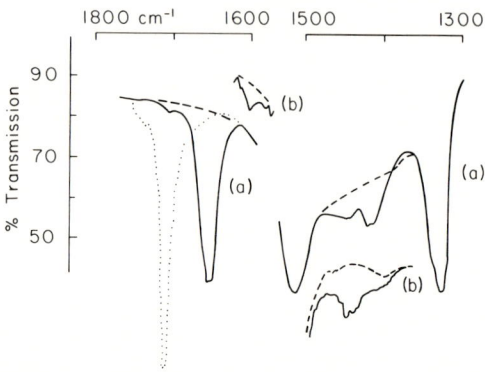

Fig. 12. Spectrum of ethylene in the presence and absence of hydrogen: (a) $P_{H_2} = 420$ mm Hg, $P_{C_2H_4} = 340$ mm Hg (dotted curve is for H_2 in helium with $P_{H_2} = 420$ mm); (b) $P_{C_2H_4} = 240$ mm Hg (displaced ordinate).

sorbed ethylene (at 240 mm) in a flow system with helium as a diluent. Although the reference cell does cancel the strong absorption due to gaseous ethylene between 2975 and 3140 cm^{-1}, the energy loss in both beams at these pressures is sufficiently great so that there is limited instrument response in this region. Nevertheless, we see once again prominent bands at roughly the same positions listed in Table III for the better-defined spectrum at lower pressures.

Figure 11(a) shows the spectrum of adsorbed species on an active catalyst in a hydrogen–ethylene stream. This spectrum appears and stabilizes within minutes after hydrogen is blended into the ethylene stream. Three new bands appear in the presence of hydrogen at 2892, 2860, and 2812 cm^{-1}. The appearance and location of these bands were verified by expanded scale spectra. Experiments at lower ethylene pressures reveal that there is an additional band at about 2940 cm^{-1} partially obscured in Fig. 11 by overlap of the ethylene spectrum. On a poisoned catalyst, which does not show the ZnH and OH bands, only the bands characteristic of chemisorbed ethylene are seen.

Figure 12 shows the spectrum for the deformation region in an ethylene–helium stream and an ethylene–hydrogen stream. In the deformation region the two bands at 1451 and 1438 cm^{-1} due to ethylene alone appear to weaken and shift slightly and a new band (or perhaps two) appears at about 1415 cm^{-1}. Figure 12 also shows that under reaction conditions the ZnH band is shifted from 1709 to 1655 cm^{-1}; the corresponding shift in the OH band is from 3490 to 3510 cm^{-1}.

Changes that occur in a reacting system are not limited to the appearance of new bands. Close scrutiny of the data in Figs. 11 and 12 reveals that the bands definitely associated with chemisorbed ethylene are weakened. The bands at 2984 and 1600 cm^{-1} are no longer evident; hence, it seems that a major portion of the chemisorbed ethylene is converted to the new species under reaction conditions. In view of this we believe that the bands centered at about 1440 cm^{-1} under reaction conditions (which are weaker than those due to chemisorbed ethylene alone) are due to the species formed under reaction conditions.

Similar studies to the above have been carried out with C_2H_4–D_2 as the reactant. These results are consistent with studies of the C_2H_4–H_2 mixture insofar as they suggest that a large portion of the chemisorbed ethylene is converted to a new species. A strong band occurs at about 2905 cm^{-1} with a somewhat weaker band at 2860 cm^{-1}; in addition a broad band seems to appear at 2955 cm^{-1}, but since runs at lower ethylene pressures were not carried out, this band is uncertain. Bands in the deformation region are found at 1445 and 1415 cm^{-1} and both of these are somewhat sharper than those found in the C_2H_4–H_2 reaction mix. It is most significant that only a *single* band appears in the paraffinic C—D stretching region at 2150 cm^{-1} which suggests that a monodeutero paraffinic species is formed.

Similar studies to the above, but more abbreviated, were also carried out with C_2D_4–H_2 as a reactant mixture. The species formed under reaction conditions yields a *single* weak band at 2890 cm^{-1} which suggests a monohydrido species. There are also several very weak bands in the CD region between 2100 and 2160 cm^{-1} and a weak band at 1289 cm^{-1}.

Figure 13(a) and (b) shows the spectrum of the supposed intermediate in the region of its strongest bands under steady state reaction conditions

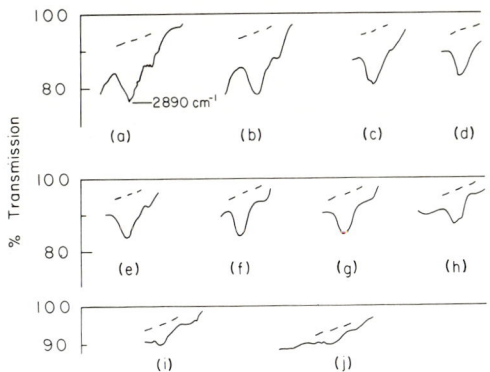

FIG. 13. Changes in 2890 cm^{-1} band with time. (a) and (b) steady state (P_{H_2} = 420 mm, $P_{C_2H_4}$ = 340 mm): (a) 20 min on stream; (b) 65 min on stream. (c)–(j) In H_2/He stream (P_{H_2} = 420 mm, P_{H_2} = 340 mm): (c) 0.4 min; (d) 1.5 min; (e) 2.8 min; (f) 4.4 min; (g) 6.9 min; (h) 13.5 min; (i) 29.5 min; (j) 35.5 min.

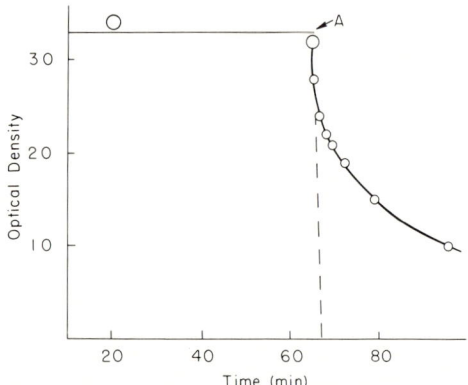

Fig. 14. Optical density (in arbitrary units) of 2890 cm^{-1} B and as a function of time. Open symbols are for steady state conversion (P_{H_2} = 420 mm, $P_{C_2H_4}$ = 340 mm). At point A the ethylene was removed from the stream but the hydrogen flow was continued. The dashed line represents the initial rate.

in a C_2H_4–H_2 stream; the spectrum, which forms immediately, is clearly stable for a prolonged period of time. Immediately after the spectrum in Fig. 13(b) was obtained, helium was substituted for ethylene in the reacting stream; the decay of the dominant 2890 cm^{-1} band in the He:H_2 stream is shown in Figure 13(c)–(j).

Figure 14 is a plot of the optical density of the 2890 cm^{-1} band center versus time for the spectra shown in Fig. 13. At the point labeled A on this graph, the H_2:C_2H_4 stream was changed to a H_2:He stream. The decrease in intensity of the spectrum due to this species (hereafter called X) is rapid initially but becomes slower as time proceeds. The initial rate of falloff (based on the first two points) is such that if this rate were maintained (dashed line), it would take about 2 min in pure hydrogen to remove all of X from the surface. Similar conclusions are reached if the integrated intensity in the 2925–2825 cm^{-1} region is plotted as a function of time. Corresponding runs were also made for C_2H_4–D_2 reactant mixture.

These results are best interpreted in terms of the proposed mechanism. The rate of the reaction in the steady state is the rate of *either* formation of the ethyl radical or its reaction with adsorbed hydrogen (steps 13 and 14). Accordingly, in the steady state, the concentration of the intermediate, I, should be constant; if the ethylene is suddenly removed but hydrogen is still present, the concentration of I should decrease, and *the initial rate of decrease of I should be equal to the steady state conversion to ethane.*

In order to estimate the rate of disappearance of X, we must relate the intensity of the IR band to the amount of X. The IR data suggest that all of the chemisorbed ethylene (0.3 cm³/gm) is converted to X. If this is indeed the amount of X, we get the decay rates shown in Table IV which also shows the steady state rate determined by analysis of the effluent. The errors are large but the agreement clearly suggests X is the intermediate.

(Attempts were made to check the amount of X corresponding to the intensity of the bands shown in Fig. 13. In one procedure a catalyst in the steady state was purged to remove ethylene and the surface species was stripped off in a hydrogen or deuterium stream. The amount of C_2H_6 or C_2H_5D stripped off provides another estimate for the amount of X. The results of this procedure were somewhat ambiguous because replacement of surface hydrogen with gaseous hydrogen or deuterium could not be achieved instantaneously. If we ignore this and combine this estimate with the other estimate, this "best value" decay rate is about 30% less than the values listed in Table IV.)

Table V lists the bands observed for X and compares them to bands observed for mercury diethyl and the resolved bands observed for diethyl zinc. The band for X at 2812 cm⁻¹ is too low to be a CH stretching fundamental and must be a reasonably strong overtone or combination. The three bands at 2860, 2892, and 2940 cm⁻¹ suggest a species which is paraffinic and contains at least three hydrogens. The occurrence of these bands in the C_2H_4–D_2 reaction mixture together with a single band at 2150 cm⁻¹, the C—D stretching region, suggests that X is a species with at least four hydrogen atoms. If we admit to the possibility that the 2940 cm⁻¹ band, which is broad, may be more than one band, the assignment given for these three bands in Table V follows.

The spectra of all metal alkyls studied by Kaesz and Stone (46) show a band at about 2750 cm⁻¹ assignable to an overtone of the symmetric methyl deformation vibration at about 1375 cm⁻¹. If we assign the band at 2812 cm⁻¹ to this overtone, the symmetric methyl deformation frequency would occur at about 1406 cm⁻¹, i.e., near the broad band centered at 1415 cm⁻¹. If we accept this assignment and take account of the fact that

TABLE IV

Comparison of Decay Rates to Steady State Conversion

Reactants	Steady state rate (molecules/sec gm)	Decay of X (molecules/sec gm)
C_2H_4–H_2	7.9×10^{16}	6.7×10^{16}
C_2H_4–D_2	3.3×10^{16}	4.5×10^{16}

TABLE V

Assignments for Intermediate and Diethyl Mercury

Bands of diethyl mercury[a,b] (cm^{-1})	Bands of diethyl zinc[c] (cm^{-1})	Assignment[a]	Intermediate bands[b] (cm^{-1})
2975 vs		CH$_3$ anti	2940 br
2930 vs		CH$_2$ anti	2940 br
2900 vs		CH$_3$ sym	2892 s
2850 sh		CH$_2$ sym	2860 m
2750 m	2735	2δ CH$_3$	2812 w
1468	1465	CH$_3$ anti def	1440 s br
1460 ssh	1465	CH$_3$ anti def	1440 s br
1430 s	1415	CH$_2$ scissor	1415 s br
1375 s	1373	CH$_3$ sym def	1415 s br

[a] These data were taken from Kaesz and Stone (*46*). Above 2800 cm^{-1} we accepted their assignment; below 2800 cm^{-1} we used the more explicit assignments of Green (*47*). Only those bands were listed above 1300 cm^{-1} that were listed as m or stronger.

[b] The abbreviation br stands for broad; all other symbols are standard.

[c] Bands in the CH stretching region are not resolved.

a band also occurs here for C$_2$H$_4$–D$_2$, we must conclude that this broad band also includes the scissor-like vibration of the CH$_2$ group attached to the zinc as in diethyl zinc. This leaves the band at 1440 cm^{-1} for the antisymmetric methyl deformation vibration. This band would be expected to persist and does for the species C$_2$H$_4$–D$_2$ since monodeuteration effectively removes the degeneracy of the antisymmetric deformation of methyl and gives rise to one band nearly unchanged in position and another at much lower frequencies (*48*). On the other hand, one would expect that for C$_2$H$_4$–D$_2$ the band due to the symmetric methyl deformation at about 1415 cm^{-1} would be absent, and, hence, the overtone at 2812 cm^{-1} would also be missing; this seems to be the case. Further support for the presence of a methyl group in X is given by the spectrum from C$_2$D$_4$–H$_2$. The band at 1289 cm^{-1} due to the deformation of a lone C—H group is in a range attributed to the C—H bending vibration in CD$_3$–CD$_2$H (*49*).

Some further discussion of the validity of these assignments is given in reference (*13*). Alternatives were considered and the only possible (not probable) alternative seemed to be a polymeric paraffinic species formed only when ethylene was in the presence of hydrogen. A careful search in the effluent from reaction of hydrogen with X failed to reveal the presence of any compounds other than ethane. Accordingly, we conclude that X is an intermediate of the form S—CH$_2$—CH$_3$ and the occurrence of the band at 2812 cm^{-1}, assigned to an overtone, is strong evidence that S is a zinc ion.

IV. Reactions with Propylene

A. Proplyene on Zinc Oxide. The π-Allyl

Propylene adsorbed on zinc oxide is, in part, weakly adsorbed and, in part, strongly adsorbed (12). Exposure to propylene followed by brief evacuation at room temperature leaves 0.37 cm^3/gm on the catalyst, and this propylene is not removed even if evacuation is continued for several hours. Degassing at 70°C for 1 hr removes roughly half of this propylene and all of it can be recovered by degassing 1.5 hr at 125°C. Analysis of the desorbed propylene reveals that it is 99.9% propylene and 0.1% propane with no dimeric products; hence, adsorption occurs without signicant chemical change. Thus, propylene is much more strongly held than ethylene and, accordingly, may be bound in a different manner. Details of the behavior of the strongly held propylene suggest it is a single species.

The spectrum of chemisorbed propylene in the CH and O—H stretching region is shown in Fig. 15. The band at 3593 cm^{-1} is clearly due to an OH frequency; hence, dissociation accompanies propylene adsorption. Careful scrutiny of the region from 1500 to 2000 cm^{-1} reveals no band assignable to a ZnH band. Since the presence of adsorbed propylene has been found to block out the infrared active hydrogen chemisorption on the ZnO pair sites, we may assume that propylene adsorption occurs as follows:

$$C_3H_6 + O\text{---}Zn\text{---}O \rightarrow \overset{\overset{\displaystyle C_3H_5}{|}}{O}\text{---}\overset{\overset{\displaystyle H}{|}}{Zn}\text{---}O.$$

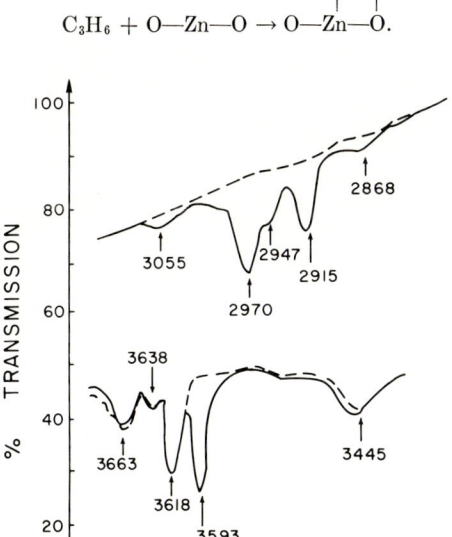

Fig. 15. Spectrum of chemisorbed propylene (CH$_3$—CH=CH$_2$).

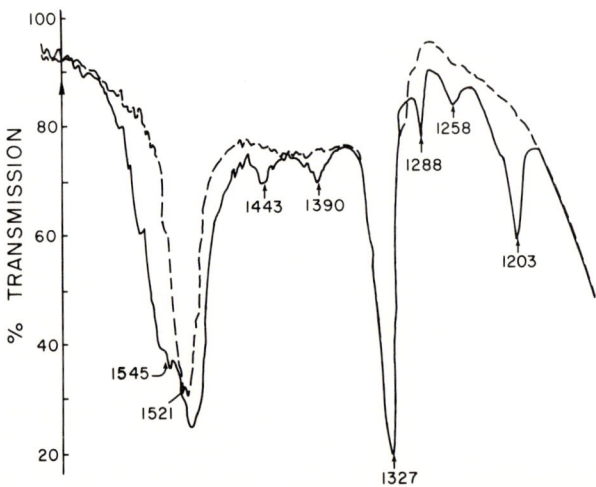

Fig. 16. Spectrum of chemisorbed propylene (CH_3—CH=CH_2).

Thus, the hydrocarbon fragment blocks out the zinc half of the active site, and the ZnH band is not observed.

Five bands are observed in the region near 3000 cm^{-1} corresponding to C—H stretching vibrations. (The band at 2947 cm^{-1}, which appears as a shoulder in Fig. 15, is seen as a separate peak when the spectrum is observed on an expanded transmission scale.) Detailed identification of these bands will be deferred, but the weak band at 3055 cm^{-1} suggests that the hydrocarbon fragment is olefinic.

Figure 16 shows the spectrum of chemisorbed propylene in the C—H deformation region. Bands between 1450 and 1200 cm^{-1} can reasonably be assigned to C—H deformation vibrations, but there are one or more bands almost on top of the background band at 1521 cm^{-1}. This band could result from broadening and enhancement of the background band, but it is most likely due to the hydrocarbon. A band centered at 1545 cm^{-1} corresponds to a carbon—carbon stretch with appreciable olefinic character. Coupled with the evidence of olefinic character in the C—H stretching region this assignment seems reasonable. In the gas phase, the double-bond stretch for propylene occurs at 1652 cm^{-1} (50); hence, interaction of the double bond with the surface has shifted the stretching frequency by 107 cm^{-1}.

If the spectrum of adsorbed propylene is observed in the presence of gaseous propylene, additional bands to those shown in Figs. 15 and 16 are observed. These additional bands are due to a more weakly bound form of propylene which is readily removed by a brief evacuation. The salient feature of the spectrum of the weakly bound propylene is a band of 1620 cm^{-1}. This band, clearly in the double-bond stretching region, is about 30 cm^{-1} less than that for gaseous propylene. Since a similar shift is observed in the spectrum of chemisorbed ethylene, it seems reasonable to conclude that the bonding of this weakly held propylene is the analog of that found for chemisorbed ethylene, i.e., it is a surface π-complex.

Figure 17 shows the spectrum for chemisorbed C_3D_6 in the OH stretching region ($\sim$3500 cm^{-1}), the OD stretching region ($\sim$2600 cm^{-1}), and the C—D stretching region. Near 3500 cm^{-1} we find only slight changes in the background hydroxyl bands which may be perturbations due to the adsorbed olefin. At 2653 cm^{-1}, however, we see a strong band (with some structure) due to an OD formed by the adsorption of propylene. Thus, these observations support the view that propylene adsorbs dissociatively.

Figure 18 shows the spectrum of C_3D_6 in the CH deformation region. We would expect the normal isotope effect to shift C—D deformations about 400 cm^{-1}, that is, completely out of this region. Thus, the observed bands are due to C—C vibrations. The band at 1473 cm^{-1} (with a shoulder at 1460 cm^{-1}) can only correspond to the 1545 cm^{-1} band in C_3H_6. The

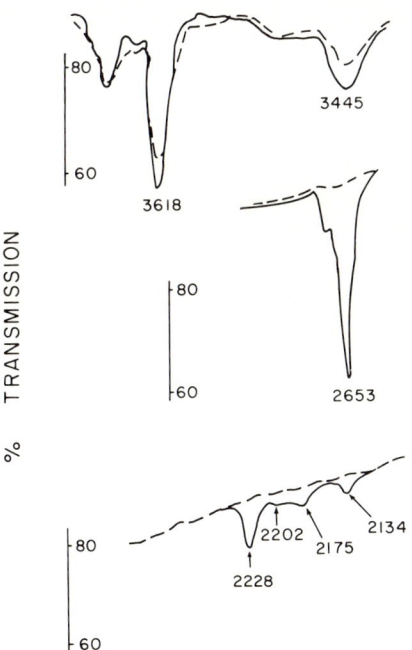

FIG. 17. Spectrum of chemisorbed propylene (CD_3—CD=CD_2).

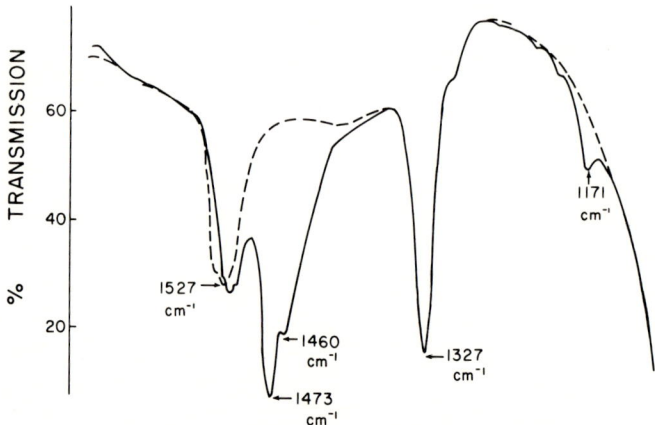

FIG. 18. Spectrum of chemisorbed propylene (CD_3—CD=CD_2).

isotope shift, 72 cm^{-1}, is very nearly the same as the isotope shift of the C=C frequency (70 cm^{-1}) for the gas phase spectra (*50*); hence, the position of this band substantiates our previous conclusion that the 1545 cm^{-1} band is the double-bond frequency shifted by adsorption.

The spectra of C_3H_6 and C_3D_6 show that chemisorption of propylene is dissociative, but they fail to identify which carbon–hydrogen bond is broken on adsorption. To this end the spectra of a number of deuterium-labeled propylenes were studied and compared. These results are summarized in abbreviated form in Table VI, which specifies the hydrogen fragment formed on adsorption; the fragment was identified as an OH if a band appeared near 3593 cm^{-1} or as an OD if a band appeared near 2653 cm^{-1}. In those cases where the spectrum changed with time the summary

TABLE VI

Spectrum of Chemisorbed Propylenes

	Compound	Surface hydrogen fragment	Spectrum
I	CH_3—CH=CH_2	O—H	Stable
II	CD_3CD=CD_2	O—D	Stable
III	CH_3CD=CH_2	O—H	Stable
IV	CH_3—CH=CD_2	O—H	Changes
V	CD_3—CH=CH_2	O—D	Changes
VI	CD_3—CH=CD_2	O—D	Stable

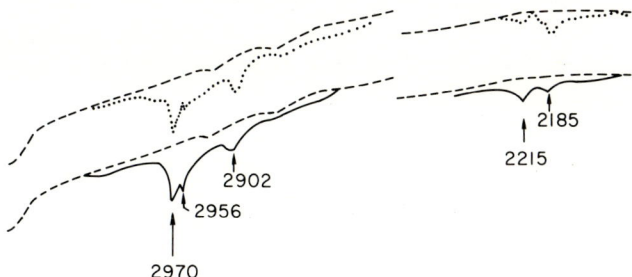

FIG. 19. Spectrum of chemisorbed propylene (CD_3—CH=CH_2 and CH_3—CH=CD_2): dotted line, chemisorbed CD_3—CH=CH_2 on zinc oxide; solid line, chemisorbed CH_3—CH=CD_2 on zinc oxide.

in Table VI applies to the initial spectrum. These results show that adsorption of propylene occurs by cleavage of a methyl carbon—hydrogen bond to form an allylic species.

Figures 19 and 20 show the initial spectra of chemisorbed CH_3—CH=CD_2 (IV) and CD_3—CH=CH_2 (V). The wavelength scale is the same but, for clarity, the transmission scales for the two spectra have been shifted. Although IV yields an OH fragment on adsorption and V yields

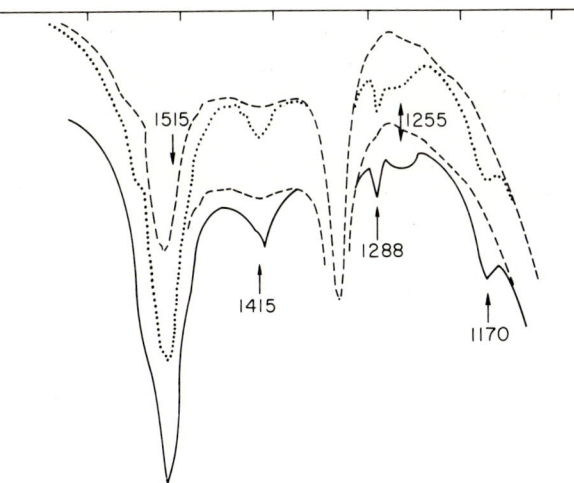

FIG. 20. Spectrum of chemisorbed propylene (CD_3—CH=CH_2 and CH_3—CH=CD_2): dotted line, chemisorbed CD_3—CH=CH_2 on zinc oxide; solid line, chemisorbed CH_3—CH=CD_2 on zinc oxide.

an OD fragment on adsorption, *the initial spectra of the surface hydrocarbons formed by these two compounds is the same within experimental error.* The coincidence of the initial hydrocarbon spectra for adsorbed IV and V strongly suggest that propylene adsorbs to form a symmetric allylic species:

$$CD_3—CH=CH_2 \xrightarrow{-D} CD_2\text{===}CH\text{===}CH_2 \xleftarrow{-H} CD_2=CH—CH_3$$

It is conceivable that IV and V both yield two rapidly equilibrating species, i.e.,

$$\underset{*}{CD_2}=CH—\underset{*}{CH_2} \rightleftarrows \underset{*}{CD_2}—CH=\underset{*}{CH_2}$$

and that Figs. 19 and 20 represent the composite spectrum, but the spectrum itself makes this interpretation unlikely. The two equilibrating species would give rise to six C—H and four C—D stretches; nature is not so vindictive that it would combine overlap and intensity of these ten bands to yield the three C—H and two C—D bonds expected for the symmetric species.

When propylene chemisorbs to form this symmetric allylic species, the double-bond frequency occurs at 1545 cm^{-1}, a value 107 cm^{-1} lower than that found for gaseous propylene; hence, by the usual criteria, the propylene is π-bonded to the surface. For such a surface π-allyl there should be gross similarities to known π-allyl complexes of transition metals. Data for allyl complexes of manganese carbonyls (*51*) show that for the σ-allyl species the double-bond frequency occurs at about 1620 cm^{-1}; formation of the π-allyl species causes a much larger double-bond frequency shift to 1505 cm^{-1}. The shift observed for adsorbed propylene is far too large to involve a simple σ-complex, but is somewhat less than that observed for transition metal π-allyls. Since simple π-complexes show a correlation of bond strength to double-bond frequency shift, it seems reasonable to suppose that the smaller shift observed for surface π-allyls implies a weaker bonding than that found for transition metal complexes.

Bonding of a π-allyl ligand to the metal can be viewed as follows:

$$H_2C\underset{M}{\overset{\overset{\displaystyle H}{C}}{\diagup \diagdown}}CH_2 \quad \longleftrightarrow \quad H_2C\underset{M}{\overset{\overset{\displaystyle H}{C}}{\diagup \diagdown}}CH_2$$

wherein the plane of the three carbons is nearly perpendicular to the metal-allyl bond axis with a slight tilt so that the end carbon atoms are closer to the metal than the central carbon atom. Although x-ray studies suggest the two carbon—carbon bonds are not quite the same length (*52*), the validity of this conclusion has been questioned (*53*). Thus, to a first approximation, the allyl ligand in π-allyls has the form:

$$\mathrm{H_2C \cdots CH \cdots CH_2}$$

(with H atoms shown on each carbon)

We believe the surface species has this form with the zinc atom of the active site playing the role of the metal. Note that this structure is symmetric as required by the IR data for adsorbed CD_3—CH=CH_2 and CH_3—CH=CD_2.

The infrared spectra of a number of π-allyl complexes have been determined (*54*), but the band positions due to the π-allyl ligand shift considerably from one metal to another; hence, further comparison of these spectra to that for adsorbed species is of little value. It is more productive at this point to proceed *ab initio*. To do so it is assumed, following Fritz (*55*), that the bonding to the surface determines the geometry, force constants, etc., of the absorbed hydrocarbon fragment; but the hydrocarbon fragment behaves vibrationally as an independent entity. This is valid because the low frequency allyl-surface vibrations couple only loosely with the hydrocarbon frequencies. This view is similar to the view used to justify the "constancy" of group frequencies (*56*); now, however, we are taking the whole adsorbed species as a complex "group frequency." On this basis, the vibrations of the surface π-allyl are those expected for a molecule with C_{2v} symmetry.

For a species with the formula C_3H_5 and C_{2v} symmetry we can expect eighteen vibrations. Of these, five are coupled C—H stretching vibrations, ten are coupled C—H bending vibrations, and three are skeletal vibrations. With the help of the six labeled propylenes, we can make a rather firm assignment for many of these vibrations. For example, the carbon—hydrogen stretch that appears for adsorbed propylene but is absent in adsorbed 2-deuteropropylene is that for the center C—H stretch. Verification of this is given by the spectrum of 1,1,3,3,3-pentadeuteropropylene which shows only one C—H stretch at a position corresponding to the missing band in adsorbed 2-deuteropropylene. Via similar procedures the eleven bands observed for propylene can be clearly assigned to a normal mode expected for a molecule with the assumed structure. The positions of the remaining modes not observed can be estimated and it is found that these bands probably fall at frequencies below the cutoff for zinc oxide, where they would be unobservable. We can further check the validity of this assignment by application of the theoretical rules giving the changes in frequency expected for isotopic substitution(*57*). These rules hold with considerable accuracy only for the correct assignment of the normal modes. When these rules were applied to the observed bands with the assignment cited above, the rules held with the same accuracy as that found for simple gas phase molecules (57). Thus, the consistency of these analyses offers strong support for the assumed structure.[3]

Further details on the structure of the adsorbed π-allyl are supplied by analysis of the carbon—carbon vibrations. We shall assume, as is often done (58), that the hydrogen atoms move in concert with the carbons in these vibrations so that the molecule is essentially a symmetric bent triatomic of the form:

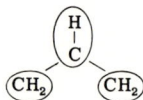

Such a molecule, with ankylosed CH bonds, will have three frequencies: symmetric stretching ν_1', bending ν_2', and asymmetric stretching ν_3'. Generally, the orders of these frequencies will be $\nu_3' > \nu_1' > \nu_2'$. These frequencies, in the valence force approximation, depend on the bond angle, the stretching force constant, and the bending force constant (58). The form of the equations is such that the frequencies ν_3' and ν_1' are sensitive to the angle and stretching force constant but not sensitive to the bending force constant. On the basis of the structural assignment, the most reasonable bond angle is 120°. We shall further assume ν_2' is about the same as that for propylene, 417 cm^{-1}. Now we can fix the stretching force constant by assigning the 1545 cm^{-1} band for adsorbed propylene to ν_3'. This permits us to compute *with no further assumptions* the positions of ν_1' for adsorbed propylene and both ν_1' and ν_3' for all other symmetric deuterium isomers. These calculations and their comparison to experiment are shown in Table VII. The agreement is clearly very good.

In order to test the sensitivity of the model, calculations for adsorbed proplyene were repeated with the assumed value of ν_2' 517 cm^{-1}. This increase of 24% in the value of ν_2' caused ν_1' to increase to 1220 cm^{-1}, a change of 1% in frequency. By way of contrast, if the angle was changed by 4%, the change in the calculated value of ν_1' was 50 cm^{-1} (4%). Thus, the calculations depend little on the assumed bending frequency but provide a sensitive criteria for the angle. The value of 120°, the ideal value for a π-allyl, provides the best fit to the data. The value of the stretching

---

[3] Details of the analysis described herein can be found in Dent and Kokes (12). This very abbreviated summary is meant to be indicative of the support available for the structural assignment.

TABLE VII

Calculated versus Observed Carbon—Carbon Frequencies

Molecule	Vibration[a]	Experimental	Calculated[b]	% Error
C_3H_6	ν_3'	1545	—	—
	ν_1'	1203	1208	0.4
C_3H_5D	ν_3'	1525	1517	0.5
	ν_1'	1203	1208	0.9
C_3D_5H	ν_3'	1502	1504	0.1
	ν_1'	1159	1162	0.3
C_3D_6	ν_3'	1473	1470	0.2
	ν_1'	1170	1147	2.0

[a] The numbering of frequencies is that of Herzberg (*40*) for a triatomic with C_{2v} symmetry.
[b] The ν_3' frequency for adsorbed C_3H_6 used to compute the stretching force constant with the assumption that the bond angle was 120° and the C—C—C bending frequency was the same as that for propylene.

force constant calculated by this procedure is 7.54×10^5 dyn/cm. This is about midway between typical values (*56*) for single and double bonds, 4.5 and 9.6×10^5 dyn/cm, respectively. The value found, close to that found for benzene, is consistent with the expected half double-bond character for a π-allyl species.

It seems quite clear that propylene forms a π-allyl species when it adsorbs on zinc oxide. The interpretation of its interaction with the proposed active sites, however, has two seemingly major flaws. First, the amount of proplyene held on active sites is threefold greater than the amount of type I hydrogen adsorption. Second, it seems unlikely that the bulky allyl group can approach the sequestered zinc close enough to form a strong bond. Neither of these apparent flaws is insuperable. The larger adsorptive capacity for propylene may be a reflection of the heterogeneity of the active sites. Bonding to the sequestered zinc may be facilitated by movement of the zinc out of the trigonal hole; such motion is the analog of a change in coordination number for a transition metal complex. Accordingly, we shall retain this picture of adsorbed propylene in the sections that follow.

B. Reactions of π-Allyls

Presumably, adsorption of propylene occurs as follows:

$$CH_3-CH=CH_2 \; + \; -O-Zn-O- \; \longrightarrow \; \underset{-O-Zn-O-}{H_2C\overset{\overset{\displaystyle H}{C}}{\diagup\diagdown}CH_2} \quad \text{H} \tag{16}$$

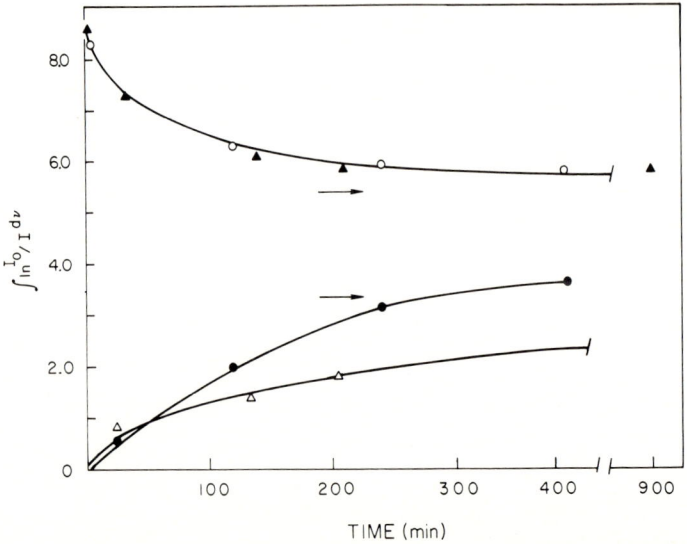

Fig. 21. Integrated intensity OH and OD bands versus time for adsorbed labeled propylene. CH_3—CH=CD_2: ○, OH; ●, OD. CD_3—CH=CH_2: △, OH; ▲, OD. The integrated intensity for OD was multiplied by 1.35, the isotopic shift, in an attempt to correct for expected differences in the integrated absorption coefficient.

Such a mode of adsorption provides a mechanism for double-bond isomerization which we can follow by use of labeled propylenes (13). For example, on the basis of the foregoing, we expect the following sequence for adsorbed CH_3—CH=CD_2:

$$CH_3-CH=CD_2 + \underset{|}{-}O-\underset{|}{Zn}-\underset{|}{O}- \rightleftarrows \underset{-O}{\overset{H_2C\overset{\overset{H}{|}}{\underset{\cdot\cdot}{C}}CD_2}{|}}\underset{Zn}{}\underset{-O-}{|} \qquad (17)$$

$$\underset{-O}{\overset{\overset{H}{|}H_2C\overset{\cdot\cdot}{\underset{\cdot\cdot}{C}}CD_2}{|}}\underset{Zn}{}\underset{-O-}{|} \rightleftarrows \underset{-O}{|}\underset{Zn}{\overset{H_2C\overset{\overset{H}{|}}{\underset{\cdot\cdot}{C}}CD_2\overset{H}{|}}{}}\underset{O}{|} \qquad (18)$$

$$\underset{-O}{|}\underset{Zn}{\overset{H_2C\overset{\overset{H}{|}}{\underset{\cdot\cdot}{C}}CD_2\overset{H}{|}}{}}\underset{O-}{|} \rightleftarrows \underset{-O}{}\underset{Zn}{\overset{H_2C\overset{\overset{H}{|}}{\underset{\cdot}{C}}CD_2H}{}}\underset{-O-}{|} \qquad (19)$$

Step (18) in the above is the analog of step (8), which is required for H_2—D_2 equilibration; it is a necessary step if we view the π-allyl as an immobile species on the surface. The products of step (19) can be viewed as propylene in the form of a loosely held π-complex which on desorption yields isomerized propylene. Readsorption of the isomerized propylene or further reaction of the π-complex would yield surface OD groups. When equilibrium is achieved, the concentration of surface OD groups should equal 40% of the initial concentration of OH groups. Figure 21 shows a plot versus time of the intensity (multiplied by a scale factor to yield concentration) of the surface OH and OD. The expected equilibrium points are indicated by arrows. Corresponding data for CD_3—CH=CH_2 are also shown. Except for the OH species from CD_3—CH=CH_2, which is a relatively weak band on the side of a surface hydroxyl, the curves approach the expected value.

Table VIII shows the results of a CH_3—CH=CD_2 isomerization run. Product analysis was carried out by IR and mass spectrographic analysis.

TABLE VIII

Product Analysis of Isomerized CH_3—CH=CD_2 Infrared Analysis

Compound	Gaseous product		Surface product	
	Observed %[a]	Equilibrium %[b]	Observed %[a]	Equilibrium %[b]
A: CD_2=CH—$CH_{3-x}D_x$	30 ± 2	11	8 ± 1	12
B: CH_2=CH—$CH_{3-x}D_x$	59 ± 3	31	64 ± 2	40
C: CHD=CH—$CH_{3-x}D_x$	10 ± 4	58	28 ± 2	48
	Mass spectroscopic analysis			
C_3H_6	0.6	8	8.9	10.8
C_3H_5D	9.9	26	27.5	30.2
$C_3H_4D_2$	79.0	35	41.6	34.0
$C_3H_3D_3$	10.0	23	17.8	19.0
$C_3H_2D_4$	0.6	8	4.1	5.4
Av # D/molecule	1.99		1.80	

[a] The products were collected after 6 hr reaction at room temperature. The total pressure was 300 mm Hg. Gaseous products were obtained by condensing the gas phase in a liquid nitrogen trap. Surface product was the strongly adsorbed propylene obtained by degassing at 125°C.

[b] These values were computed from the experimentally determined deuterium content with the assumption that the distribution of deuterium between the end carbons was random.

TABLE IX
Products from Addition of Deuterium to Propylene[a,b]

	Hydrogen	Propane	Propylene	Propylene[c]
% d_0	1.9	12.2 (16)	84.4 (84)	49.9 (44)
% d_1	6.1	20.9 (33)	14.1 (15)	28.9 (39)
% d_2	92.0	55.6 (30)	1.6 (1)	15.5 (14)
% d_3		11.3 (16)	—	5.7 (2.5)
% d_4		—	—	—
Av # D/molecule	1.90	1.66	0.173	0.770

[a] A 90-min run at 25°C with $C_3H_6:D_2 = 122$ mm: 608 mm gave 18.5% conversion.

[b] Values in parentheses are values expected for random distribution of deuterium with the stated overall deuterium content.

[c] This is the chemisorbed propylene after the run which is removed from the catalyst by degassing 1 hr at 125°C.

Consider the first three components listed for the gaseous product. *Random statistics favor C as a product; despite this, B is the major product.* This suggests that equilibrium is achieved via the pathway, $A \rightleftarrows B \rightleftarrows C$. This is the expected pathway if isomerization occurs via the 1,3-hydrogen transfer depicted in Eqs. (17)–(19); on this basis, $CH_2\!\!=\!\!CH\!\!-\!\!CD_2H$, corresponding to B, should be the initial product. The "surface product," which had more opportunity for reaction, is, as expected, somewhat closer to the statistical equilibrium.

If the hydrogen from the adsorbed propylene is mobile, we would expect not only intramolecular hydrogen migration but also intermolecular hydrogen migration. This means that in the reaction of the dideuteropropylene randomization of deuteriums should occur, i.e., $C_3H_4D_2$ should react to form products of the type C_3H_5D, $C_3H_3D_3$, etc. Data in Table VIII shows this reaction does occur, but it is slower than the intramolecular hydrogen migration. As before, the surface product more nearly attains the statistical equilibrium.

The proposed mechanism of the exchange suggests that the intermolecular exchange should not involve the center CH bond. On the other hand, if the exchange were random among the six hydrogen positions, one would expect about one out of three molecules to have a deuterium on the central carbon at equilibrium. Analysis by IR shows no evidence for a deuterium at this position even for the nearly equilibrated surface products. Thus, this result also conforms to expectations based on the proposed mechanism.

If the π-allyl species is the reactive species in hydrogenation, one would expect behavior dramatically different from that for ethylene, which adsorbs as a π-complex. For the π-allyl, exchange of hydrocarbon with deuterium is possible via the hydrogen formed from dissociation of propylene. Accordingly, we would expect addition of deuterium to propylene to result in the appearance of HD in the gas phase, exchange of deuterium with the propylene, and an isotopic smear in the product propane. None of these effects shows up in the addition of deuterium to ethylene. Table IX shows typical data for addition of deuterium to propylene. The expected reactions are quite evident. Moreover, as expected, the distribution of deuterium in the propylene corresponds roughly to what one would find for one-at-a-time exchange of five of the hydrogens.

The reactions described above are those expected if a π-allyl species is functioning as an intermediate. Thus, it appears that not only is the π-allyl formed on zinc oxide but it is an important intermediate in the reactions of propylene.

V. Reactions of Butene

A. STEREOCHEMICAL CONSIDERATIONS

Isomerization of butene via a π-allyl species introduces an added dimension to the stereochemistry. The π-allyl species from propylene is presumed to be planar with its plane approximately parallel to the surface. Since it is attached to the electropositive zinc, it may have considerable carbanion character. A corresponding structure for adsorbed butene would lead to two isomeric forms, viz:

$$\underset{\text{"syn"}}{\overset{}{\underset{}{\text{H}\diagdown\underset{H}{\overset{H}{\underset{|}{C}}}\text{::}\underset{H}{\overset{|}{C}}\text{::}C\diagup\text{CH}_3}}} \quad \text{and} \quad \underset{\text{"anti"}}{\overset{}{\underset{}{\text{H}\diagdown\underset{H}{\overset{H}{\underset{|}{C}}}\text{::}\underset{H}{\overset{|}{C}}\text{::}C\diagup\text{H}\atop\text{CH}_3}}}$$

Cis-butene should lead initially to the anti form; trans-butene should lead initially to the syn form and 1-butene should give rise initially to both. The equilibrium distribution of syn and anti forms usually differs greatly from the equilibrium distribution of cis- and trans-butene; for cobalt complexes (59, 60) the syn form, precursor of trans-butene, is by far the most stable. By way of contrast for the corresponding carbanion, the cis anion seems by far the more stable. This preference for the cis carbanion is presumed to be the source of the high initial cis-to-trans ratio in the initial products of base catalyzed isomerization. In the base catalyzed isomerization of more complex cis-olefins (cis-S-methyl-stilbene), the ions corresponding to syn and anti are not interconvertible and cis–trans isomerization involves the α-olefin as an intermediate *(16)*; for the simpler *cis*-olefins (*cis*-butene) heterogeneous base catalyzed cis–trans isomerization is direct and the α-olefin is not an intermediate *(17)*. Isomerization via π-allyl ligands of transition metal complexes may be important in some cases *(61)*, but it has not been established if the syn–anti conversion is direct. The observation *(62, 63)* that many π-allyls are "dynamic" and undergo rapid σ-to-π-allyl interconversion suggests the possibility of a mechanism for cis-to-trans conversion which does not involve butene-1 as an intermediate:

$$\underset{H}{\overset{H_3C}{>}}C=C\underset{H}{\overset{CH_3}{<}} \underset{+H}{\overset{-H}{\rightleftarrows}} \underset{H}{\overset{H_2C}{>}}\overset{*}{C}\cdots C\underset{H}{\overset{CH_3}{<}}$$

$$\updownarrow$$

$$\underset{H_2C^*}{\overset{H}{>}}C\cdots C\underset{H}{\overset{CH_3}{<}} \rightleftarrows \underset{H}{\overset{H_2C}{>}}\overset{*}{C}\cdots C\underset{H}{\overset{CH_3}{<}}$$

$$+H \updownarrow -H$$

$$\underset{H_3C}{\overset{H}{>}}C=C\underset{H}{\overset{CH_3}{<}}$$

where * represents the complexed atom. Thus, in the isomerization of *cis*-butene over oxide catalysts wherein surface π-allyls are intermediates, one can expect either the sequential pathway,

cis-butene ⇌ 1-butene ⇌ *trans*-butene,

or the simultaneous conversion of *cis*-butene to 1- and *trans*-butene.

B. Butene on Zinc Oxide

Figure 22 shows the spectrum in the OH region for zinc oxide after admission of butene-1 at a pressure of about 8 mm *(14)*. Spectrum (a), taken after 8 min exposures, shows two features: (1) the strong surface hydroxyl band at 3615 cm^{-1} is shifted about 5 cm^{-1} to lower frequencies; (2) a new band appears at 3587 cm^{-1}. This new band, clearly an OH, appears to arise from dissociation of the adsorbed butene. Spectrum (b) shows the same region after exposure to the gas phase for 1 hr. It is clear that the OH band formed from butene grows with time; detailed studies, however, reveal that there is little change after the first 20 min. Spectrum (c) was taken after 20 min evacuation. Two features are evident: (1) in the absence of the gas phase the hydroxyl band of the zinc oxide has shifted back to its previous position; (2) the OH band formed from butene is reduced somewhat in intensity. Spectrum (d) was taken after degassing for 90 min;

further reduction in the OH band is evident. When the sample was degassed a total of 16 hr, this band intensity decreased by roughly an additional 20%.

These results are similar to those with propylene insofar as they indicate dissociative adsorption of the olefin. The hydrogen that yields the hydroxyl has not been identified but it seems reasonable to suppose that, once again, the allylic hydrogen is lost. Results with butene, however, do differ from those with propylene in two respects: first, the dissociation (as evidenced by the OH band) is rapid but not instantaneous as found for propylene; second, dissociatively adsorbed butene is more easily removed by room temperature evacuation than dissociatively adsorbed propylene. These facts suggests that steric effects are present; hence, the kinetic behavior of these two species may be quite different.

Figure 23a shows the C—H region of the spectrum about 10 min (solid line) and 60 min (dotted line) after admission of 8 mm of butene-1 to a sample of zinc oxide. These spectra, which are primarily due to physically adsorbed and gas-phase butenes, show sizeable changes as a function of time. The region above 3000 cm^{-1} is particularly clear cut. Initially, a band is observed only at 3082 cm^{-1}; this corresponds closely to the 3086 cm^{-1} band for gaseous butene-1 (*64*). After 1 hr the band at 3082 cm^{-1} is gone and two new bands (above 3000 cm^{-1}) have appeared at 3035 and 3018 cm^{-1} which correspond within the experimental uncertainty to the expected bonds for *cis*-butene (*64*) (3030 cm^{-1}) and *trans*-butene (*64*) (3021 cm^{-1}). Other bands are consistent with these changes. Thus, it is evident that double-bond isomerization has occurred.

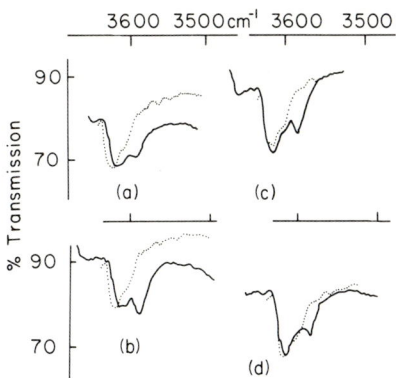

Fig. 22. Spectra of zinc oxide in the OH region in butene-1: (a) 8 min after exposure to butene-1 at 8 mm; (b) 60 min after exposure to butene-1 at 8 mm; (c) 70 min exposure to butene-1 followed by 20 min degassing; (d) shows (d) after 90 min degassing.

Figure 23 b and c, show the spectrum in the double-bond region after 20 min and 70 min exposure to the gas phase, respectively. Bands due to adsorbed species dominate this region of the spectrum. The initial spectrum (b) shows at least four bands at about 1650, 1630, 1610, and 1550–1570 cm^{-1}. [Later work (65) shows the broad band in the 1550–1570 cm^{-1} region is actually two bands.] Of these only the band at about 1650 cm^{-1} can be assigned to the gas phase and the most reasonable assignment is to gaseous butene-1 (1645 cm^{-1}) (64). In the spectrum after 70 min exposure the bands at about 1650 and 1610 cm^{-1} are no longer prominent and no bands assignable to gaseous species are observed. This is not unexpected if the composition of the gas phase has approached equilibrium. At equilibrium the dominant species is *trans*-butene (77%) (66) which shows no IR active C=C band and the *cis*-butene (20% at equilibrium) (66) has a much weaker C=C band than the butene-1 (67, 68) (3% at equilibrium) (66).

Firm assignments for these C=C bands require more detailed experiments but a tentative assignment can be made. The bands at 1550–1570 cm^{-1} are probably due to a π-allyl species; the shift from the double-bond region for butenes is about 100 cm^{-1} compared to the shift of 107 cm^{-1} observed for the π-allyl formed from propylene, but the butene is less firmly held. With propylene we observed a "π-complex" in which the shift in C=C stretch was about 30 cm^{-1}. We believe the band at 1610 cm^{-1}

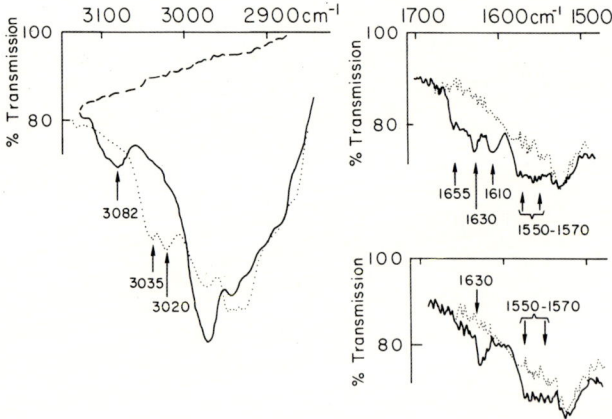

FIG. 23. Spectra of zinc oxide in the presence of butene-1: (a) solid line, after 10 min exposure to butene-1 (8 mm); dotted line, after 60 min exposure to butene-1 (8 mm); (b) After 20 min exposure to butene-1 (8 mm); (c) After 70 min exposure to butene-1 (8 mm). Arrows mark positions of peaks referred to in text.

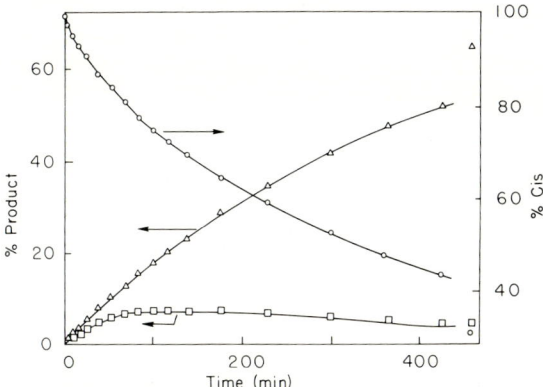

FIG. 24. cis-Butene isomerization over zinc oxide: ○, cis-butene (right-hand ordinate); △, trans-butene; □, butene-1.

seen in the initial spectrum represents a butene-1 "π-complex"; disappearance of this band in time is consistent with this assignment. Similarly, the band at 1630 cm^{-1}, which persists and may intensify after 70 min, may be tentatively assigned to π-complexed butene-2 (cis and/or trans).

Some support for the above assignments is offered by the behavior of the bands at 1630 and 1550–1570 cm^{-1} on degassing. Both bands are still present after a brief degassing. In time, however, both bands decrease. The band assigned to the π-complex decreases on degassing much faster than the band assigned to the more strongly held π-allyl species. After 90 min degassing, the π-complex band is gone but the π-allyl band persists and is still evident at about half its initial intensity after overnight degassing. This decrease in the 1550–1570 cm^{-1} band on degassing is qualitatively similar to that observed for the OH band formed on adsorption; hence, it appears that both this band and the OH band stem from the same species, a π-allyl species.

C. Isomerization of cis-Butene

Figure 24 shows the cis-butene isomerization over zinc oxide as a function of time at room temperature (14). On a per unit area basis the initial rate at room temperature is 4×10^{10} molecules/sec cm^2, a rate roughly one third that reported for alumina (69). Since the activation energy for alumina is less than that found for zinc oxide, this means that zinc oxide is comparable (on a per unit area basis) to alumina as an isomerization catalyst at slightly higher temperatures.

Extrapolation of the rate data in Fig. 24 to zero conversion shows that the initial ratio of butene-1 to *trans*-butene formation is about unity. Thus, butene-1 is not an intermediate in the cis–trans isomerization and direct cis–trans isomerization occurs. Similar results are found for the heterogeneous base catalyzed isomerization over sodium on alumina (*17*).

Results with butene are not as extensive as those with propylene. Nevertheless, on the basis of the ground work laid by the more extensive propylene studies, we are able to apply similar criteria to the more limited data for butene and conclude that a π-allyl species forms. Some preliminary studies suggest that two π-allyl species form from 1-butene (*65*), corresponding to the syn and anti forms. The results for propylene, the fact that π-allyl species form from butene, and the fact that zinc oxide is an effective catalyst for butene isomerization strongly suggest that these π-allyls are intermediates in the isomerization reaction.

VI. Reactions of Acetylenes

Acetylenes on Zinc Oxide

The hydrogen on acetylene is far more acidic than the allyl hydrogen of olefins (*70*). If zinc oxide is sufficiently basic to abstract a proton from propylene to form an allyl species with considerable carbanion character, one would expect it to react even more readily with acetylene to abstract a proton and form an acetylide. Adsorption of acetylene on zinc oxide is rapid and irreversible (*65*). The species formed is not removable even by degassing at 300°C. Infrared data suggests that a surface reaction occurs at a rapid rate in the presence of gaseous acetylene, which we have tentatively ascribed to polymerization. If only enough acetylene is added to the system to form a partial monolayer, the surface reaction is considerably slower. Under these conditions if the OH region is scanned shortly after admission of the acetylene, a new OH band appears. This, of course, suggests that adsorption of acetylene occurs with the loss of a proton to the oxygen part of the active site and the formation of an acetylide bound to the zinc half of the active site. Data for perdeuteroacetylene are consistent with this view. Unfortunately, the occurrence of the surface reaction makes it difficult to make a more complete analysis of spectra.

The infrared spectrum in the OH region for the adsorption of methyl acetylene is completely analogous to that for acetylene. The spectrum of CH_3—C≡C—D, however, introduces some new features (*65*). Initially, an OH band appears; in time, however, an OD band appears. By analogy with base catalyzed reactions of acetylene (*71*) we believe the methyl hydrogen is lost to form a propargyl ion. The propargyl ion has two contributing forms to a resonance hybrid:

$$\bar{C}H_2-C\equiv C-H \longleftrightarrow CH_2=C=\bar{C}-H \quad \text{or} \quad [CH_2\mathrel{\overset{...}{=}}C\equiv CH]^-.$$

This ion plays the role in base catalyzed reactions of acetylenes that allyl carbonions play in base catalyzed reactions of olefins. For a base catalyzed isomerization of methyl acetylene the following is consistent with the accepted pathways:

$$CH_3-C\equiv C-D + Zn-O \rightleftarrows \begin{matrix}[CH_2\mathrel{\overset{...}{=}}C\equiv CD]^- \\ Zn-O\end{matrix}\diagup H^+$$

$$\updownarrow$$

$$\begin{matrix}[CH_2\mathrel{\overset{...}{=}}C\equiv CH]^- \\ Zn-O\end{matrix}\diagup D^+ \rightleftarrows \begin{matrix}CH_2=C=CDH \\ Zn-O\end{matrix}$$

The initial appearance of an OH band followed by the later appearance of an OD band agrees with this scheme. Also in agreement with this scheme, we find that adsorbed allene yields the same spectrum as methyl acetylene (65).

Studies of the interaction of acetylenes with zinc oxide clearly provide a very interesting avenue for more detailed study. Results to date, though still very fragmentary, suggest that the view that reactions of unsaturated hydrocarbons over zinc oxide occur via proton abstraction to form a species with considerable anionic character has considerable merit.

VII. Concluding Remarks

This review has been concerned largely with interactions and reactions of unsaturated hydrocarbons with zinc oxide. The picture of the active site as a metal oxide pair capable of heterolytic fission of an acidic C—H bond provides a consistent framework for discussion of these results. We believe this view may be generally applicable. In its application, however, we must keep in mind that zinc oxide may be much more effective for heterolytic cleavage (i.e., more basic) than oxides such as, say, alumina.[4]

---

[4] The active site is viewed as an acid-base, cation-anion *pair*, hence, the "basicity" of the catalyst depends not only on the proton affinity of the oxide ion but also on the carbanion affinity of the cation. Thus, the *acidity* of the cation may determine the "basicity" of the catalyst. Specific interactions, i.e., effects of ion structure on the strength of the interaction, are likely to be evident when the carbanions differ radically in structure; when this is likely the concept of catalyst "basicity" should be used with caution.

Thus, the concentration of π-allyl from propylene on zinc oxide is high enough to be seen in the IR, but such species are not evident on alumina (*65*). The existence of sites capable of heterolytic cleavage, however, is evident in the IR spectra when more acidic adsorbates such as water (*72*), alcohol (*73*), ammonia (*74*), and (perhaps) acetylene (*75*) are adsorbed on alumina. Given these facts it seems reasonable to suppose that π-allyl species may be present on alumina at concentrations high enough to function as intermediates but not high enough to be seen in the IR. The fact that butene isomerization (*69*) on alumina appears to involve a large element of intramolecular hydrogen transfer is reminiscent of the data obtained for zinc oxide. When this is coupled with the fact that 1-butene isomerization yields high *cis–trans* ratios approaching those found for base catalyzed reactions and the reaction is enhanced by removal of surface water, which poisons the active sites on zinc oxide, this view becomes increasingly attractive. Clearly, such sites as we describe play no role over acidic oxides such as amorphous silica—alumina or molecular sieves, but we feel their role on other oxides, including oxidation catalysts, may prove to be an important one.

Acknowledgment

Acknowledgement is made to the donors of the Petroleum Research Fund, administered by the American Chemical Society, for support of this research. Preparation of this manuscript was aided by funds from the NSF under grant GP-22830.

References

1. Bond, G. C., "Catalysis by Metals," Academic Press, New York, 1962.
2. Bond, G. C., and Wells, P. B., *Advan. Catal. Relat. Subj.* **15**, 257 (1964).
3. Burwell, R. L., Jr., Haller, G. L., Taylor, K. C., and Read, J. F., *Advan. Catal. Relat. Subj.* **20**, 1 (1969).
4. Harrison, D. L., Nicholls, D., and Steiner, H., *J. Catal.* **7**, 359 (1967).
5. Bartek, J., Ph.D. Thesis, Johns Hopkins Univ., Baltimore, Maryland, 1970.
6. Hamilton, W. M., and Burwell, R. L., Jr., *Actes 2nd Congr. Int. Catal., Paris, 1960* **1**, 987 (1961).
7. Rooney, J. J., and Webb, G., *J. Catal.* **3**, 488 (1964).
8. Conner, W. C., Innes, R. A., and Kokes, R. J., *J. Amer. Chem. Soc.* **90**, 6858 (1968).
9. Conner, W. C., and Kokes, R. J., *J. Phys. Chem.* **73**, 2436 (1969).
10. Dent, A. L., and Kokes, R. J., *J. Phys. Chem.* **73**, 3772, 3781 (1969).
11. Dent, A. L., and Kokes, R. J., *J. Amer. Chem. Soc.* **91**, 7207 (1969).
12. Dent, A. L., and Kokes, R. J., *J. Amer. Chem. Soc.* **92**, 1092, 6718, 6709, (1970).
13. Dent, A. L., and Kokes, R. J., *J. Phys. Chem.* **74**, 3653 (1970).
14. Dent, A. L., and Kokes, R. J., *J. Phys. Chem.* **75**, 487 (1971).
15. Bank, S., Schriesheim, A., and Rowe, C. A., Jr., *J. Amer. Chem. Soc.* **87**, 3244 (1965).
16. Hunter, D. H., and Cram, D. J., *J. Amer. Chem. Soc.* **86**, 5477 (1964).

17. Haag, W. O., and Pines, H., *J. Amer. Chem. Soc.* **82**, 387 (1960).
18. Mckee, D. W., *J. Amer. Chem. Soc.* **84**, 1109 (1952).
19. Eischens, R. P., and Pliskin, W. A., *Advan. Catal. Relat. Subj.* **10**, 1 (1958).
20. Woodman, J. F., and Taylor, H. S., *J. Amer. Chem. Soc.* **62**, 1393 (1940).
21. Taylor, E. H., and Wethington, J. A., *J. Amer. Chem. Soc.* **76**, 971 (1954).
22. Aigueperse, J., and Teichner, S. J., *Ann. Chim. (Paris)* **7**, 13 (1962).
23. Aigueperse, J., and Teichner, S. J., *J. Catal.* **2**, 359 (1968).
24. Bozon-Verduraz, F., Arghiropoulos, B., and Teichner, S. J., *Bull. Soc. Chim. Fr.* p. 2854 (1967).
25. Bozon-Verduraz, F., and Teichner, S. J., *J. Catal.* **11**, 7 (1968).
26. Molinari, E., and Parravano, G., *J. Amer. Chem. Soc.* **75**, 5233 (1953).
27. Parravano, G., and Boudart, M., *Advan. Catal. Relat. Subj.* **7**, 47 (1955).
28. Taylor, H. S., and Strother, C. O., *J. Amer. Chem. Soc.* **56**, 586 (1934).
29. Kesavulu, V., and Taylor, H. A., *J. Phys. Chem.* **64**, 1124 (1960).
30. Low, M. J. D., *J. Amer. Chem. Soc.* **87**, 7 (1965).
31. Kubokawa, Y., and Toyama, O., *J. Phys. Chem.* **60**, 833 (1956).
32. Eischens, R. P., Pliskin, W. A., and Low, M. J. D., *J. Catal.* **1**, 180 (1962).
33. Glemza, R., and Kokes, R. J., *J. Phys. Chem.* **69**, 3254 (1965).
34. Thomas, D. G., *J. Phys. Chem. Solids* **10**, 47 (1959).
35. Kokes, R. J., *J. Phys. Chem.* **60**, 99 (1962).
36. Gregg, S. J., and Sing, K. S. W., "Adsorption, Surface Area and Porosity," p. 82. Academic Press, New York, 1967.
37. Twigg, G. H., *Discuss. Faraday Soc.* **8**, 152 (1950).
38. Meyer, E. F., and Burwell, R. L., Jr., *J. Amer. Chem. Soc.* **85**, 2877 (1963).
39. MacRae, A. U., *Surface Sci.* **4**, 2476 (1966).
40. Herzberg, G., "Molecular Spectra and Molecular Structures," p. 326. Van Nostrand-Reinhold, Princeton, New Jersey, 1945.
41. Powell, D. B., and Sheppard, N., *Spectrochim. Acta* **16**, 69 (1958).
42. Stoicheff, B. P., *J. Chem. Phys.* **21**, 755 (1956).
43. Pradilla-Sorzana, J., and Lacker, J. P., Jr., *J. Mol. Spectrosc.* **22**, 180 (1967).
44. Quinn, H. W., and Glev, D. N., *Can. J. Chem.* **40**, 1103 (1962).
45. Carter, J. L., Yates, D. J. C., Lucchesi, P. J., Elliot, J. J., and Kovorkian, V., *J. Phys. Chem.* **70**, 1126 (1966).
46. Kaesz, H. D., and Stone, F. G. A., *Spectrochim. Acta* **15**, 360 (1959).
47. Green, J. H. S., *Spectrochim. Acta, Part A* **24**, 863 (1968).
48. Riter, J. R., Jr., and Eggers, D. F., Jr., *J. Chem. Phys.* **44**, 745 (1966).
49. van Riet, R., *Ann. Soc. Sci. Bruxelles, Ser. 1* **71**, 102 (1957).
50. Lord, R. C., and Venkataswarlu, P., *J. Opt. Soc. Amer.* **43**, 1079 (1953).
51. McClellan, W. R., Hoehn, H. H., Cripps, H. N., Muetterties, E. L., and Howk, B. W., *J. Amer. Chem. Soc.* **83**, 1601 (1961).
52. Mason, R., and Russell, D. R., *Chem. Commun.* p. 26 (1966).
53. Cotton, F. A., Faller, J. W., and Musco, A., *Inorg. Chem.* **6**, 179 (1967).
54. Fischer, E. O., and Werner, H., "Metal π-Complexes," pp. 182–183. Elsevier, Amsterdam, 1966.
55. Fritz, H. P., *Chem. Ber.* **94**, 1217 (1961).
56. ref. 40 p. 192–201.
57. Decius, J. C., and Wilson, E. B., Jr., *J. Chem. Phys.* **11**, 1409 (1951).
58. ref. 40 p. 168–175.
59. Aldridge, C. L., Jonassen, H. B., and Pulkkinen, E., *Chem. Ind. (London)* p. 374 (1960).

60. Moore, D. W., Jonassen, H. B., Joyner, T. B., and Bertrand, A. J., *Chem. Ind. (London)* p. 1304 (1960).
61. Harrod, J. F., and Chalk, A. J., *J. Amer. Chem. Soc.* **88,** 3491 (1966).
62. van Leeuwen, P. W. N. M., and Praat, A. P., *Chem. Commun.* p. 365 (1970).
63. Wilke, G., Bogdanovic, B., Hardt, P., Heimbach, P., Keim, W., Kröner, M., Oberkirch, W., Tanaka, K., Steinrücke, E., Walter, D., and Zimmerman, H., *Angew. Chem. Int. Ed. Engl.* **5,** 151 (1966).
64. Sheppard, N., and Simpson, D. M., *Quart. Rev. Chem. Soc.* **6,** 1 (1962).
65. Chang, C. C., and Kokes, R. J., unpublished observations; see also *J. Amer. Chem. Sec.* **92,** 7517 (1970).
66. Golden, D. M., Eggers, K. W., and Benson, S. W., *J. Amer. Chem. Soc.* **86,** 5416 (1964).
67. Bellamy, L. J., "The Infrared Spectra of Complex Molecules," p. 09. Wiley, New York, 1958.
68. Conley, R. T., "Infrared Spectroscopy," p. 97. Allyn & Bacon, Boston, Massachusetts, 1966.
69. Hightower, J. W., and Hall, W. K., *J. Amer. Chem. Soc.* **89,** 778 (1967).
70. Kosower, E. M., "Physical Organic Chemistry," p. 27. Wiley, New York, 1968.
71. Iwai, I., *in* "Mechanisms of Molecular Migrations" (B. S. Thyagarajan, ed.), Vol. 2, p. 73. Wiley, New York, 1969.
72. Peri, J. B., *J. Phys. Chem.* **69,** 220 (1965).
73. Deo, A. V., and Dalla Lana, F. G., *J. Phys. Chem.* **73,** 716 (1969).
74. Peri, J. B., *J. Phys. Chem.* **69,** 231 (1965).
75. Yates, D. J. C., and Lucchesi, P. J., *J. Chem. Phys.* **35,** 243 (1961).

Chemisorption Complexes and Their Role in Catalytic Reactions on Transition Metals

Z. KNOR

Institute of Physical Chemistry
Czechoslovak Academy of Sciences
Prague, Czechoslovakia

I. Introduction	51
II. General Formulation of the Problem	52
III. Chemisorption Complexes	57
IV. Characterization of the Metal Surfaces	65
V. Concluding Remarks	71
References	71

I. Introduction

In any particular field of research a very important problem exists, viz, the problem of rationalization of all the various experimental results. There are two basic approaches for solving this problem: either a semiempirical correlation of the results, using some arbitrary parameter[1] (which can be considered only as a first step in the solution of this problem), or the purely theoretical approach (which sometimes results from the former one). Examples for both of these approaches can be found in catalysis too. Their analysis was extensively done elsewhere (1).

The complete theory of catalysis, which would start with the isolated reaction participants, was not available until now because of the lack of adequate knowledge of the participants themselves (even the complete theory of the isolated participants, starting from the first principles, is still lacking). However, in analogy with the homogeneous chemical reactions one can expect that the quantum chemical approach, based on the semiempirical quantum mechanical methods, could be a prospective one.

---

[1] The choice of a particular correlating parameter, of course, follows from more or less exactly formulated theory.

We can expect that in future it might probably enable us to characterize the reactivity of all reaction participants, including the reaction components and the catalyst itself, in terms of their electronic structure. The quantum chemical methods for approximate description of the polyatomic molecules (reaction components) have already been worked out. However, a very important problem arises here, one which has to be studied carefully, namely, the representation of the catalyst in the frame of this theoretical approach.

When we are working with such complicated objects like polyatomic molecules and metals, we are not able to describe completely the real system, and, consequently, we are forced to construct its model, the properties of which should satisfy the following conditions: (1) it should be tractable by the contemporary theoretical methods; (2) it has to simulate as much as possible the behavior of the real system; and (3) it must not contradict any experimental results. Thus, the model is the point where the theory and experiments meet their mutual requirements, and where they directly influence each other (*2*).

In the past the theoretical model of the metal was constructed according to the above-mentioned rules, taking into account mainly the experimental results of the study of bulk properties (in the very beginning only electrical and heat conductivity were considered as typical properties of the metallic state). This model (one-, two-, or three-dimensional), represented by the electron gas in a constant or periodic potential, where additionally the influence of exchange and correlation has been taken into account, is still used even in the surface studies. This model was particularly successful in explaining the bulk properties of metals. However, the question still persists whether this model is applicable also for the case where the chemical reactivity of the transition metal surface has to be considered.

In this article this question will be discussed together with the problems of model construction of the reactants on the metal surfaces. The experimental methods and their results, which can be used for this purpose, will be outlined.

II. General Formulation of the Problem

The energy of the metallic bond in transition metal crystals is in the range of 5–10 eV. The bond energy in most of the simple molecules is of the same order of magnitude, and the energy of the chemisorption bond ranges from 2 to 10 eV. Consequently, from the point of view of pure energy, it is difficult to decide the way of theoretical treatment of the surface interactions. One can either describe this interaction as a perturbation of the crystal, caused by the approaching gas particle, or as a surface complex ("surface molecule") (*3*, *4*) which is perturbed by the presence of the rest of the crystal. An unambiguous answer probably does not exist to this general question. It seems reasonable to use the latter concept (viz, the concept of surface complex) in all cases where there exists the final product, or at least an intermediate (even for a short time interval), which can be regarded as a "chemical molecule" or "transition complex," respectively (e.g., in chemisorption, field ionization near the metal surfaces, molecular scattering processes on surfaces). Consequently, we can state quite generally that all the surface interactions of particles, which are in principle distinguishable from those forming the crystal (e.g., by the use of isotopes), can be treated in the way discussed above. On the other hand, the direct interaction of particles essentially undistinguishable from those in the interior of the crystal (as are electrons, photons, and phonons), should be theoretically described as the interaction of that particle with the whole crystal (*2*).

After all, even in the first case we deal with the interaction of an electron belonging to the gas particle with all the electrons of the crystal. However, this formulation of the problem already represents a second step in the successive approximations of the surface interaction. It seems that this more or less exact formulation will have to be considered until the theoretical methods are available to describe the behavior both of the polyatomic molecules and the metal crystal separately, starting from the first principles. In other words, a crude model of the metal, as described earlier, constructed without taking into account the chemical reactivity of the surface, would be in this general approach (in the contemporary state of matter) combined with a relatively precise model of the polyatomic molecule (the adequacy of which has been proved in the reactivity calculations of the homogeneous reactions).

A close analogy to the localized surface interaction can be found in the field of chemical kinetics, namely, in the spectator stripping mechanism (*5*, *6*) of the gas reactions, as evidenced by the recent crossed-molecular-beams experiments. Here the projectile seems to meet with only a part of the target molecule (that one to be transferred), while the rest of the target behaves as a "spectator," in a sense not taking part in the reaction.

Any heterogeneous catalytic reaction can be represented by a simplified scheme:

$$\begin{array}{ccc} A & \to & B & \to & C \\ \text{(reaction components} & & \text{(chemisorption} & & \text{(catalytic reaction)} \\ \text{+ metal catalyst)} & & \text{complex)} & & \end{array}$$

Passing in this scheme from the left to the right side we can formulate several problems which have to be studied: (1) the transition from state A to B; (2) the identification of the chemisorption complexes; (3) the reactivity of these complexes; (4) their role in the particular catalytic reaction (e.g., blocking of the surface, their mutual interaction); (5) the mechanism of the reaction (the transition from state B to C) (e.g., does the reaction proceed in the chemisorbed layer or can some components react directly from the gas phase, impinging on the chemisorbed species?) ; and (6) the liberation of the reaction products from the surface into the gas phase or their stability in the surface. We consider as a *fundamental problem the identification of those chemisorption complexes that are responsible for the reaction in the desired direction.*

If we are now interested in the experimental study of the problems listed above we have to decide which particular system we will choose for this study. By a particular system we mean both the type of the reaction (e.g., hydrogen + oxygen, hydrogen + cyclopropane) and the type of the catalyst (e.g., evaporated metal film, wire, single crystal, supported metal). It is well known that the mere chemisorption of simple gases like hydrogen, nitrogen (7–9) on one metal only results in a spectrum of chemisorbed species [this is true even for one crystallographic plane of a given metal (10, 11)]. The problem is, of course, even more complicated when the interaction of polyatomic molecules is studied. Moreover, all these effects occur on the surface of a catalyst, which itself is not a chemically inert participant of the interaction. Understandably, most of the above-mentioned problems cannot be solved using an arbitrary catalytic reaction with an industrial catalyst. In this case the system is much too complicated (contamination problem, problems of mass and heat transfer, the effect of promotors, etc.). As a consequence, most of the experimental results obtained in such systems have no unambiguous interpretation. *The decision for the choice of a particular system is determined by the type of problem we are interested in.* The procedure of finding an appropriate system for the experimental study of the heterogeneous catalytic reaction is very much similar to the construction of a theoretical model. Again, we have to fulfill analogical requirements, viz, (1) the experimental method applied to this system should supply us with the results, which could be unambiguously interpreted; and (2) the basic features of the "industrial reaction" should be preserved in this "model system" (e.g., the transfer of hydrogen, oxidative splitting of a particular bond). In the first requirement an intricate problem is hidden, namely, that the more sophisticated the experimental technique used for obtaining an insight into the problem, the less direct is the information obtained. One has to use the theory for the interpretation of the results and, consequently, the preciseness of the information is determined not only by the experimental error, but also (and sometimes even to greater extent) by the preciseness of the theoretical model used.

The first requirement for the choice of an appropriate model system can be respected by using as simple a system as possible, e.g., diatomic molecules as reaction components and a well-defined surface of a catalyst (well-defined both from the point of view of its cleanliness and its crystallographic orientation). The second part of the preceding statement requires some explanation: the definition of a "well-defined surface" strongly depends on the state of the experimental technique. Both terms, cleanliness and crystallographic perfection of the surface, have no absolute meaning (12). The degree of detectable contamination and crystallographic perfection does not depend only on the sensitivity of the experimental method that was used for its determination, but also on some characteristic features of that particular experimental method. For example, ellipsometry was recently shown to be able to detect contamination of the surface during the first 100 min at the pressure of 4×10^{-10} Torr, whereas low energy electron diffraction (LEED) detected no change of the diffraction pattern (13). In spite of that, both of these experimental methods have comparable sensitivity, each of them being able to detect a given surface layer in a particular state only: LEED provides information about well-ordered systems and elipsometry detects even the amorphous layer (13). Therefore, it is understandable that in the last time several authors combine even more than two different experimental methods for surface studies within one vacuum system, e.g., (12). Consequently, it is important to always examine the level of contamination and of crystallographic perfection, detectable by a particular experimental method (in the sense of how the results can be influenced by these two factors), and to specify exactly the conditions of the experiment. From this point of view the ideal system is the clean surface of one crystallographic plane (which would be atomically flat) and diatomic molecules of two kinds only. Most of the experimental techniques suitable for these types of studies (e.g., UHV technique and mass spectrometers, which are able to detect and analyze the quality of several particles only) have become available only recently. However, even in this case there are still further difficulties. One can work either with a macroscopic single crystal plane, the cleanliness of which is well defined (UHV technique and Auger spectroscopical analysis of the surface being applied), or with a microscopic crystal plane [using a field electron or field ion emission microscope (FEM and FIM, respectively)]. In the first case the crystallographic perfection can be checked by LEED only, and there are severe limitations of the ability of LEED to detect the roughness of the surface in the atomic dimensions (*14*, *15*). On the other hand, one can easily obtain atomically flat surfaces in FEM or FIM experiments [e.g., combined with mass spectrometric analysis of the field desorbed products (*16*, *17*)], however, in this case the presence of a high electric field might complicate the interpretation of the results. Accordingly, even from the point of view of the correct interpretation of a particular result, it is useful to take into account the results of more than one experimental method.

With these types of simple systems and contemporary experimental techniques, one can study any of the basic problems of the heterogeneous catalysis listed in the preceding text (see p. 54). However, one fundamental problem appears now if the results obtained in these studies—the rate-determining step, the active chemisorption complex, etc.—are to be applicable for the "real" (contaminated and polycrystalline) surface. The change from the model system to the "real" one must be a quantitative change only (e.g., the area of the catalytically active surface is changed, the number of the active chemisorption complexes is changed) not a qualitative one (the type of adsorbed complexes is different, the reaction intermediates are formed as a result of the interaction between one reaction component and the contaminant molecule, etc.). This problem represents, in fact, another formulation of the second requirement which has to be respected in the choice of a model system (see p. 54) and can be regarded as a feedback check. Consequently, it is important to perform experiments both with the well-defined surfaces and the industrial catalyst. The final goal of these studies should be to prove whether the basic mechanism of the catalytic reaction is in principle the same one in both cases. Finally, one might doubt the reasonableness of those experimental studies, where the basic requirements for the choice of a "simple system" are only partially respected (evaporated metal films, polycrystalline wires, etc., where the cleanliness of the surface is sufficient and the influence of the crystallographic orientation can be respected mostly in a speculative way), because no experimental technique is yet available that would supply the direct information about the properties of the chemisorbed species on the surfaces without the averaging effect. However, there are two basic reasons for using these types of systems. (1) In spite of the fact that the interpretation of these results is not straightforward, one can successfully study with these systems those special questions (sometimes even easier than on better-defined surfaces) that are essentially more chemical than physical in character. (2) The experimental equipment for the study of these systems is usually less expensive. Examples will be outlined in the following text.

III. Chemisorption Complexes

We will discuss the reaction of hydrogen and oxygen on transition metals first. This reaction has been extensively studied in our laboratory (*18–32*) using evaporated metal films as a catalyst. From our previous considerations it follows that as a consequence of the choice of this particular system we must restrict ourselves to certain problems only. We cannot identify the surface species (we can indirectly indicate only some of them) nor understand completely their role in the reaction. Because of the polycrystalline character of the film, all the experimental results are "averaged" over all the surface. Several new problems thus arise, such as grain boundaries, and, consequently, the exact physical interpretation of these results is almost impossible; it is more or less a speculative one. However, we can still get some valuable information concerning the chemical nature of the active chemisorption complex. The experimental method and the considerations will be shown in full detail for nickel only. For other metals studied in our laboratory, only the general conclusions will be presented here.

The following experimental methods were used for these studies: volumetric measurements of the amount of gases (*18–28*) consumed during a particular interaction; measurements of the electrical resistance (*24–31*) and work function changes of the film (*22, 32*); Hall effect measurements (*33, 34*), and adsorption calorimetry (*35, 36*). All the results were obtained in bakeable, all-glass apparatus working in the pressure range 10^{-9} to 10 Torr. Experimental details were published elsewhere (*37–39*). All these methods give us only indirect information about the adsorbed layer. Moreover, the results of most of them have rather complicated interpretations, depending on several parameters not now accessible to direct estimation. Consequently, all these physical methods were used only to indicate the various processes on the catalyst surfaces (chemisorption, reaction) and for distinguishing between them.

First, the chemisorption of the individual reaction components (hydrogen and oxygen) was investigated on nickel. The influence of preadsorbed oxygen on the following adsorption of hydrogen was studied at 78° and 300°K (Fig. 1) (*20, 32*). At 78°K the amount of hydrogen consumed decreased with an increasing amount of oxygen, indicating that the surface covered with oxygen does not take part in the subsequent interaction with hydrogen. On the other hand, at 300°K the amount of hydrogen consumed reached a value several times higher than on the clean surface. This is true until oxygen adsorption is complete. As soon as the complete layer of oxygen is formed, almost no hydrogen is consumed when admitted. There are two possible explanations for the large amount of hydrogen consumed

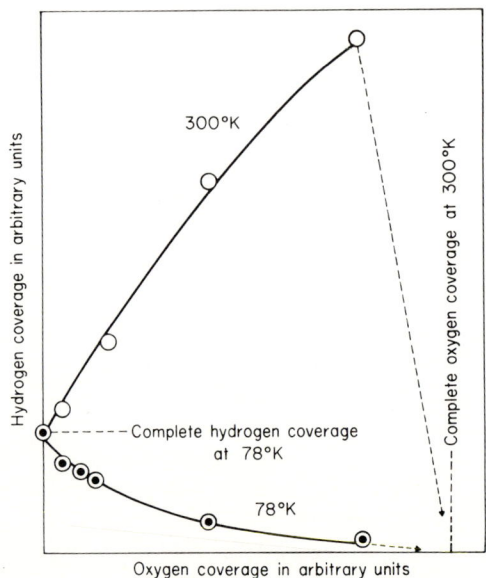

Fig. 1. Hydrogen adsorption at 78° and 300°K as a function of preadsorbed amount of oxygen on nickel.

at 300°K. (1) The surface layer of the metal is reconstructed during the oxygen adsorption in such a way that more "space" for hydrogen arises on the surface. (2) Some kind of chemical interaction between hydrogen and preadsorbed oxygen (or more exactly, the chemisorbed complex) occurs. In the first case one would not expect such an abrupt decrease in the ability of the reconstructed surface to consume hydrogen when the complete oxygen coverage is attained. Therefore, the second possibility seems to be the more probable one.

Further, the electrical resistance changes of the nickel films were studied. It was observed that regardless of the temperature (78° or 300°K), the resistance of a nickel film after chemisorption of either hydrogen or oxygen alone took place was always higher than the original value. The final values of the relative resistance changes were of the order of magnitude $(\Delta R/R)_{O_2} \sim 10^{-2}$, $(\Delta R/R)_{H_2} \sim 10^{-3}$. However, if hydrogen was admitted to the nickel surface that was partially covered with oxygen, at 300°K the film resistance decreased with time (Fig. 2a). This effect was not observed at 78°K. When oxygen was admitted to the chemisorbed hydrogen layer at 300°K, the resistance after the initial rise decreased with time (Fig. 2b). The initial rise of the film resistance indicated that oxygen chemisorption was the first step of this interaction. The decrease of the resistance with time continued even when there was no oxygen or hydrogen in the gas phase. At 78°K oxygen caused only a resistance increase. When this composite layer of adsorbed hydrogen and oxygen was heated to 300°K the resistance decreased. The explanation for the resistance decrease might be: (1) the change in the mechanism of the electrical conductivity in the presence of the two gases competitively chemisorbed on the surface, or (2) an interaction between hydrogen and oxygen on the surface. The first explanation seems to be less probable than the second one, because one can hardly expect that the mere change of temperature (from 78° to 300°K) would cause a qualitative change of the conductivity mechanism of this composite layer, without taking into account a possible chemical reaction.

Several complementary experiments were done with work function measurements (using retarding field diodes) in this system. Again, keeping in mind all the difficulties of the interpretation, we have used these experiments for qualitative indication and for distinguishing between the processes of competitive chemisorption and chemical reaction. Both oxygen and hydrogen individually caused an increase of the work function of the nickel film ($\Delta\varphi_{O_2} = -1.6$ eV, $\Delta\varphi_{H_2} = -0.3$ eV). When the composite layer was formed by adsorbing hydrogen into the incomplete oxygen layer, at 78°K the value $\Delta\varphi = -0.6$ eV resulted and at 273°K the value $\Delta\varphi = +0.8$ eV was observed. The decrease of the work function at higher tem-

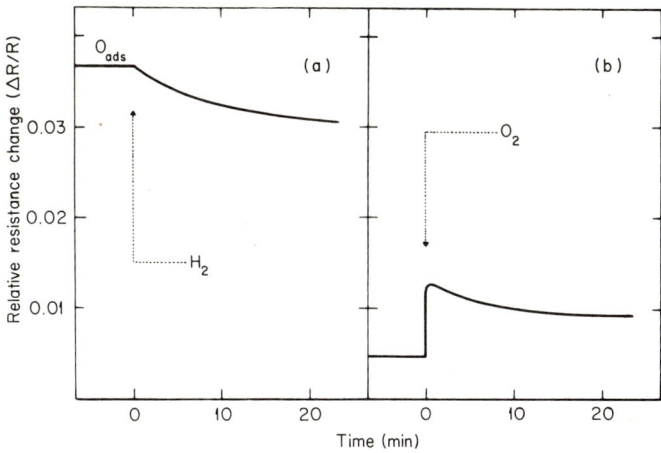

FIG. 2. Typical curves of the relative changes of the electrical resistance of nickel films as a function of time: (a) adsorption of one dose of hydrogen on the surface, partially covered by preadsorbed oxygen; (b) adsorption of one dose of oxygen on the surface, covered by preadsorbed hydrogen (both at 300°K).

perature again indicated qualitatively different process, most probably the interaction between hydrogen and oxygen on the surface.

Taking into account the results of all the above-listed experiments, it is possible to conclude that these experimental methods indeed indicate the interaction of the two gases on the surface of nickel at temperatures close to room temperature.

The only conclusions that can be drawn from our experiments concerning *the role of oxygen* in these interactions are (1) oxygen chemisorption is faster than the subsequent interaction, and (2) oxygen can react with hydrogen on the surface of nickel in the adsorbed layer at room temperature.

The role of hydrogen in these interactions can be further clarified by means of additional experiments using hydrogen atomized in the gas phase (*27*). We have seen already that hydrogen was able to react at 300°K with preadsorbed oxygen when there was some "space" left on the surface for hydrogen chemisorption (most probably a dissociative one). When the surface was completely covered by oxygen, no chemical interaction with molecular hydrogen was indicated. However, the hydrogen atoms produced in the gas phase were able to react with complete oxygen layer on the nickel surface at 300°K both when they were supplied directly from the gas phase at this temperature, and when they were trapped on the surface at 78°K and the composite layer was afterward heated to 300°K. When a small amount of nickel was additionally evaporated onto the surface completely covered with oxygen, the reaction between molecular hydrogen and oxygen proceeded. The question now arises whether the role of the bare nickel atoms in the surface lies in the dissociation of hydrogen molecules only.

When hydrogen, atomized in the gas phase, interacted at 78°K with the complete layer of chemisorbed hydrogen (prepared by the chemisorption of hydrogen molecules at 78°K) (*27*), the amount additionally trapped on this surface was several times higher than the complete coverage resulting from ordinary chemisorption.[2] The interaction of this layer with oxygen was studied at 78°K and no reaction was indicated. However, when this composite layer was heated to 273°K, the amount of hydrogen, approximately equivalent to the amount trapped previously as gas atoms, was released to the gas phase. No oxygen was desorbed during this rise of temperature. Finally, it was observed that with nickel it is not possible to repeat any of the above-described processes on the same surface.

[2] It is interesting that a similar effect, viz, the considerable increase of the consumed amount of hydrogen, is also caused by the incomplete layer of oxygen. This could suggest that chemisorbed oxygen facilitates the splitting of the bond in the hydrogen molecule and reacts with one part of it only [see the later discussion of the propylene self-hydrogenation on nickel, partially covered with oxygen (p. 64).]

From all these experiments one can conclude that (1) the active hydrogen particles on the surface are probably chemisorbed hydrogen atoms; (2) the necessary prerequisite for the reaction is a certain configuration of oxygen and hydrogen atoms on the surface that is different for the various origins of hydrogen atoms on the surface; (3) the reaction itself needs some activation energy; and (4) the products of the reaction are stable on the surface at the conditions of these experiments, i.e., they do not desorb.

The same type of analysis was applied to palladium, rhodium, molybdenum (*22, 25, 27, 31, 32*). Without going into details, it is possible to say that the reaction between hydrogen and oxygen proceeds easily at room temperature on palladium and rhodium, and it does not take place on molybdenum unless the hydrogen atoms are supplied to the chemisorbed oxygen layer directly from the gas phase. The reaction can be repeated several times on the same surface of palladium and rhodium (the products of the reaction are desorbable), in contrast to nickel and molybdenum, where the products are firmly bound to the surface. Consequently, either the products of the reaction are the same on all these metals, with different bond energy to the metal, or, more probably, the products are different (H_2O, desorbable; OH, firmly bound). Again it follows from these experiments that the active surface complex contains atomic hydrogen, and that both the energetics and the chemical specificity of the metals, together with the specific configuration of the reactants, play important roles. The energetical effects both of mere chemisorption and of the chemical reaction on the surface can be studied directly by the adsorption calorimetry. In our laboratory the modified Beeck-type calorimeter was used (*35, 36*) and consequently the processes where no gaseous products resulted could be studied.

Similar analysis was also applied to the study of the forms of nitrogen adsorbed on iron and their possible roles in ammonia synthesis (*40*). The experiments with atomized hydrogen revealed one interesting result here, viz, that hydrogen atoms were able to react with preadsorbed nitrogen only when they were supplied directly from the gas phase at 273°K. Analogical behavior has already been mentioned in the earlier discussion of the hydrogen–oxygen interaction on molybdenum.

Several additional conclusions concerning the nature of the chemisorbed layer can be drawn from the Hall effect measurements (*33, 34*). The chemisorbed species, together with the surface metal atoms, represent complexes analogical to the ordinary chemical compounds and, consequently, one might expect that the metal atoms involved in these complexes will contribute to lesser extent or not at all to the bulk properties of the metal. Then we should speak about the demetallized surface layer (*41*). When the Hall voltage was measured as a function of the evaporated film thickness

(*33, 34*), the effect resulting from the change of about 100 Å was of the same order of magnitude as was the effect of chemisorbed gases (e.g., hydrogen). Thus it is possible to conclude that the role of the chemisorption complex is not the demetallization only, i.e., not only the effective thickness of the metallic layer is decreased in chemisorption. This statement is supported by the fact that the film resistance changes caused by either chemisorbed hydrogen or oxygen are considerably different from each other; however, in the case of Hall voltage measurements the changes of the same order of magnitude resulted for both of these gases (the role of a particular chemisorption complex might be different in both of these processes). Similar conclusions can be drawn from the measurement of the Hall voltage at different temperatures; e.g., hydrogen on nickel causes the increase of this voltage at 78°K and the decrease at 273°K.

One important restriction of the applicability of all the above-mentioned conclusions should be always kept in mind—that only the properties of *strongly bound particles* were studied. Always when the second gas was introduced, it reacted with the chemisorbed layer of the first one, the gas phase being pumped off beforehand. This need not be a serious restriction with the hydrogen layer at 78°K and the oxygen layers both at 78° and 300°K, where only a few percent are desorbed during evacuation. However, in the case of hydrogen at room temperature, as much as approximately 25% of the adsorbed amount can be desorbed by mere pumping off the gas phase (*19*).

When the problem of the choice of an appropriate model system for a catalytic study was discussed previously, it was automatically accepted that the simplest reaction components should be the diatomic molecules (e.g., hydrogen, oxygen, nitrogen). This is true from the point of view of the system entering the catalytic reaction. However, it need not be necessarily true from the point of view of the reaction mechanism. It might easily happen that a bigger molecule will be less sensitive to all the physical and chemical inhomogeneities of the surface and, consequently, the number of the active chemisorption complexes might be reduced to only one.

Therefore, it seems worthwhile to examine this idea by studying the interaction of some more complicated molecules (e.g., unsaturated hydrocarbons, such as ethylene, propylene, cyclopropane) with the metal surfaces. This type of study might also be of some practical importance.

The interaction of cyclopropane with clean surfaces of several metals was studied in our laboratory (*42–46*), and it was shown that at 273°K self-hydrogenation occurred, resulting in products strongly dependent on the nature of the metal. This conclusion followed from the indirect estimation of the average composition of the chemisorbed layer, based on the mass spectrometric analysis of the gas phase. Furthermore, the products of the reaction in a gaseous mixture of cyclopropane and hydrogen on the catalyst, precovered either with chemisorbed hydrogen or cyclopropane, were studied mass spectrometrically. The displacement of chemisorbed species by means of carbon monoxide chemisorption, and the hydrogen–deuterium exchange with the composite layer were also studied. Simultaneously, electrical resistance changes were measured. In some cases the infrared spectra of the chemisorbed species both on the evaporated metal films and on the supported metal catalysts were recorded (*46*), and the heat of adsorption was calorimetrically determined.

On platinum and palladium, the surface complexes containing three carbon atoms result from cyclopropane chemisorption (*43*). These complexes are partially dehydrogenated. A part of hydrogen atoms on the surface can recombine with the dehydrogenated residues, forming mostly propane or propylene (depending on the degree of the surface coverage), desorbing into the gas phase when carbon monoxide is added to this layer on platinum. On nickel, iron, and molybdenum, negligible amounts of hydrocarbons are displaced by carbon monoxide, and hydrogen appears in the gas phase. In addition to propane, methane and ethane are also produced by cyclopropane self-hydrogenation on these metals, thus indicating that the original cyclopropane molecule cracked on their surfaces to residues containing one or two carbon atoms, respectively. The difference between platinum, on which no cracking of cyclopropane occurs, and molybdenum, which gives rise to cyclopropane cracking, is reflected in the different amount of heat evolved in the interaction of cyclopropane with these metals: on platinum the initial heat is about 65 kcal/mole, whereas on molybdenum the value of about 140 kcal/mole results (*35a*).

Close parallelism was found between the products of cyclopropane self-hydrogenation and the products of the hydrogenation resulting when a mixture of cyclopropane and hydrogen is admitted onto the film (*43*). The observed small differences were attributed to the different stationary composition of the surface layers attained when cyclopropane interacted with the virgin surface or when the reaction mixture was added. It was shown that the mass spectrometric analysis of the gas phase and the measurement of the electrical resistance of the films are sensitive enough to indicate the different composition of the surface layers in each one of the following cases: (1) cyclopropane interaction with the clean surface, (2) cyclopropane interaction with the hydrogen layer on the surface, and (3) interaction of the cyclopropane–hydrogen mixture with the chemisorbed layer of cyclopropane.

From all these examples, based mainly on the "chemical" analysis of the surface layer, one can conclude that the chemisorption of cyclopropane on the clean surfaces of metals is again, unfortunately, a rather complicated process, resulting in several adsorbed species: hydrocarbon residues with a varying content of carbon atoms and hydrogen atoms. Platinum, being the most thoroughly studied, was shown to be the exception, where the C_3 complexes are preserved without cracking into smaller residues (*43*). However, even in the case of the complicated dissociative chemisorption of cyclopropane on other metals, important conclusions can be drawn from the mass spectrometric study of methane, ethane, and propane production, namely, the conclusions concerning the role of C_1, C_2, and C_3 residues in a particular reaction, as shown in (*43–45*).

In the preceding section, the problem of the general applicability of the results obtained with well-defined surfaces was briefly mentioned. An important contribution to the solution of this problem might be the study of the artificially contaminated (poisoned) surfaces. The most common poison for hydrogenation reactions is oxygen. Consequently, the study of the influence of preadsorbed oxygen on the catalytic hydrogenation can more or less represent a bridge between the clean and contaminated surface of the model and industrial catalyst, respectively. The role of one particular poison can be checked in this way. The problem of this additional participant (either a chemically inert or reactive one) can be also formulated as a problem of the modification of the catalyst surface properties. From this point of view, the problem promises to be an important one for the study of the catalyst selectivity.

In our laboratory the influence of preadsorbed oxygen on the following self-hydrogenation of cyclopropane on platinum and of propylene on nickel were studied (*47*). First, the reactivity of the preadsorbed oxygen with the hydrocarbon molecules had to be checked. It was shown that oxygen probably does not react with the hydrocarbon complex on the surface of nickel (compare with the hydrogen–oxygen interaction, as discussed previously), whereas on platinum a part of the preadsorbed oxygen is able to split oxidatively the hydrocarbon molecules. Preadsorbed oxygen represents a third reaction participant (cyclopropane residues, oxygen, and the catalyst), which competes with the others for hydrogen atoms on the surface. It is possible to conclude that a small amount of preadsorbed oxygen effectively increases the "space" for hydrogen on the surfaces of both nickel and platinum and facilitates the transfer of hydrogen atoms from one hydrocarbon molecule to another. This resembles the previously mentioned case, where oxygen preadsorbed on nickel increased several times the consumed amount of hydrogen at 300°K.

In this section we have discussed several chemisorption complexes from the point of view of the chemical nature of one part of this complex—the one resulting from the gas molecule. However, it follows from the above-mentioned results that the metal atoms, constituting the second part of each particular chemisorption complex, do also influence its reactivity. Consequently, when treating theoretically the chemisorption complex, one has to characterize not only its "chemically active" constituent, but also the metallic surface.

IV. Characterization of the Metal Surfaces

In the Introduction the problem of construction of a theoretical model of the metal surface was briefly discussed. If a model that would permit the theoretical description of the chemisorption complex is to be constructed, one must decide which type of the theoretical description of the metal should be used. Two basic approaches exist in the theory of transition metals (48). The first one is based on the assumption that the d-electrons are localized either on atoms or in bonds (which is particularly attractive for the discussion of the surface problems). The other is the itinerant approach, based on the collective model of metals (which was particularly successful in explaining the bulk properties of metals). The choice between these two is not easy. Even in contemporary solid state literature the possibility of d-electron localization is still being discussed (49–51). Examples can be found in the literature that discuss the following problems: high cohesion energy of transition metals (52), their crystallographic structure (53), magnetic moments of the constituent atoms in alloys (54), optical and photoemission properties (48, 49), and plasma oscillation losses (55).

In the frame of the itinerant model, the surface is represented by a potential barrier of various origins and shapes, in most cases treated as one-dimensional problem (e.g., 56–60), without taking into account the potential variation in the plane of the surface[3] [with the exception of (61) where this effect is qualitatively discussed in connection with the field ionization probability]. Obviously, the nonlocalized model is suitable and often used for the theoretical interpretation of the changes of the bulk properties of the metals caused by the surface effects (the changes of the electrical resistance, magnetic properties, galvanomagnetic effects, etc.).

The other extreme of the theoretical treatment of the metal surfaces is the approximation of the surface by a single atom or by a simple array of two or more atoms (e.g., 62–65): one can use for the description of these atoms either the electronic structure of the individual atoms [sometimes even simplified, e.g. (63)] or fractional occupancy of their atomic orbitals

---

[3] Recently the potential variation in the plane of the surface was taken into account in the calculation of the field ionization probability (85).

corresponding to the bulk metal value, e.g., (*64*, *66*). In other words, this model of the surface is based on the assumption that the surface atoms preserve to some extent their individual character, viz, that *in the moment of the interaction* between a gas atom and the surface of a crystal the wavefunctions, describing the states of electrons in the metal surface, can be approximated in the close neighborhood of the surface atom by the wavefunctions of an isolated atom of the same metal (*2*). As a consequence, when the gas particle approaches the surface, one can describe the behavior of the system during a short time interval by means of some atomic-like orbitals or their hybrids, corresponding to the symmetry around the surface atom (which is the same one both for the array of the atomic orbitals and for their hybrids), which interact with the orbitals of the gas particle. When speaking about "atomic" orbitals in the metal surfaces, one has to specify their type, number, and spatial arrangement. This problem is usually handled by assuming that the direction and the type of the orbitals emerging into the free space are not too much different from those involved in the metal–metal bonds inside the crystal, consequently having the direction toward the nearest missing neighbors. Figure 3 represents the schematic view of these directions for an atom of the fcc lattice, showing the situation on various crystallographic planes. Evidently, the crystallographic planes do not differ only in the number and geometrical arrangement of the surface atoms, but also in the number and spatial orientation of the orbitals emerging from one surface atom.

The model of the metal surface, as described above, has been used recently by several authors for the interpretation of their results of chemisorption and catalytic studies (*67–72*), mostly without stating explicitly that the atomic orbitals are probably hybridized and can be approximated by purely atomic orbitals during the interaction only. The hybridization

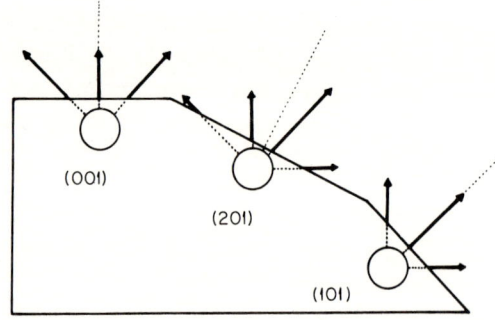

FIG. 3. The scheme of the directions toward the nearest missing neighbors, cut by the (001), (201), and (101) planes for an atom of the fcc lattice.

Fig. 4. Platinum image (helium, liquid hydrogen temperature).

of the atomic orbitals in clean metal surfaces has been discussed in the literature (72–76), mostly in connection with the strengthening of the surface metal–metal bonds and the rearrangement of the surface layer (in comparison with the equivalent layer inside the crystal).

The above-mentioned view of the metal surfaces can be used for the

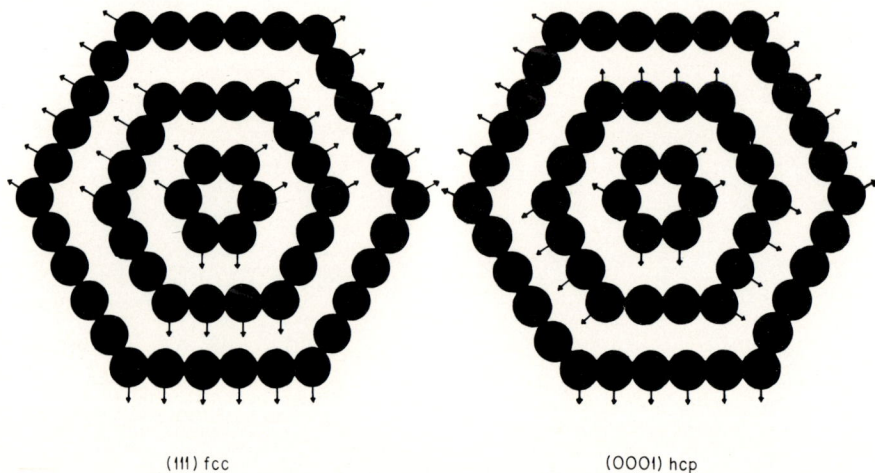

(111) fcc   (0001) hcp

FIG. 5. The scheme of the arrangement of atoms around the (111) plane of the fcc lattice and around the (0001) plane of the hcp lattice. The arrows represent those directions toward the nearest missing neighbors, which are sticking out perpendicularly to a given plane. No other directions are indicated. Only the atoms on edges are reproduced here.

interpretation of some fine details of the field ion images (77), making essentially one assumption only: the field ionization proceeds preferentially in those regions where the fully occupied orbitals of the image gas atoms can overlap with those exposed and only partially occupied orbitals of the surface metal atoms. As an example, we will discuss the platinum surface image (Fig. 4). Figure 3 shows the directions of the orbitals emerging from the (001), (201), and (101) planes. In the case of the (001) and (101) planes, they form a completely symmetrical pattern around the normal to the plane considered. On the (201) plane [and the same holds true for (301) plane], these directions form an asymmetric pattern. On the latter planes the favorable orbitals, which can considerably overlap with the orbitals of the image gas atoms, are those sticking out on one side of the edge of these planes, namely in the direction toward the (101) plane (compare Figs. 3 and 4). In this way it is possible to understand the difference in the brightness of otherwise equivalent rows of atoms on both sides of these net planes. Analogically the regional brightness around the (111) plane can be explained by comparing Figs. 4 and 5 (where only those directional orbitals are indicated that emerge perpendicularly to the particular plane).

The model of the (111) plane of the fcc lattice can be easily adapted for the (0001) plane of the hcp lattice because the atomic arrangement in both of these planes is the same one, only the sequence of successive layers differs (Fig. 5). Using the model for the (0001) plane, one can understand the alternating visibility of the atomic rows in certain directions around this plane (compare Figs. 5 and 6). These types of considerations can be applied to other fcc and hcp metals as well (*77–80*). The specific behavior of various metals of the same lattice type can be understood in terms of the different degree of occupation of the surface orbitals and of the varying polarization by the applied high electric field.

The basic assumption used for the interpretation of the field ion images, viz, the overlapping mechanism, was independently confirmed by the recent atom-probe experiments which have supplied evidence for the formation of a quasichemical complex between an inert gas atom and a metal

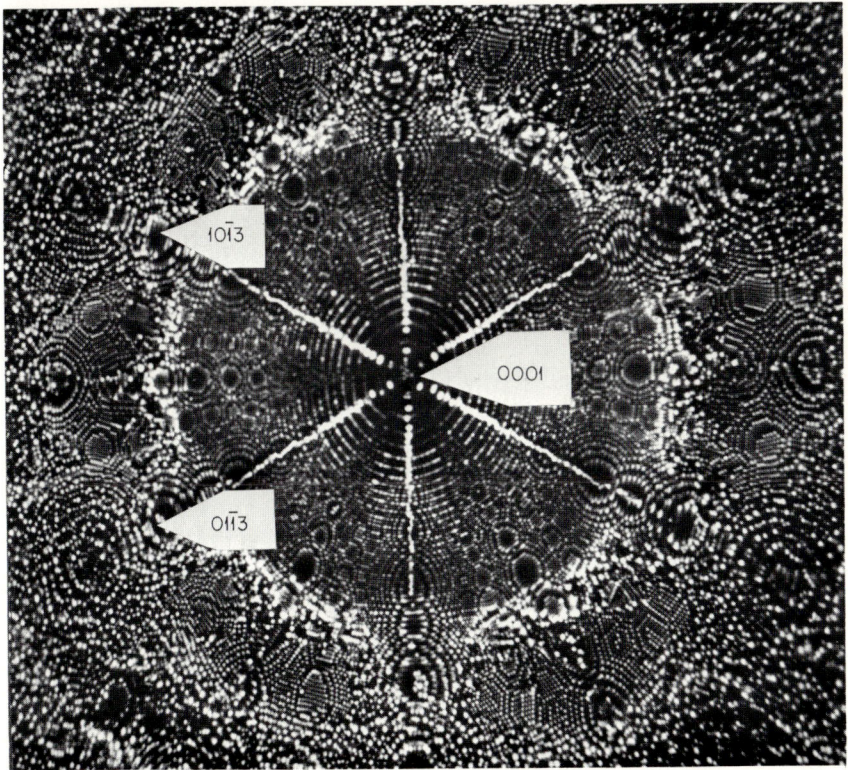

Fig. 6. Rhenium image (helium, liquid hydrogen temperature).

atom in the process of field ionization (e.g., WHe^{3+}, $RhHe^{2+}$) *(16, 80)*. The formation of a surface complex, even of a transient one, probably decreases its bond strength with the rest of the crystal and in this way the promoted field desorption can be explained *(81, 82)*.

The molecular orbital view of the metal surfaces permits the specification of all the effects that can be included under the term "surface rearrangement" *(77)*: (1) the change of the orientation of the surface orbitals, (2) the change of their occupancy by the electrons, and (3) the change of the position of a metal atom in the surface. The surface rearrangement in this general sense can explain both the so-called hydrogen promotion effect [the increased resolving power of the field ion microscope in the presence of low partial pressure of hydrogen *(77, 80)*] and the contrast reduction of the regional brightness in the presence of the chemisorbed layer *(77)*.

Further arguments supporting the idea of the localized character of the interaction of gas atoms with the metal surfaces can be found in the literature. (1) Using the field electron emission microscope, it was shown that the kinetics of oxygen chemisorption on tungsten did not depend on the degree of coverage of tungsten by germanium *(83)*. The authors themselves explained this finding by assuming the localized character of this interaction between oxygen and tungsten atoms. (2) In the study of hydrogen chemisorption on copper–nickel alloys it was shown that hydrogen is able to distinguish copper and nickel atoms from each other *(41)*. (3) The work function difference between (111) and (110) planes of tungsten as measured by the retarding potential technique was $\Delta\varphi = 1.6$ eV, however, $\Delta\varphi = 0$ resulted from the Auger electron ejection studies, where excited xenon atoms were used *(84)*. This result was explained as a consequence of the strong influence of the xenon ion, formed from the excited atom immediately adjacent to the metal surface. The strong influence of the ion causes great reduction of the effect of the surface topography and, consequently, the effective work function difference between these two planes decreases in this case almost to zero. Thus, the gas particle again interacts with its closest neighborhood only, without recognizing the rest of the crystal. (4) From the ion neutralization spectroscopy and the LEED study of the system X + nickel (X = O, S, Se) *(3)*, it was concluded that complexes of the type Ni_2X were formed on the Ni (100) plane. The energies, corresponding to the characteristic peaks on the energy distribution curves of the Auger electrons, were correlated with the orbital energies of the free X atoms, thus supplying evidence for the validity of the assumption that the surface complex can be approximated by the individual molecule, perturbed by the presence of the crystal.

V. Concluding Remarks

Limited numbers of metals are known as active catalysts, in contrast with the great variety of inorganic and organic compounds that might be active in the heterogeneous catalytic reactions. The number of active metals, interesting from the industrial point of view, is further decreased because some of them are too expensive for large-scale use. Of course, one cannot exclude the fact that a new technology will someday appear and several of these rare metals will become less expensive; nevertheless, for the present time the search for the rules of rational selection of a metallic catalyst does not seem to be of utmost importance.

It was shown in the preceding text that even in the simplest systems many different chemisorbed particles originate on the surface during the catalytic reaction. In principle most of them can interact with each other and probably with gaseous reaction components as well. As a consequence, any catalytic reaction represents a system of simultaneous reactions, and the problem is how to influence the course of a particular reaction—in other words, it is essentially the selectivity problem. Thus in catalysis by metals, probably the modification of the surface properties (by forming the alloys, stable surface complexes, or by the addition of promotors, etc.) seems to be the most promising direction of the further fundamental research.

ACKNOWLEDGMENT

The author would like to express his gratitude to his co-workers at the Institute of Physical Chemistry of the Czechoslovak Academy of Science, namely, to Drs. S. Černý, R. Merta, and Z. Bastl, all of whom read the manuscript and whose many suggestions have markedly improved the article.

REFERENCES

1. Ponec, V., and Černý, S., *Rozpr. Cesk. Akad. Ved, Rada Prirod. Ved.* **75**, No. 5 (1965).
2. Knor, Z., *J. Vac. Sci. Technol.* **8**, 57 (1971).
3. Hagstrum, H. D., *Phys. Rev. Lett.* **22**, 1054 (1969).
4. Grimley, T. B., *in* "Molecular Processes on Solid Surfaces" (E. Drauglis, R. D. Gretz and R. I. Jaffee, eds.), p. 299. McGraw-Hill, New York, 1969.
5. Henglein, A., Lacman, K., and Jacobs, G., *Ber. Bunsenges. Phys. Chem.* **69**, 279 (1965).
6. Herman, Z., Kerstter, J., Rose, T., and Wolfgang, R., *Discuss. Faraday Soc.* **44**, 123 (1967); Polanyi, J. C., *Discuss. Faraday Soc.* **44**, 293 (1967).
7. Becker, J. A., and Hartman, C. D., *J. Phys. Chem.* **57**, 157 (1953).
8. Hickmott, T. W., and Ehrlich, G., *J. Phys. Chem. Solids* **5**, 47 (1958).

9. Redhead, P. A., *Proc. Symp. Electron Vac. Phys., Balatonfoldvar, Hung., 1962* p. 89 (1963).
10. Delchar, T. A., and Ehrlich, G., *J. Chem. Phys.* **42,** 2686 (1965).
11. Tamm, P. W., and Schmidt, L. D., *J. Chem. Phys.* **51,** 5352 (1969).
12. Melmed, A. I., *in* "Molecular Processes on Solid Surfaces" (E. Drauglis, R. D. Gretz, and R. I. Jaffee, eds.), p. 105. McGraw-Hill, New York, 1969.
13. Muller, R. H., Steiger, R. F., Somorjai, G. A., and Morabito, J. M., *Surface Sci.* **16,** 234 (1969).
14. Heckingbottom, G., *Surface Sci.* **17,** 394 (1968).
15. Feinstein, L. G., and Macrakis, M. S., *Surface Sci.* **18,** 277 (1969).
16. Müller, E. W., McLane, S. B., and Panitz, J. A., *Surface Sci.* **17,** 430 (1969).
17. Block, J., *Z. Phys. Chem. (Frankfurt am Main)* **39,** 169 (1963).
18. Ponec, V., and Knor, Z., *Collect. Czech. Chem. Commun.* **26,** 29 (1961).
19. Knor, Z., and Ponec, V., *Collect. Czech. Chem. Commun.* **26,** 37 (1961).
20. Knor, Z., and Ponec, V., *Collect. Czech. Chem. Commun.* **26,** 961 (1961).
21. Knor, Z., and Ponec, V., *Collect. Czech. Chem. Commun.* **26,** 579 (1961).
22. Ponec, V., Knor, Z., and Černý, S., *Discuss. Faraday Soc.* **41,** 149 (1966).
23. Ponec, V., and Knor, Z., *Collect. Czech. Chem. Commun.* **27,** 1091 (1962).
24. Ponec, V., and Knor, Z., *Actes 2nd Congr. Int. Catal., Paris, 1960* p. 195 (1961).
25. Ponec, V., Knor, Z., and Černý, S., *Proc. 3rd Int. Congr. Catal., Amsterdam, 1964* p. 353 (1965).
26. Ponec, V., Knor, Z., and Černý, S., *Collect. Czech. Chem. Commun.* **29,** 3031 (1964).
27. Ponec, V., Knor, Z., and Černý, S., *J. Catal.* **4,** 485 (1965).
28. Ponec, V., Knor, Z., and Černý, S., *Collect. Czech. Chem. Commun.* **30,** 208 (1965).
29. Ponec, V., and Knor, Z., *Collect. Czech. Chem. Commun.* **25,** 2913 (1960).
30. Ponec, V., and Knor, Z., *Collect. Czech. Chem. Commun.* **27,** 1443 (1962).
31. Knor, Z., Ponec, V., and Černý, S., *Kinet. Katal.* **4,** 437 (1963).
32. Knor, Z., and Ponec, V., *Collect. Czech. Chem. Commun.* **31,** 1172 (1966).
33. Bastl, Z., *Collect. Czech. Chem. Commun.* **33,** 4133 (1968).
34. Bastl, Z., *Surface Sci.* **22,** 465 (1970).
35. Černý, S., Ponec, V., and Hládek, L., *J. Catal.* **5,** 27 (1966).
35a. Černý, S., and Cuřínová, A., unpublished observations (1971).
36. Hládek, L., *J. Sci. Instrum.* **42,** 198 (1965).
37. Knor, Z., *Catal. Rev.* **1,** 257 (1967).
38. Knor, Z., *Rev. Sci. Instrum.* **31,** 351 (1960).
39. Knor, Z., *Czech. J. Phys.* **13,** 302 (1963).
40. Ponec, V., and Knor, Z., *J. Catal.* **10,** 73 (1968).
41. Sachtler, W. M. H., and van der Plank, P., *Surface Sci.* **18,** 62 (1969).
42. Knor, Z., Ponec, V., Herman, Z., Dolejšek, Z., and Černý, S., *J. Catal.* **2,** 299 (1963).
43. Merta, R., and Ponec, V., *Proc. 4th Int. Congr. Catal., Moscow 1968.*
44. Merta, R., and Ponec, V., *J. Catal.* **17,** 79 (1969).
45. Merta, R., *Collect. Czech. Chem. Commun.* **36,** 1504 (1971).
46. Cukr, M., Thesis, Inst. of Phys. Chem., Czech. Acad. of Sci., Prague, 1969.
47. Kadlecová, H., Kadlec, V., and Knor, Z., *Collect. Czech. Chem. Commun.* **36,** 1205 (1971).
48. Phillips, J. C., *Phys. Rev. A* **140,** 1254 (1965).
49. Ehrenreich, H., *Opt. Prop. Electron. Struct. Metals Alloys, Proc. Int. Colloq., Paris, 1965* p. 109 (1966).
50. Harrison, W. A., *Phys. Rev.* **181,** 1036 (1969).
51. Friedel, J., *J. Phys. (Paris) Cl 1* **31,** 85 (1970).

52. Wannier, G. H., Misner, C., and Schay, G., *Phys. Rev.* **185**, 983 (1969).
53. Pettifor, D. G., *Proc. Phys. Soc., London (Solid State Phys.)* **3**, 367 (1970).
54. Brooks, H., *in* "Electronic Structure and Alloy Chemistry of Transition Metals" (P. A. Beck, ed.). Wiley, New York, 1963. [Russian transl. p. 9.]
55. Bakulin, E. A., Balabanova, L. A., and Bredov, M. M., *Fiz. Tverd. Tela* **12**, 72 (1970).
56. Gadzuk, J. W., *in* "The Structure and Chemistry of Solid Surfaces" (G. A. Somorjai, ed.), p. 43-1. Wiley, New York, 1969.
57. Nagy, D., and Cutler, P. H., *Phys. Rev.* **186**, 651 (1969).
58. Gadzuk, J. W., *Surface Sci.* **18**, 193 (1969).
59. Duke, C. B., and Alferieff, M. E., *J. Chem. Phys.* **46**, 923 (1967).
60. Horiuti, J., and Toya, T., *in* "Solid State Surface Science" (M. Green, ed.), p. 1. Dekker, New York, 1969.
61. Holscher, A. A., Thesis, Univ. of Leiden, Leiden, 1967.
62. Blyholder, G., and Coulson, C. A., *Trans. Faraday Soc.* **63**, 1782 (1967).
63. van der Avoird, A., *Surface Sci.* **18**, 159 (1969).
64. Zacharov, I. I., and Sutula, V. D., *Kinet. Katal.* **10**, 632 (1969).
65. Dunken, M. H., and Opitz, G., *Proc. 4th Int. Congr. Catal., Moscow 1968*.
66. Grimley, T. B., quoted in Sachtler and van der Plank (*41*), p. 75.
67. Bond, G. C., *Surface Sci.* **18**, 11 (1969).
68. Dowden, D. A., *J. Res. Inst. Catal., Hokkaido Univ.* **4**, 1 (1966).
69. Anderson, J. R., and Avery, N. R., *J. Catal.* **7**, 315 (1967).
70. Vasko, N. P., Ptushinskii, Y. G., and Chuikov, B. A., *Surface Sci.* **14**, 448 (1969).
71. Morgan, A. E., and Somorjai, G. A., *J. Chem. Phys.* **51**, 3309 (1969).
72. Tamm, P. W., and Schmidt, L. D., *J. Chem. Phys.* **51**, 5352 (1969).
73. Richman, M. H., *Jap. J. Appl. Phys.* **8**, 1273 (1969).
74. Rhodin, T. N., Palmberg, P. W., and Plummer, E. W., *in* "The Structure and Chemistry of Solid Surfaces (G. A. Somorjai, ed.), p. 22-1. Wiley, New York, 1969.
75. Palmberg, P. W., and Rhodin, T. N., *J. Chem. Phys.* **49**, 134 (1968).
76. Plummer, E. W., and Rhodin, T. N., *J. Chem. Phys.* **49**, 3479 (1968).
77. Knor, Z., and Müller, E. W., *Surface Sci.* **10**, 21 (1968).
78. Rendulic, K. D., and Knor, Z., *Surface Sci.* **7**, 205 (1967).
79. Reisner, T., Nishikawa, O., and Müller, E. W., *Surface Sci.* **20**, 163 (1970).
80. Müller, E. W., *Quart. Rev. Chem. Soc.* **23**, 177 (1969).
81. Tsong, T. T., and Müller, E. W., *Phys. Status Solidi* (a) **1**, 513 (1970).
82. Bell, A. E., Swanson, L. W., and Reed, D., *Surface Sci.* **17**, 418 (1959).
83. Sokolskaya, I. L., and Mileshkina, N. V., *Surface Sci.* **15**, 109 (1969).
84. Mac Lennan, D. A., and Delchar, T. A., *J. Chem. Phys.* **50**, 1772 (1969).
85. Fonash, S. J., and Schrenk, G. L., *Surface Sci.* **23**, 30 (1970).

Influence of Metal Particle Size in Nickel-on-Aerosil Catalysts on Surface Site Distribution, Catalytic Activity, and Selectivity

R. VAN HARDEVELD and F. HARTOG

Catalysis Department
Central Laboratory DSM
Geleen, The Netherlands

I. Introduction	75
II. Statistics of Surface Atoms and Surface Sites	77
A. General Remarks on Model Studies	77
B. Statistics of Surface Atoms	79
C. Statistics of Surface Sites	84
III. Infrared Studies on the Adsorption of N_2, CO, and CO_2	86
A. Experimental Procedure	86
B. Adsorption of N_2, CO, and CO_2 on Nickel Catalysts	87
C. Adsorption of N_2 and CO on Palladium and Iridium Catalysts	96
IV. Deuteration and Exchange of Benzene	100
A. Experimental Procedure	100
B. Reaction of Benzene with Deuterium on Nickel Catalysts	103
C. Reaction of Benzene with Deuterium on Iridium Catalysts	107
V. Conclusions	110
VI. Preparation and Characterization of Catalysts	110
References	112

I. Introduction

In the last few years remarkable progress has been made in the preparation of supported metal catalysts. Entirely new methods have been developed, comprising precipitation of the metal as an insoluble salt or hydroxide on the support under controlled conditions, or loading the support with the metal by means of ion exchange. A feature of catalysts prepared according to the former method (*1, 2*) is that, after reduction, they have a high metal content (50% by weight, or more), while the metal crystals are still small (20–40 Å) and distributed very uniformly over the support. The latter approach yields catalysts with metal crystallites of approximately 10 Å; however, the metal content is rather low [about 2% (*3–5*)].

These advances in catalyst preparation techniques have certainly stimulated the already growing interest in the relations between the catalytic and sorptive properties of catalysts and their mode of preparation. Many authors have studied the dependence of specific reaction rate upon particle size, mainly in hydrogenation, dehydrogenation, and hydrogenolysis reactions. The results of this work have recently been compiled by Schlosser (*6*).

In quite a number of cases the particle size was not found to have an effect on the specific reaction rate, whereas in some others this effect was clearly observed. Boudart *et al.* (*7*) coined the term "facile" for reactions in which the specific activity does not depend on the particle size, using the term "demanding" in referring to those cases where such a dependence exists.

The first time the present authors found a clear indication of an effect of the metal particle size on the properties of a metal-on-carrier catalyst was when one of them (v.H.) tried to confirm a statement by Eischens and Jacknow (*8*) to the effect that nitrogen is adsorbed on nickel at room temperature, as is evident from the appearance of an infrared absorption band at 2202 cm^{-1}. At first Eischens' results could only be duplicated by working in strict conformity to his method of sample preparation, i.e., impregnation of the support with a solution of $Ni(NO_3)_2$, drying, and reducing with hydrogen. When the sample was calcined before reduction no infrared band due to adsorption of nitrogen was observed. This induced van Hardeveld and van Montfoort (*9*) to extend the investigation to cover various nickel, palladium, and platinum catalysts. They found that a measurable amount of the infrared-active form of adsorbed nitrogen occurs only on metal crystallites with diameters in the range between 15 and approximately 70 Å. In studies on crystal models they noted that crystallites within this size range exhibit a relatively large number of a special type of sites, and also that these sites are much less prevalent on crystallites outside this range.

There exist two geometrically different varieties of these sites, which are referred to as B_5 sites because both can be made to accommodate a nitrogen molecule, which is then coordinated by five atoms. They occur at steps on the (100) and (111) planes, and particularly on (110), (311), and other high-index planes. A later paper by van Hardeveld and van Montfoort (*10*) contains additional evidence showing that the B_5 sites are indeed responsible for the infrared-active form of nitrogen adsorption, and also that the number of B_5 sites in the sample can be estimated with fair accuracy from the intensity of the 2200 cm^{-1} band. This means that infrared study of nitrogen adsorption can give valuable information about the structure of the surface of metal particles.

The outcome of this research naturally raised the question whether the catalysts investigated differed also in chemisorptive and catalytic properties. We therefore examined how the adsorption of CO and CO_2, and the reaction of benzene with deuterium proceed on these catalysts. This study forms the subject of the present article.

II. Statistics of Surface Atoms and Surface Sites

A. General Remarks on Model Studies

To be capable of relating the adsorptive and catalytic properties of metal crystals to the crystal size, one must know the structure of the crystal surface on an atomic scale. As a rule, this information is obtained from studies on crystal models. A description of infinitely extended crystal planes has been given by Nicholas (11), while finite crystals have been treated by various other authors. Poltorak and Boronin (12) analyzed the surface of a fcc octahedral crystal, while Schlosser (6) calculated the number of free valencies and the various types of surface atoms of a cubo-octahedral crystal. The latter authors also investigated other, less symmetric crystals bounded by (111) and (100) planes, as well as the effect of progressive rounding of the corners and edges of the symmetrical cubo-octahedron. Van Hardeveld and Hartog (13) published an extensive calculation of the number of different surface atoms and surface sites on various fcc, bcc, and hcp crystals. All these studies were undertaken on crystals free of lattice defects; the surface structures of defects in crystals have been established by Jaeger and Sanders (14).

Application of models always raises the question of whether the models chosen are close enough to reality to warrant that the conclusions drawn from the model studies will be valid. Determining the shape of very small metal crystals by electron microscopy is not possible owing to the limited resolving power of the electron microscope.

Derivation of the shapes from theoretical considerations also presents nearly insurmountable difficulties, the first being that every geometrically defined shape can exist only at certain discrete values of the total number of atoms N_T (in the case of an octahedron: 6, 19, 44, 85, etc.). Generally, the last step in the preparation of supported metal crystals is a reduction process in which the metal crystals are formed. When this reduction has gone to completion, the over-all process of crystal growth stops. From this moment onward, the number of atoms in every crystal does not change anymore, unless metal atoms can migrate over the support from one particle to another. However, conditions allowing this migration to occur give rise to sintering and, therefore, will generally be avoided in catalyst preparation.

Hence, the number of atoms of which a crystal is built up equals the number it has happened to acquire in the reduction process; so there is no ground for expecting that only crystals with certain discrete values of N_T should occur. Comparison of differently shaped crystals is a valid procedure only if the crystals contain equal numbers of atoms. If there are indications suggesting that the crystals are so shaped as to have a minimum of free energy, the best one can do is to determine the equilibrium shape for a number of crystals with randomly chosen values of N_T. In our earlier paper (*13*) we gave an example of such a calculation for a fcc crystal consisting of 683 atoms, using the number of free valencies as an approximate stability criterion. We arrived at the conclusion that at this number of atoms a nearly spherical shape possesses the highest stability. Alpress and Sanders (*15*) calculated the energy for crystals of various shapes containing up to 2000 atoms. They assumed that the total energy is the sum of independent, pairwise interactions that can be described by means of a Morse or Mie potential. From the energy versus N_T plot these workers concluded that the stablest shape achievable in most cases is that of a multiply twinned icosahedral crystal. Crystals showing this configuration had already been observed by them (*16*) in deposits of gold and nickel evaporated onto mica.

Second best in their view is the octahedron. However, Alpress and Sanders did not make mention of the cubo-octahedron, which, owing to its smaller surface-to-volume ratio, is probably stabler than the octahedron. Work along similar lines has been done by Fukano and Wayman (*17*). These authors included only the interactions between nearest neighbors in their considerations; however, they also made allowance for the interaction with the substrate by assuming that an atom contacting the substrate is joined to it by two bonds.

There exists, however, some doubt as to whether catalysts made under the usual conditions always consist of crystals of equilibrium shape. For example, the electron micrographs of our catalyst B (see Fig. 14) show that the crystal shapes are not very uniform and rather asymmetric, which certainly does not suggest that they are in a state of thermodynamic equilibrium. Displacement of individual surface atoms will ultimately yield a crystal in which no surface atom can find a site with a binding energy exceeding that of the site occupied by it. However, this situation need not prove that the crystal has assumed its equilibrium shape; a metastable state is much more likely. For the equilibrium shape to be attained, it is necessary not only that the surface atoms be able to migrate, but also that the interior atoms have a certain degree of mobility. The temperatures at which these conditions are fulfilled are rather high, especially for the more refractory metals, and are usually not attained during catalyst preparation. Drechsler and Nicholas (18) have shown by means of a calculation that field emitter tips are in thermodynamic equilibrium, but these tips are annealed at temperatures around 2000°C, whereas the temperatures employed in catalyst preparation usually fall in the range between 400–500°C or lower.

In conclusion, it can be said that there is always some uncertainty as to the correctness of the model(s) chosen. However, as will be pointed out in the next section, studies on several plausible crystal shapes enable us to draw some conclusions that are independent of the actual crystal shapes chosen.

B. Statistics of Surface Atoms

A surface atom has, by definition, a smaller number of nearest neighbors than an atom in the interior of the crystal. An interior atom in the metals with which we are concerned here, and which are all fcc, has twelve nearest neighbors. On every metal crystal there will be found various types of surface atoms that differ in the number j and in the arrangement of their

TABLE I

Surface Atoms of fcc Crystals

Coordination number	Notation sources				Occurrence
	van Hardeveld and Hartog (13)	Nicholas (11)	Bond (19)	Perry and Brandon (20)	
4	$C_4^{9,10}$				Vertex of octahedron
6	C_6^6	C_6	C_6	6.3; 6.4^a	Vertex of cubo-octahedron
7	C_7^5	C_7	E_7^1	7.3a; 7.4a	Intersection of (100) and (111) planes; (311) plane
7	C_7^9	C_7†	E_7	7.3b; 7.4b	Intersection of (111) planes; (110) plane
8	$C_8^{4,5}$	C_8	P_8		(100) plane
9	C_9^3	C_9†	P_9		(111) plane
10	C_{10}	C_{10}	P_{10}		(311) plane; steps at (100) and (111) planes (B_5 site)
11	C_{11}	C_{11}	P_{11}		(110) plane; steps at (111) plane (B_5 site)

^{a} The second digit gives the number of next-nearest neighbors.

nearest neighbors. We denote such a surface atom by the symbol $C_j^{p,q,r,\ldots}$ where the lower index gives the number of nearest neighbors, and the upper indices the arrangement of these neighboring atoms. How these upper indices are obtained is fully set forth in our earlier paper (*13*). A basic assumption made in our work is that the chemisorptive and catalytic properties of a surface atom vary with the number and arrangement of its nearest neighbors, which implies that the influence of more remote atoms can, to a first approximation, be neglected. Table I lists the types of surface atoms that are observed most frequently on fcc crystals, their notation as given by various authors, and the places where they are found on different crystals.

In what follows we shall denote the number of $C_j^{p,q,r,\ldots}$ atoms in a crystal by $N(C_j^{p,q,r,\ldots})$, and the total number of surface atoms by N_s. The crystal size will be expressed in terms of the relative diameter d_{rel}. This dimensionless parameter is equal to the ratio of the diameter of a sphere with a volume equal to N_T times the volume of the unit cell divided by the number of atoms in this cell, to the diameter of one atom. Hence, this parameter does not depend upon the lattice constant, and its use relieves us of the need to define a diameter for every crystal shape.

With fcc crystals we have $d_{\text{rel}} = 1.105 \sqrt[3]{N_T}$. According as a metal crystal is reduced in diameter with retention of its shape, a growing fraction

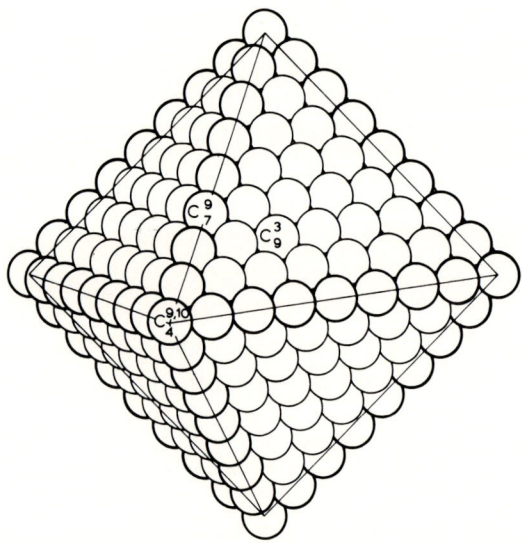

Fig. 1. Face-centered cubic (fcc) octahedron, $m = 9$ (m is the number of atoms along an edge).

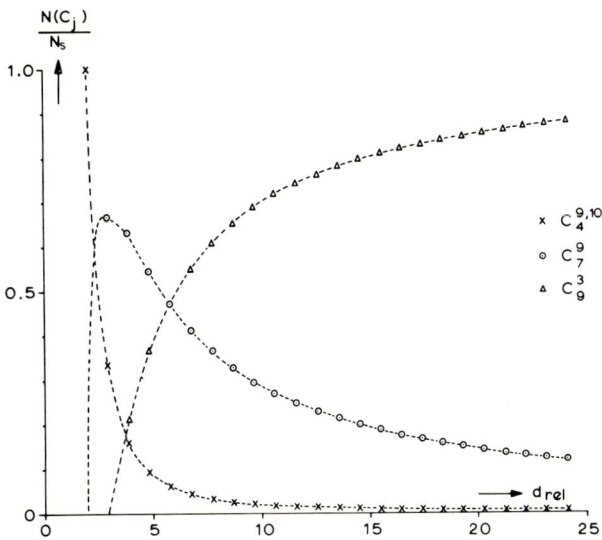

Fig. 2. Plots of $N(C_j)/N_s$ versus d_{rel} for a fcc octahedron.

of the total number of surface atoms will come to lie near the vertices and edges of the crystal. Let us consider, for example, the regular octahedron depicted in Fig. 1.

At the crystal corners $C_4^{9,10}$ atoms are located; their number is six, irrespective of the crystal size. The number of C_7^9 atoms, occurring along the edges, varies linearly with the crystallite size. The faces are built up of C_9^3 atoms, whose number is a quadratic function of the crystallite size. Hence, the surface of a very large octahedron consists almost exclusively of C_9^3 atoms, whereas in a very small crystallite it is made up to a substantial degree of $C_4^{9,10}$ atoms. One may reasonably assume that somewhere in between these extremes there will be a region of crystallite diameters where a considerable fraction of the surface atoms are C_7^9 atoms. For quantitative confirmation we refer to Fig. 2, in which the quantities $N(C_j^{p,q,r,\ldots})/N_s$ for the various atom types are plotted versus d_{rel}. (In this and other similar plots the lines drawn through the points merely serve as an illustration, and do not refer to any actual crystals.) The main conclusions that can be drawn from Fig. 2 are:

(1) The most drastic changes in the composition of the surface take place in the region of d_{rel} values below approximately 15, i.e., in particles smaller than ~40 Å. This has already been pointed out by Poltorak and Boronin (*12*).

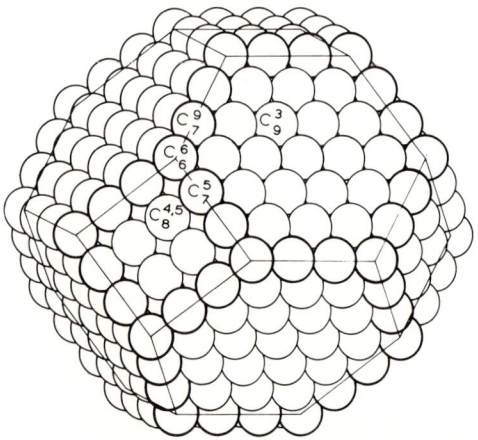

Fig. 3. Face-centered cubic cubo-octahedron, $m = 4$.

(2) No great variations in surface composition with crystal diameter occur when the particle diameter exceeds ~100 Å. The surface of these crystals is built up almost entirely of C_9^3 atoms.

(3) Edge atoms predominate in the region around $d_{rel} \sim 5$, and only at still smaller diameters does the fraction of corner atoms become very large.

That the validity of these conclusions is not restricted to the model chosen will appear from the following discussion on several other models. Let us first consider the cubo-octahedron depicted in Fig. 3. This highly symmetric crystal is bounded by (111) and (100) planes. It has altogether five types of surface atoms: the (100) and (111) planes contain $C_8^{4,5}$ and C_9^3 atoms, respectively; where two (111) planes intersect C_7^9 atoms are found, and at the intersection of a (111) plane with a (100) plane C_7^5 atoms occur, while a C_6^6 atom is present at each corner. The C_7^9 and C_7^5 atoms differ only in the geometrical arrangement of their nearest neighbors. When we compare Fig. 4 with Fig. 2 it becomes evident that the proportions of face atoms, edge atoms, and corner atoms to the total number of surface atoms of the cubo-octahedron change with particle size in essentially the same way as they do in the octahedron. A minor complication is that on the cubo-octahedron there occur two types of face atoms ($C_8^{4,5}$ and C_9^3) and edge atoms (C_7^5 and C_7^9). Another relatively unimportant difference is that owing to the hexagonal outline of the (111) planes even the smallest cubo-octahedron ($N_T = 38$; $d_{rel} = 3.86$) contains some C_9^3 atoms.

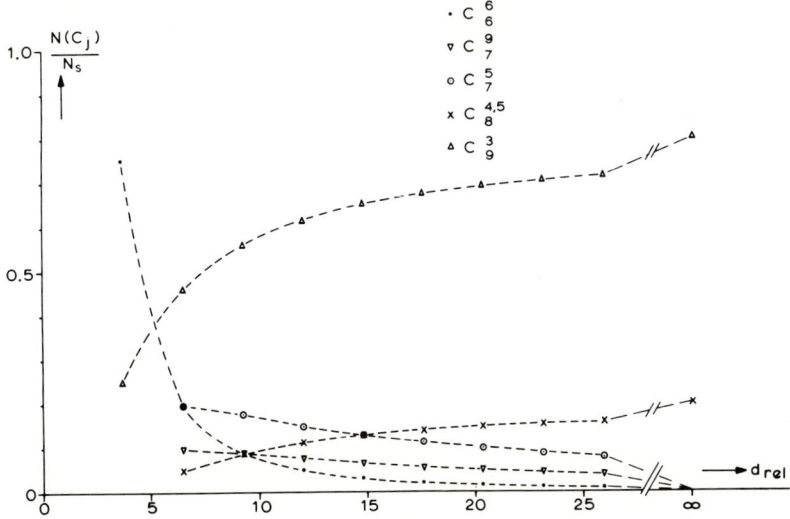

Fig. 4. Plots of $N(C_j)/N_s$ versus d_{rel} for a fcc cubo-octahedron.

Both the octahedron and cubo-octahedron are "complete" crystals, i.e., they do not contain incomplete atom layers in their planes and, hence, are devoid of B_5 sites. We know, however, that also "incomplete" crystals occur in our catalyst samples. An example of this type is the octahedron-

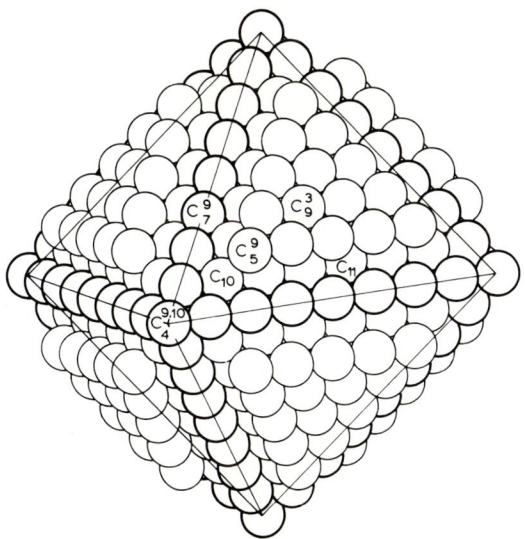

Fig. 5. Face-centered cubic octahedron-max B_5.

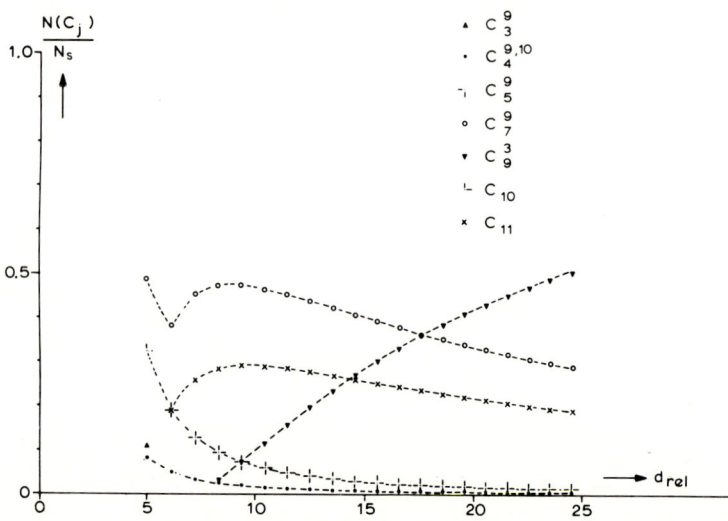

Fig. 6. Plots of $N(C_j)/N_s$ versus d_{rel} for a fcc octahedron-max B_5.

max B_5. This shape (Fig. 5) is derived from an ordinary octahedron by placing an incomplete layer of atoms on every (111) plane, the size of these layers being so chosen that the largest possible number of B_5 sites will be generated along their edges. In addition to the various types of surface atoms present on the octahedron, this crystal also contains C_5^9, C_{10}, and C_{11} atoms. The first two may be regarded as corner atoms in the sense that their numbers are independent of the crystal diameter. The C_{11} atoms are found along the edges and their number varies linearly with the diameter. Plots of $N(C_j)/N_s$ versus d_{rel} are shown in Fig. 6; the similarity with Figs. 2 and 4 is evident. The principal difference is that the region where the most profound changes in surface composition occur is slightly extended toward the side of larger diameters. Apart from this, the conclusions drawn with regard to the octahedron apply also to this representative of the class of incomplete crystals.

C. Statistics of Surface Sites

Up to now only individual surface atoms have been considered. In many cases, however, a sorptive process or a catalytic reaction requires the combined action of a number of surface atoms. In these cases it is the number of sites active in the process under consideration that is of paramount importance. In dealing with reactions requiring the presence of B_5 sites, we are in the fortunate circumstance that the number of these sites can be derived by IR measurement of the amount of nitrogen adsorbed under suitable conditions. In these reactions the catalytic activity should show a strong correlation with the number of B_5 sites. In other cases we must again resort to crystal models. By way of an example we shall determine the number of B_3 sites on a regular octahedron, this site being defined as an array of three surface atoms lying at the vertices of an equilateral triangle with sides equal to the atomic diameter. The properties of such a site will depend upon the coordination numbers of the atoms constituting it. As can be seen in Fig. 1, there are four types of sites that show the above-mentioned geometry; they differ, however, in the coordination number of the constituent atoms, and are referred to as $B_3^{4,7,7}$, $B_3^{7,7,9}$, $B_3^{7,9,9}$, and $B_3^{9,9,9}$, where the upper indices denote the coordination numbers of the three atoms of the site in question. Only the smallest octahedra with N_T values of 6 and 19, respectively, contain other B_3 sites, viz, $B_3^{4,4,4}$ and $B_3^{7,7,7}$. For the details of the calculation, which is rather straightforward, the reader is referred to our earlier paper (13); only the results will be represented here. The total number of B_3 sites, $N(B_3)_T$ turns out to be equal to $8(m-1)^2$, where m is the number of atoms along an edge, whereas $N_s = 4(m-1)^2 + 2$. It follows therefore that from $m = 6$ onward, $N(B_3)_T/N_s$ is practically constant. This value of m corresponds to a crystal diameter of 15 Å, a finding that can be generalized by stating that if a given type of site is not confined to special parts of the surface (as are, e.g., B_5 sites), the number of these sites will be proportional to the number of surface atoms (and, hence, to the surface area), unless the dimensions of the site are comparable to the diameter of the crystal.

Though the number of B_3 sites per unit of metal surface area is practically independent of the crystallite size, the catalytic activity may yet vary owing to the dependence of the catalytic activity of a site upon the coordination numbers of its constituent atoms. A plot of $N(B_3^{i,j,k})/N(B_3)_T$ for the various $B_3^{i,j,k}$ sites versus $d_{\rm rel}$, shown in Fig. 7, is in all respects analogous to the plot of $N(C_j)/N_s$ in Fig. 2. The cause of this analogy is that while surface atoms can be divided into corner atoms, edge atoms, and atoms at the faces of the crystal a similar distinct can be made for the various $B_3^{i,j,k}$ sites as the $B_3^{4,7,7}$ sites occur only at the corners, the $B_3^{7,7,9}$ and $B_3^{7,9,9}$ sites at the edges, and the $B_3^{9,9,9}$ sites at the faces of the crystal. This similarity of Figs. 2 and 7 leads to the important conclusion that the relations between specific activity and metal particle size in reactions proceeding on individual atoms will be of the same nature as those in reactions taking place on sites of a not too complicated structure.

From a consideration of the Figs. 2, 4, 6, and 7 some conclusions can be drawn. The specific activity in the range upward of approximately 70 Å will not be very sensitive to changes in metal particle size. The greatest

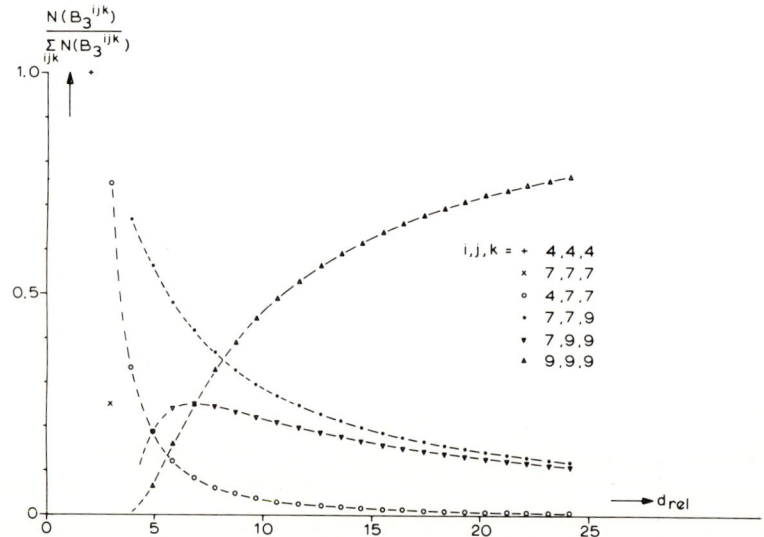

FIG. 7. Plot of $N(B_3^{i,j,k})/N(B_3)_T$ for various $B_3^{i,j,k}$ sites versus d_{rel} for a fcc octahedron.

variations in specific activity will be found in particles smaller than 40 Å. When the faces of the crystal are more active than the corners and the edges, the specific activity will increase with the crystallite diameter. This increase will rarely exceed a factor of ~2. When the edges of the crystal are far more active than the corners and the faces, the greatest specific activity will be found in metal crystals of about 15–20 Å diameter. When, finally, by far the greatest activity resides in the corners of the crystal, the specific activity will decrease with increasing metal particle diameter. Even when the differences in catalytic activity among corner, edge, and face atoms (or sites) are considerable, the specific catalytic activity of a number of catalyst samples will not show any appreciable variation unless certain conditions are fulfilled. All samples should preferably have a narrow particle size distribution, while the mean particle sizes of the sample should vary between 100 and 10 Å or less. It is not surprising therefore that only a few examples of a large particle size effect have been described in the literature.

III. Infrared Studies on the Adsorption of N_2, CO, and CO_2

A. Experimental Procedure

The infrared spectra were recorded with a Grubb–Parson single-beam grating spectrometer provided with a GS 4-type monochromator (resolution about 2 cm^{-1}). The whole apparatus was purged with dry nitrogen. The design of the infrared cell is practically identical to that described by Peri and Hannan (*21*). The infrared part of the cell can be cooled to $-100°$C by passing a stream of cooled nitrogen gas through the space between the double walls of the cell. Evacuating the cell could be done by connecting it to a conventional vacuum system comprising a two-stage diffusion pump. Dynamic vacuums better than 10^{-6} Torr could be reached in this way. The hydrogen for reducing the catalysts was purified by passing it over a deoxo-catalyst (Baker) and through a trap cooled in liquid nitrogen. The nitrogen was obtained from Air Liquide (type A 47; purity, >99.997%) and used without further purification. Carbon monoxide was obtained from the same source (type N 45; purity, >99.995%) and used as received. The carbon dioxide was purified by repeated distillation with retention of the middle fractions only. No impurities could be detected in the pure carbon dioxide by mass spectrometry.

The catalyst powders were compressed to thin disks under a pressure of about 50 kg/cm^2, with the exception of the alumina-supported catalysts which required a pressure of 1500 kg/cm^2 to obtain reasonable transmittance. The samples were reduced in a stream of hydrogen supplied at a rate of 10 l hr^{-1} (SV ~30,000 hr^{-1}). The temperatures of reduction were 350°–450°C for the nickel samples, 475°C for the palladium samples, and 425°C for the iridium catalysts.

Next, the adsorbed hydrogen was removed by evacuation at elevated temperatures. The choice of the evacuation and reduction temperatures was based on the following considerations. Depending on the type of metal, removal of adsorbed hydrogen within a reasonable time calls for temperatures in the range 350–450°C (*22, 23*), while the reduction should be carried out at some 20–30°C above the evacuation temperature since, otherwise, water given off by the support during evacuation will poison the catalyst. Hence, the somewhat unusually high reduction temperatures employed with the noble metal catalysts.

The time of evacuation was 4 hours for the nickel and iridium samples and 16 hours for palladium. The ultimate vacuums were about 10^{-6} Torr.

B. Adsorption of N_2, CO, and CO_2 on Nickel Catalysts

The analytical data of the catalysts used are given in Table II. The number of B_5 sites was determined from infrared measurements and nitrogen adsorption isotherms in the way outlined by van Hardeveld and Montfoort (*10*). The values found are higher than those mentioned in an earlier paper (*24*), owing partly to an improvement of the method for determining the extinction coefficient per molecule of nitrogen adsorbed,

TABLE II

Analytical Data of Nickel Catalyst

	Reduction		Metal content after reduction (%)	$\bar{d}_w$, from x-ray line broadening (Å)	$\bar{d}_{vs}$, from H$_2$ chemisorption at room temp. (Å)	Metal particle size distribution from electron micrographs	Number of B$_5$ sites per mg of Ni	Number of B$_5$ sites per cm^2 of Ni surface
	Time (hr)	Temp. (°C)						
A$_f$	16	350	7.4	70	48	Broad, with a large fraction of particles of d <70 Å	20 × 10^{16}	14 × 10^{13} [a] 21 × 10^{13} [b]
A$_0$	16	350	6.7	70	59	Broad, but fewer small particles than in A$_f$	5 × 10^{16}	4.2 × 10^{13} [a] 4.0 × 10^{13} [b]
B	16	425	6.8	185	214	Very broad	<10^{16}	<3 × 10^{13} [a] <2 × 10^{13} [b]
C	65	450	23.3	20	53	Very sharp	40 × 10^{16}	34 × 10^{13} [a] 13 × 10^{13} [b]
D	340	425	25.3	21	44	Very sharp	40 × 10^{16}	25 × 10^{13} [a] 12 × 10^{13} [b]

[a] The surface area per gram of nickel was calculated from $\bar{d}_{vs}$.
[b] The surface area per gram of nickel was calculated from $\bar{d}_w$.

partly to the allowance now made for the fact that at full coverage only half of the B_5 sites are occupied. Proof for this partial coverage was obtained from a study on the intermolecular interaction of the adsorbed N_2 molecules by means of nitrogen isotopes (*10*).

1. *Adsorption of* CO

Figure 8 shows the spectra of CO adsorbed on our samples; they were taken at a CO pressure of about 1 torr. To facilitate comparison, the extinction per square meter of nickel surface, as calculated from the spectra and the analytical data of the samples, has been plotted versus the wavenumber. It should be borne in mind that in the calculation of the extinction values, use has been made of the experimentally determined nickel surface areas. Hence, all inaccuracies in the surface area measurements will be reflected in the values of extinction per square meter (E/m^2).

The spectra are arranged in the order B, A_o, A_f, C, D, i.e., in the order

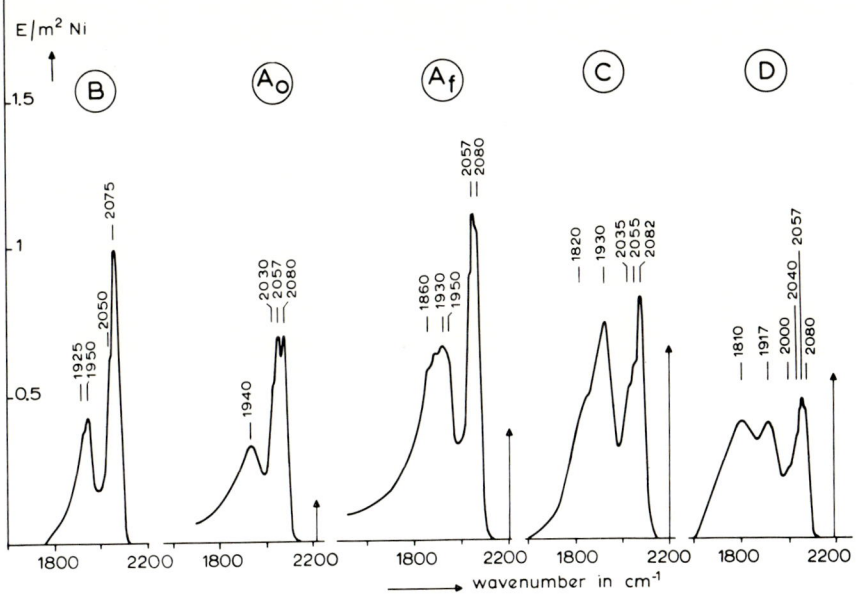

FIG. 8. Spectra of CO adsorbed on nickel catalysts B, A_o, A_f, C, and D taken at 1 Torr CO pressure and room temperature. The extinction per square meter of Ni surface area is plotted along the ordinate. The specific metal surface used in the calculation of the extinction per square meter of Ni surface area was determined from $\bar{d}_{vs}$ for the catalysts B, A_o, A_f, and from $\bar{d}_w$ for the catalysts C and D. The length of the arrow is a measure of the number of B_5 sites on the surface (see text).

of decreasing particle size. Turning now to the spectra themselves, we note that the CO adsorbed on catalyst B (no measurable amount of B_5 sites) produces a prominent peak at 2075 cm^{-1} with a shoulder at 2050 cm^{-1}, and another peak at 1950 cm^{-1} with a shoulder at 1925 cm^{-1}. In the A_0 sample these latter two bands have coalesced into a single one with a maximum at 1940 cm^{-1}. Obviously, the 1925 cm^{-1} band has distinctly increased in intensity relative to that at 1950 cm^{-1}. Another feature of the low-frequency part of this spectrum is the enhanced extinction in the 1700–1900 cm^{-1} region. The high-frequency part now contains two bands of approximately equal intensities centered at 2057 and 2080 cm^{-1}, while a new band makes its appearance as a shoulder at 2030 cm^{-1}. Passing over now to the catalysts with still smaller nickel particles, we note that the low-frequency part of the spectrum (i.e., below 2000 cm^{-1}) becomes even more pronounced owing to the appearance of bands around 1920 cm^{-1} and 1810 cm^{-1}; in the spectrum of sample D these are so strong as to completely obscure the 1950 cm^{-1} band that was so prominent in sample B. Important bands are observed also in the high-frequency part of the spectrum. A disturbing factor is the 2057 cm^{-1} band, which must be ascribed to $Ni(CO)_4$, present either in the gas phase or adsorbed on the support. Obviously, the infrared spectrum of CO adsorbed on nickel varies strongly with the particle size, the main difference between the spectra of CO adsorbed on large and on small crystallites being a relatively greater intensity of the bands around 1800 cm^{-1} in the latter case. Before discussing these results in detail we shall briefly review the data mentioned in the literature. A summary listing the frequencies of the bands observed by various authors is shown in Table III.

According to Eischens *et al.* (*25*), the adsorption bands above 2000 cm^{-1} must be ascribed to the stretching vibration of linearly bonded CO, while the bands below this frequency are due to bridge-bonded CO. This assignment was based upon the IR spectrum of $Fe_2(CO)_9$. Blyholder (*31*) attacked this view, stating that the intensities of these bands do not seem to be related to each other in a systematic manner. Moreover, he calculated that the bridge structure must also have an adsorption band in the 700–1000 cm^{-1} region. This band, however, was not seen even in those cases where the 1940 cm^{-1} band was rather strong. In a later paper, Blyholder (*32*) demonstrated by means of molecular orbital calculations that if allowance is made for the surface heterogeneity, all the experimental data available can equally well be explained by assuming that the CO is adsorbed in the linear form only. He argues that the adsorbed CO will be bonded more strongly on the more exposed metal atoms, with the result that the CO stretching frequency will shift toward lower values, so, quite conceivably, the infrared absorption band of CO adsorbed on edges, ridges, and corners

TABLE III

Wavenumbers (cm^{-1}) of CO-Absorption Bands Observed By Various Authors

Eischens et al. (25) Ni/SiO$_2$	Garland (26) Ni/Al$_2$O$_3$	Yates and Garland (27) Ni/SiO$_2$	O'Neill and Yates (28) Ni/SiO$_2$	O'Neill and Yates (28) Ni/Al$_2$O$_3$	O'Neill and Yates (28) Ni/TiO$_2$	Garland et al. (29) Ni film evaporated in CO of 2 Torr pressure	Garland et al. (29) Ni film evaporated in CO of 12 Torr pressure	Bradshaw and Pritchard (30) Ni film evaporated in vacuo at −160°C	Bradshaw and Pritchard (30) Ni film evaporated in vacuo at −120°C	Bradshaw and Pritchard (30) Ni film evaporated in CO of 2 Torr pressure at 25°C
2075	2075	2082 2057	2080	2075	2080	2083 2058	2058	2070	2070	2050
2030 –2040	2045	2035	2025 –2050	2025 –2030	2030	2025 –2030				
		1963			1995					
	1960			1950 –1960		1960?		1710 –1960	1850 –1960	1720 –1970

of the crystal will come to lie in the 1800–1900 cm^{-1} region. It is generally agreed that the infrared spectra of adsorbed CO furnish strong evidence for the heterogeneity of the surface. Eischens et al. (*25*) and Eischens and Pliskin (*33, 34*) are of the opinion that the variations in the intensity ratios of the various bands as a function of surface coverage is due to the presence of different crystal faces which are homogeneous themselves. O'Neill and Yates (*28*) ascribe the spectral differences observed upon variation of the type of support to an effect of the support on the chemisorption properties of the metal. This view was opposed by Yates and Garland (*27*), who studied the spectra of CO chemisorbed on three nickel-on-silica samples differing in nickel content only. They noted marked differences in the intensity ratios of the absorption bands obtained with their samples, and even found that certain absorption bands were entirely absent in some cases. They state that the bands at 2035, 2057, and 2082 cm^{-1} are due to CO linearly adsorbed on crystalline, semicrystalline, and dispersed nickel sites, respectively, and that the band at 1915 cm^{-1} is caused by CO adsorbed in the bridge form. Finally, the band at 1930 cm^{-1} should, according to the same authors, be assigned to the stretching vibration of CO molecules adsorbed in the structure:

$$\begin{array}{c} O \\ \parallel \\ OC \diagup C \diagdown CO \\ \mid \mid \\ -Ni\!-\!\!-\!\!-\!Ni- \end{array}$$

Garland et al. (*29*) studied the spectra of CO chemisorbed on nickel films and explained the differences observed between them after changes of the evaporation conditions along the same lines as did Yates and Garland (*27*) for supported catalysts. We hold that there is no conclusive evidence in favor of the existence of bridge-bonded CO in nickel and share Blyholder's opinion (*32*) that the infrared spectra can be explained on the assumption that CO is bonded in linear form only.

In our discussion of the spectra, the 2057 cm^{-1} band will be left out of consideration, because this band is due to small amounts of Ni(CO)$_4$. For the sake of convenience, the origin of the 2050–2080 band will be dealt with later. From our treatment of the statistics of surface atoms we know that the catalyst with the largest metal particle size, i.e., catalyst B, has a surface consisting predominantly of C_8 and C_9 atoms. Accepting Blyholder's view (*32*) that CO adsorbed on more exposed metal atoms is bound more strongly and has its stretching frequency shifted to lower wavelengths, we may safely assign the band around 1950 cm^{-1} and that at 2030–2075 cm^{-1} to CO bonded to C_8 and C_9 atoms, respectively. A decrease of the metal particle size leads to an increase in the fraction of surface atoms with lower coordination numbers, i.e., C_6 and C_7 atoms. Looking at the spectra we see that catalysts with smaller nickel particles, e.g., C and D, produce relatively strong bands at 1810 and 1920 cm^{-1} and also that these bands are either absent or much less prominent in catalyst B. We therefore ascribe the 1920 band to CO attached to a C_7 atom, and the 1810 band to CO attached to a C_6 atom.

This brings us to the discussion of the 2080 cm^{-1} band. This band appears only at relatively high CO pressures and, as found by Eischens *et al.* (*25*), is the first to vanish upon desorption of the CO. In our view this band owes its existence to species like:

$$\text{Ni}\begin{smallmatrix}\diagup\text{CO}\\\diagdown\text{CO}\end{smallmatrix} \quad \text{or perhaps even} \quad \text{Ni}\begin{smallmatrix}\diagup\text{CO}\\-\text{CO}\\\diagdown\text{CO}\end{smallmatrix}$$

These structures can, for steric reasons, be formed only when the coordination number of the nickel atom is 7 or less. We call this species the subcarbonyl species as it is an intermediate in the formation of $Ni(CO)_4$. Further evidence on this point will be given below.

2. *Dissociative Adsorption of* CO_2

From thermodynamic considerations it is evident that bulk nickel cannot be oxidized by CO_2. However, it is not justified to conclude from this that dissociative chemisorption of CO_2 will not occur. Consider, for example, the chemisorption of oxygen or hydrogen which on several metals takes place under conditions where bulk oxides or hydrides are not at all thermodynamically stable. Dissociative adsorption of CO_2 has indeed been observed by Eischens and Pliskin (*35*).

Upon exposure of our samples at room temperature to a CO_2 pressure of 50 Torr, no adsorption was observed on sample B; on sample A_0 some reaction took place, but the resulting infrared adsorption bands were too weak to permit proper identification. Distinct absorption spectra were obtained on the other samples; these are shown in Fig. 9.

The low-frequency band in the spectrum of sample A_f was superposed on a rather intense high background absorption band; its exact shape could not be determined. The conclusion that these bands are due to the presence of CO at the surface follows from their close similarity to those found upon adsorption of CO on the same samples. Hence, dissociative chemisorption of CO_2 does indeed take place. However, the CO bands observed upon dissociative adsorption of CO_2 are much weaker than those produced by CO directly adsorbed as such. This becomes evident if we compare them with the nitrogen absorption bands in Figs. 8 and 9 where

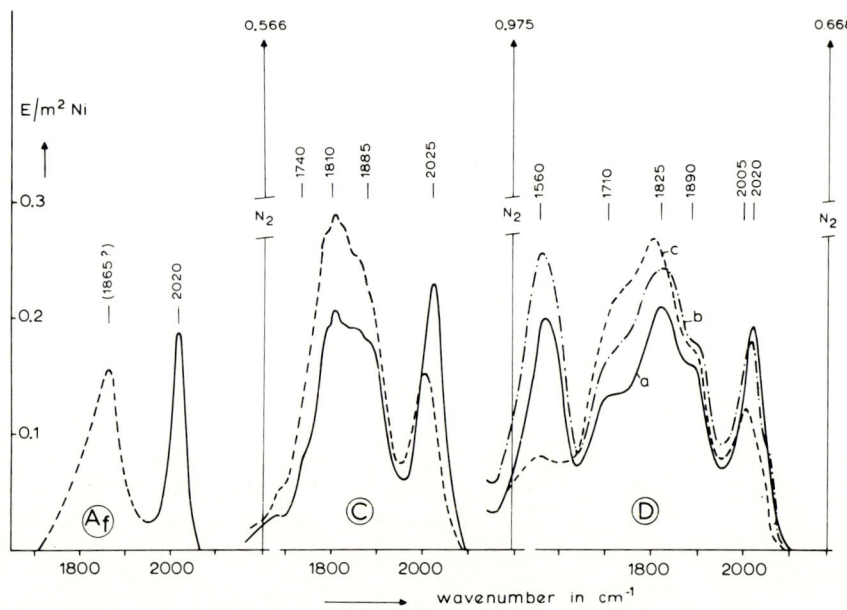

Fig. 9. Absorption spectra obtained on the authors' catalyst samples. A_f: - - - and —, 5 cm CO_2. C: —, 5 cm CO_2; - - -, after 16 hr, 45 sec evacuation. D: —, 5 cm CO_2; - . -, after 16 hr with or without 30 min cryopumping at $-196°C$ in the cell; - - - after 16 hr, 45 sec evacuation.

the intensities are indicated by the length of the vertical arrows. A particularly clear example is catalyst A_f. Even when considering that one chemisorbed oxygen atom per CO molecule is present on the surface, we must conclude that a considerable part of the surface is still bare. Notwithstanding this we found that when a catalyst had chemisorbed CO_2, it had lost completely its ability to adsorb nitrogen in an infrared active form, but was still able to chemisorb CO. The infrared bands resulting in the latter case show a slightly weaker total intensity than those obtained upon chemisorption of CO on the clean catalyst. In catalyst C (particle size 20 Å) this difference in intensity was 25–30%, whereas for catalyst A_f it amounted to about 10%. Hence the chemisorption of CO_2 takes place on the B_5 sites and results in partial coverage of the surface with CO and O.

Upon closer examination of the spectra obtained by adsorption of CO_2 and CO, it appears that the CO_2 bands lie at a slightly lower frequency than those produced by CO. The shift, $\Delta\nu$, is greater according as the frequency of the band is lower, as can be seen in Table IV. In all probability these shifts are caused by oxygen present on the surface. Upon dissociation of CO_2, the CO and O will be bonded to adjacent nickel atoms. The oxygen atom will cause the nickel atom to which it is bonded, as well as its neighbors, to become positively charged. As a result, the adsorbed CO molecule will get polarized, and more electrons will be forced into the Ni—C bond. This will have a weakening effect on the C—O bond and, in consequence, shift its stretching frequency to lower values. As a nickel atom will tend to transfer part of its charge to its neighbors, it seems likely that the formal charge thus imposed on an atom will be greater according as the coordination number of the nickel atom is lower. An alternative explanation is that the nickel atom to which the oxygen atom is bonded is, in effect, removed from the lattice, with the result that the coordination number of its neighbor bearing the CO molecule decreases by one.

The spectrum taken after adsorption of CO_2 in sample D showed a band at 1560 cm^{-1} which had no counterpart in the spectrum obtained after adsorption of CO. It must be pointed out that catalyst D is much more difficult to reduce than the other samples, which makes it likely that this band is caused by CO_2 adsorbed on remnants of unreduced nickel oxide.

After leaving our samples in contact with CO_2 for about 16 hr, we found that the bands below 1950 cm^{-1} had increased in intensity, while the band at 2030 cm^{-1} had remained essentially constant. Obviously, some of the CO had migrated to other parts of the surface, while, in addition, some of the oxygen may have migrated, or even diffused into the bulk of the metal, whereupon some further chemisorption has taken place on the B_5 sites, which have thus become vacated. When the CO_2 is frozen out by cooling part of the cell to the temperature of liquid nitrogen, no changes in the spectra are observed. Evacuation of the cell with a diffusion pump for a

TABLE IV

Assignment of the Absorption Bands of CO

Type of surface atom	Surface species	ν (cm^{-1}) CO	ν (cm^{-1}) CO from CO_2	$\Delta\nu$
C_0	$Ni(CO)_4$ gas phase	2057	—	—
C_6, C_7	Ni(CO)(CO), Ni(CO)(CO)(CO)	2050–2080	2020–2025	25–60
C_9	Ni—CO	2030–2075	2005	25–70
C_8	Ni—CO	1950	1880–1890	60–70
C_7	Ni—CO	1920	1810–1825	95–110
C_6	Ni—CO	1810	1710–1740	70–100

very short time (45 sec), however, markedly lowers the intensity of the 2050 cm^{-1} band and increases that of the other ones. The difference between these two treatments is that the freezing-out procedure removes the CO_2 only, while evacuation also eliminates any CO that is present in the gas phase in equilibrium with the CO adsorbed on the catalyst. The long period during which the catalyst was left in contact with CO_2 before evacuation ensured that the distribution of CO over the surface had attained equilibrium. Hence, the only effect of pumping can be removal of the most weakly bonded CO molecules. From the change observed in the spectra it appears that pumping removes one of the CO molecules of the subcarbonyl species. The $Ni\begin{smallmatrix}\diagup CO \\ \diagdown CO\end{smallmatrix}$ species, with its infrared absorption band at about 2050 cm^{-1}, is thus converted into a Ni—CO species which absorbs in the region below 1950 cm^{-1}. These findings support the assignment of the 2080 cm^{-1} band in the CO spectra to the subcarbonyl species. Recent results obtained with a sample containing still smaller nickel particles lend support to the conclusions drawn in the present section. This sample (catalyst F) having a mean metal particle size of about 10 Å, we expected it to contain only a very small number of B_5: sites. This was confirmed by the finding that, after being contacted with nitrogen, the catalyst did not produce a band at 2200 cm^{-1}, and was incapable of chemisorbing carbon dioxide. The spectrum obtained after adsorption of CO on this catalyst was quite remarkable. When the CO pressure was about 10^{-1} Torr, the spectrum consisted of a strong subcarbonyl band at 2080 cm^{-1}, and some weak absorption bands in the range 1700–2000 cm^{-1}. The band at 2080 cm^{-1} appeared already at CO pressures of 10^{-4} Torr, whereas on the other catalysts it became noticeable only at pressures in the range above 10^{-2}–10^{-1} Torr. When the CO pressure was increased to 1 Torr, formation of $Ni(CO)_4$ took place, and within a few hours the nickel was completely converted to carbonyl. Obviously, this catalyst has a surface with relatively large amounts of C_6 and C_7 atoms, but hardly any B_5 sites. Hence, the results obtained with it (after exposure to nitrogen and carbon dioxide) support our view that B_5 sites, and not only C_6 and C_7 atoms, are needed to adsorb nitrogen at room temperature, and to chemisorb carbon dioxide.

C. Adsorption of N_2 and CO on Palladium and Iridium Catalysts

In order to ascertain in what measure the conclusions drawn from our experiments on nickel hold also for the other metals in group VIII, we undertook some experiments with palladium and iridium catalysts. Five palladium and three iridium catalysts were available for our purpose. The analytical data for these catalysts are given in Table V.

1. *Adsorption of N_2 and CO on Palladium*

The nitrogen adsorption experiment was carried out with catalysts Pd-15, Pd-45, and Pd-105 only. As shown in Fig. 10, the 2260 cm^{-1} band due to adsorbed nitrogen was observed on all of them.

The strongest band was seen in the spectrum of catalyst Pd-15, which was to be expected since the mean particle size of this catalyst lies near the value where, according to van Hardeveld and Montfoort (9), the number of B_5 sites per unit of metal surface area is maximum. The CO spectra of these three catalysts taken at 1 Torr CO pressure are shown in Fig. 10. They are much less detailed than those obtained for nickel which renders their interpretation rather difficult. However, some information can be derived from the differences among them. Compared with the others, the band in the region 2086–2094 cm^{-1} region shows a relative increase in intensity with decreasing particle size. This behavior is analogous to that of the 2050–2080 band obtained on the nickel catalysts; hence, we are inclined to ascribe this 2086–2094 cm^{-1} band to the subcarbonyl species.

TABLE V

Analytical Data of Palladium and Iridium Catalysts

Catalyst	Type of support	Metal content by weight (%)	d_w, from x-ray line broadening (Å)	d_{vs}, from H_2 chemisorption at room temp. (Å)
Pd-15	Aerosil	2.7	≤15	nd[a]
Pd-45	Alumina P 110	5	45	nd
Pd-105	Aerosil	10	105	nd
Pd-115	Alumina P 110	5	115	nd
Pd-190	Aerosil	10	190	nd
Ir-8	Aerosil	2.01	≤10	8
Ir-37	Aerosil	2.53	45	37
Ir-100	Aerosil	10	nd	nd[b]
Ir-600	None	100	215	590

[a] nd, Not determined.
[b] Particular size around 100 Å

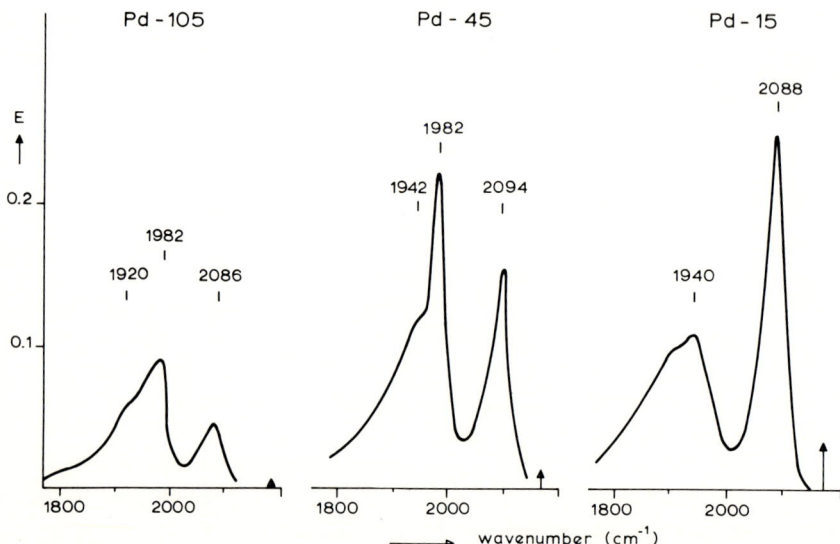

Fig. 10. Spectra of CO adsorbed on palladium samples Pd-105, Pd-45, and Pd-15. The height of the arrow gives the intensity of the band at 2260 cm^{-1} after adsorption of N_2 on the samples.

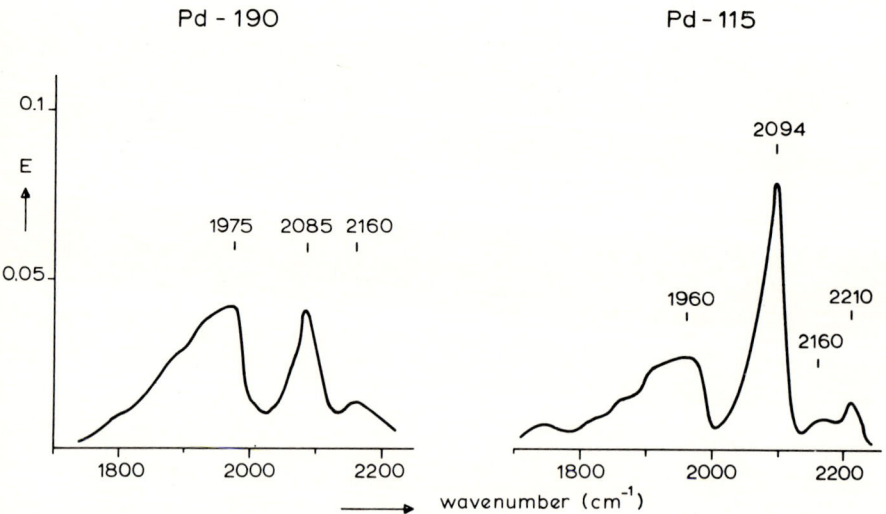

Fig. 11. Spectra of CO adsorbed on palladium catalysts Pd-190 and Pd-115.

Much less can be said about the other bands. The 1982 cm^{-1} band is very pronounced in the spectrum of catalyst Pd-45, barely discernible in that of catalyst Pd-105, and completely absent in the spectrum of CO adsorbed on catalyst Pd-15. We recently made two palladium catalysts, Pd-115 and Pd-190, following the same procedures as for Pd-45 and Pd-105, respectively, the only difference being that unlike with the latter two, we started from the acetate and not from the chloride. The spectra of CO adsorbed on catalysts Pd-190 and Pd-115 are given in Fig. 11.

Comparing the spectra of Pd-45 and Pd-105 with those of Pd-115 and Pd-190, respectively, we see that in the samples prepared from palladium acetate the 1982 cm^{-1} band is absent, or at any rate strongly reduced in intensity. Hence, it is not impossible that the 1982 cm^{-1} band is due to CO adsorbed on a palladium surface that is partially covered with chlorine.

On the other hand, additional bands at 2160 and 2210 cm^{-1} are present in the spectra of the acetate-based samples. Clearly, further experimental work is necessary to enable more definite statements to be made on the origin of the various bands.

2. *Adsorption of N_2 and CO on Iridium*

When nitrogen was adsorbed at room temperature on the catalysts Ir-8, Ir-37, and Ir-100, an infrared absorption band due to adsorbed N_2 was observed only on catalyst Ir-37. This was to be expected since the

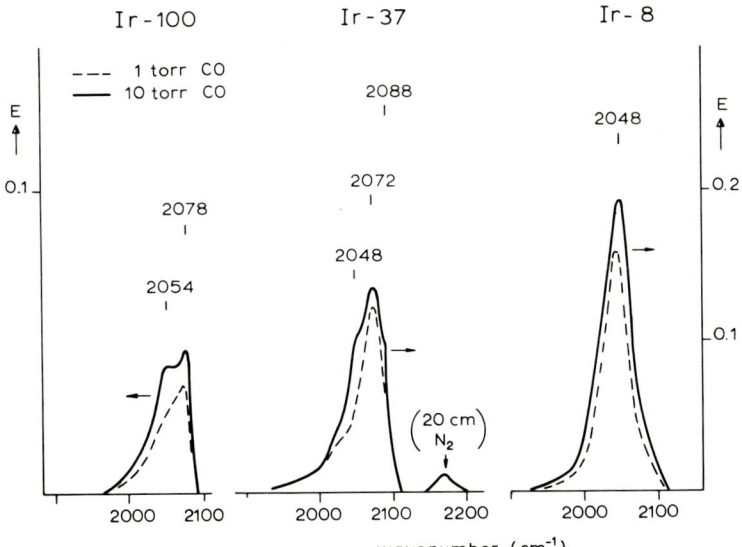

Fig. 12. Infrared spectra of CO adsorbed on IR-100, IR-37, and IR-8.

mean particle sizes of the other catalysts lie outside the range where the surface contains a noticeable number of B_5 sites. The infrared spectra of CO adsorbed on these catalysts are represented in Fig. 12.

These spectra show even fewer features than those of the palladium catalysts. Absorption takes place almost exclusively in the region 2000–2100 cm^{-1}. There are some weak bands below 2000 cm^{-1}, but our experimental method did not allow us to determine their frequencies with reasonable accuracy. It is clear that also with the iridium catalysts the particle size has an effect on the spectra. The spectrum of Ir-8 shows only one intense band at 2048 cm^{-1}, whereas the other two have additional bands at higher frequencies. There is also a marked dependence of the intensity of the 2048 cm^{-1} band on the CO pressure, especially in the case of Ir-37 and Ir-100. We shall not try to interpret the CO spectra of the iridium samples, as we consider the data available insufficient for the purpose.

However, it will at any rate be clear now that the palladium, nickel, and iridium catalysts used in our experiments differ widely in surface characteristics, as is evident from the variations in chemisorptive behavior. An obvious question that may be asked now is whether the catalysts differ also in catalytic behavior. This induced us to study the reaction of benzene with deuterium on the nickel and iridium catalysts.

IV. Deuteration and Exchange of Benzene

A. Experimental Procedure

The reaction of benzene with deuterium is an excellent means for studying the influence of the particle size on the catalytic activity. Working with a relatively simple mixture of reactants, one can study a number of reactions at the same time.

The first reaction yields a mixture of all deuterobenzenes, the second one, the deuteration reaction, a mixture of deuterocyclohexanes, mainly $C_6H_6D_6$ up to C_6D_{12}. The rates of these reactions can be measured with fair accuracy, and additional information can be derived from the isotopic distributions of the reaction products. Schrage and Burwell (*36*) already demonstrated the usefulness of exchange reactions for studying the heterogeneity of catalyst surfaces.

The experimental procedure is outlined schematically in Fig. 13; a detailed description was given by Hartog *et al.* (*37*). Benzene vapor and deuterium gas, in the molar ratio of 1:18, were passed through a catalyst bed and then through a cold trap immersed in liquid nitrogen in which the hydrocarbons were frozen out. The temperature of the catalyst bed was

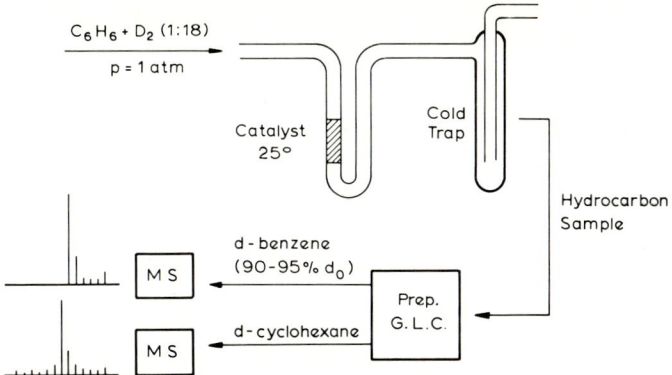

Fig. 13. Schema for the experimental procedure of the reaction of benzene with deuterium.

maintained at 25°C, and the total pressure was 1 atm. The flow rate of the reactants and the amount of catalyst in the reactor were so adjusted that the benzene fraction of the reaction product still contained more than 90% of benzene-d_0. In this way, the effect of further reactions of the deuterobenzenes formed could be largely suppressed. Under these conditions more than 95% of the reaction products are primary products, i.e., they are formed directly from benzene-d_0. Separate experiments proved that under our experimental conditions cyclohexane does not react with deuterium, which is in accordance with the findings of Anderson and Kemball (43) on evaporated films of platinum, palladium and nickel.

At the end of the experiment the hydrocarbons were removed from the cold trap and analyzed. The cyclohexane content of the sample was determined by gas chromatography on a Perkin–Elmer F11 gas chromatograph. The remainder of the sample was separated into a benzene and a cyclohexane fraction on a preparative gas chromatograph (Varian Autoprep).

The isotopic composition of the two fractions was then determined on an AEI MS-10 mass spectrometer. From the analysis of the reaction products and the known flow rate of the reactants, the rates of exchange (R_E) and deuteration (R_H) could then be calculated.

In the derivation of the expressions for R_E and R_H we have used the following symbols:

R_E, R_H represent the rates of exchange and the rate of deuteration, respectively (moles per unit time and per unit metal surface area);

X, Y, Z are mole fractions of benzene-d_0, deuterobenzene, and deuterocyclohexane in the hydrocarbons leaving the reactor ($X + Y + Z = 1$);

$X(l)$, $Y(l)$, $Z(l)$ are mole fractions of benzene-d_0, deuterobenzene, and deuterocyclohexane in the hydrocarbons at a distance l from the beginning of the catalyst bed;

S is the metal surface area of catalyst charged into the reactor;

L is the length of catalyst bed;

f is the amount of benzene entering the reactor per unit time.

Bearing in mind that both reactions are zero order, or nearly zero order, in benzene [see Bond (38)], and that the deuterium was present in excess, we arrive at:

$$\frac{dY(l)}{dl} = \frac{S}{fL} R_E \frac{X(l)}{X(l) + Y(l)} - \frac{S}{fL} R_H \frac{Y(l)}{X(l) + Y(l)}, \qquad (1)$$

and

$$\frac{dZ(l)}{dl} = \frac{S}{fL} R_H. \qquad (2)$$

The second term in Eq. (1) makes allowance for the fact that part of the benzene exchanged is removed by deuteration. This correction is important when $R_H \gg R_E$.

Substituting $p = (S/fL)R_E$ and $q = (S/fL)R_H$, and considering that

$$X(l)/[X(l) + Y(l)] = 1 - Y(l)/[1 - Z(l)],$$

we get:
$$dY(l)/dl = p - (p + q)Y/(1 - Z), \qquad (3)$$

$$dZ(l)/dl = q. \qquad (4)$$

On the condition that $Z(l) = 0$ for $l = 0$ and $Z(l) = Z$ for $l = L$, integration of Eq. (4) gives:

$$Z = qL = (S/f)R_H, \qquad R_H = (f/S)Z.$$

Integration of Eq. (3), on the condition that $X(l) = 1$ for $l = 0$ and $X(l) = X$, $Y(l) = Y$, and $Z(l) = Z$ for $l = L$, yields:

$$Y = (1 - qL) - (1 - qL)^{(p+q)/q}$$

or

$$Y = (1 - Z) - (1 - Z)^{(R_E + R_H)/R_H},$$

which gives

$$(X + Y)^{(R_E + R_H)/R_H} = X,$$

and, finally,

$$R_E = R_H[\log X - \log(X + Y)]/\log(X + Y).$$

B. Reaction of Benzene with Deuterium on Nickel Catalysts

The catalysts used in these experiments included those already employed in the infrared measurements in addition to some others. The results are presented in Tables VI and VII along with some older measurements on Raney-nickel and a nickel-on-kieselguhr catalyst. These older measurements are slightly less accurate because the cyclohexane content of the reaction product was determined by mass spectrometry. The surface area of catalyst E was not determined; hence, its reaction rates per unit of surface area could not be calculated.

The catalysts are listed in the order of decreasing mean metal particle size. The most striking fact emerging from the results is the tremendous increase in the specific rate of exchange with decreasing particle size. In contrast to this, the rate of deuteration (R_H) hardly depends on the crystallite size. This latter finding agrees with the results on the rate of benzene hydrogenation obtained by Ljubarskii (39), Nikolajenko et al. (40), Nikolajenko et al. (41), and Aben et al. (42).

Notwithstanding the large variations in exchange rate, the isotopic distributions of the deuterobenzenes formed are practically the same in all catalysts, and as shown in Table VII, this holds true also for the isotopic distributions of the cyclohexane formed.

There is some diversity of opinion in the literature on the issue as to whether or not exchange and deuteration proceed on the same sites. Anderson and Kemball (43) hold that these reactions take place on different sites,

TABLE VI

Rates of Exchange and Deuteration on Nickel

Catalyst	$\bar{d}_w$ (Å)	$\bar{d}_{vs}$ (Å)	Crystallite size distribution[a]	R_E/R_H	$R_E \times 10^4$ (moles hr^{-1} m^{-2})	$R_H \times 10^4$ (moles hr^{-1} m^{-2})
Ni Guhr		~380	br	~30		
Raney-Ni		~200	br	~50		
B	185	214	br	19	8.57	0.44
A_0	70	59	br	4.9	2.11	0.43
A_f	70	48	br	1.9	1.86	1.00
Y-G		67	fn	0.94	0.75	0.80
C	20	53	n	0.30	0.59	1.97
D	20	44	n	0.07	0.09	1.18
E	18		n	0.10		

[a] br, broad; fn, fairly narrow; n, narrow.

TABLE VII

Isotopic Product Distributions (mole %) on Nickel

Catalysts	Ni-Guhr	Raney-Ni	B	A_0	A_f	Y-G	C	D	E
Benzene-d_1	71.2	68.7	72.9	74.6	68.8	71.6	69.1	82.5	70.8
d_2	15.6	15.4	16.5	14.7	17.0	15.6	15.6	11.3	14.2
d_3	6.8	7.2	5.8	5.4	6.7	6.0	6.4	2.6	6.1
d_4	3.1	3.9	2.5	2.5	3.5	3.0	3.6	1.1	3.6
d_5	1.8	2.7	1.3	1.5	2.2	2.0	2.6	1.0	2.6
d_6	1.5	2.2	0.9	1.2	1.8	1.8	2.6	1.4	2.6
Cyclohexane-$d_{<5}$	1.8	nd[a]	0.0	0.0	0.6	nd	0.0	0.6	nd
d_5	8.3		3.6	3.6	4.0		4.6	6.9	
d_6	57.9		64.6	64.2	62.7		60.2	58.4	
d_7	18.7		20.4	19.4	19.5		20.3	20.8	
d_8	7.2		7.5	8.0	7.9		9.0	8.3	
d_9	2.9		2.5	3.0	3.0		3.4	3.1	
d_{10}	1.7		1.0	1.4	1.5		1.5	1.3	
d_{11}	0.8		0.3	0.4	0.6		0.6	0.4	
d_{12}	0.7		0.2	0.2	0.3		0.3	0.2	

[a] No data.

while Hartog, Tebben and Weterings, from a comparison of the isotopic distribution of the deuterobenzene and deuterocyclohexane, arrive at the opposite conclusion. From the present results it follows that a nickel surface must at any rate contain sites with R_E/R_H values higher than 20 in addition to sites for which R_E/R_H is smaller than 0.1. The small variations in the isotopic distribution of benzene suggest that the former sites account for practically all deuterobenzene formed on all catalysts, and that the cyclohexane must come from the other type of sites. The nature of the sites with the high R_E/R_H value is rather puzzling. Schrage and Burwell (44) and Bond (19) argue that the "topside" addition of deuterium needed for exchange of benzene and other cyclic systems via an additive mechanism takes place on step sites at the surface. If this is true one would expect the exchange activity to be highest in catalysts with the largest number of B_5 sites, because these sites occur on surface steps only. However, just the opposite is observed; hence, Burwell and Bond's view must be rejected.

From the relation between R_E/R_H and particle size it follows that the density of sites with the high R_E/R_H value is about a factor 100 higher on crystals of approximately 200 Å than on crystals of approximately 20 Å. If the crystal is indeed free of defects, we cannot imagine any type of

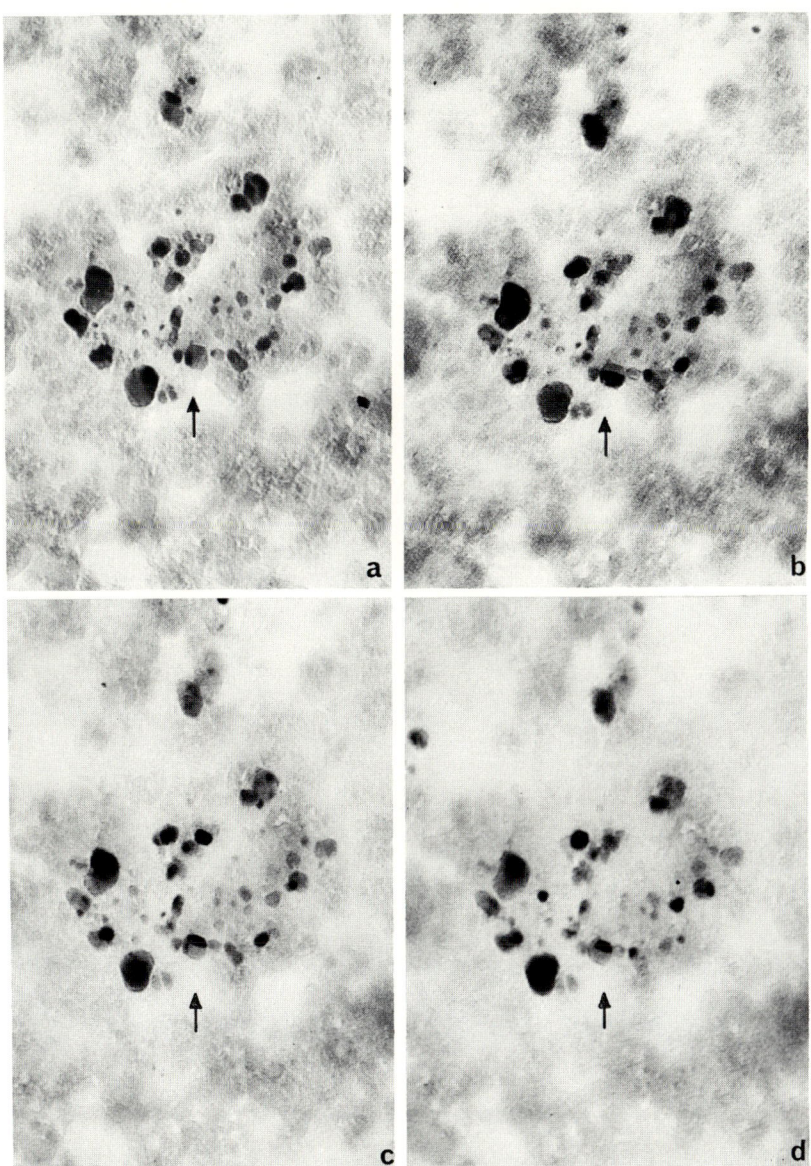

Fig. 14. Influence of specimen orientation (catalyst B) relative to the electron beam on the appearance of the electron-micrograph. Angle of tilt: (a) 0°20′; (b) 4°30′; (c) 7°20′;(d) 16°.

site to be present on it showing this relation between the rate of occurrence and the particle size.

For the time being, we shall not go further than tentatively assuming that the sites with the high R_E/R_H value(s) are found where stacking faults, for instance, twin boundaries, intersect the surface. That twin boundaries occur in our catalysts is evident from electron micrographs of catalyst B. Figure 14 shows a series of electron micrographs taken of the same specimen by means of a goniometer stage. Between the exposures the preparation was tilted through the angles indicated.

It can be concluded from these micrographs that the crystal marked by an arrow is made up of at least three parts differing in the orientation of the crystal axes. During tilting of the specimen these parts are successively so orientated as to favor a strong scattering of the electron beam, with the result that they appear as black spots on the micrograph. Probably, twin boundaries are formed when during catalyst reduction two or more crystals coalesce; hence, they will be more frequent on larger crystals. The atomic arrangement at twin boundaries and other stacking faults is fundamentally different from that of the remainder of the crystal [see Jaeger and Sanders (14)]. Hence, they give rise to sites showing a geometry that cannot occur in those parts of the crystal that have the normal fcc structure.

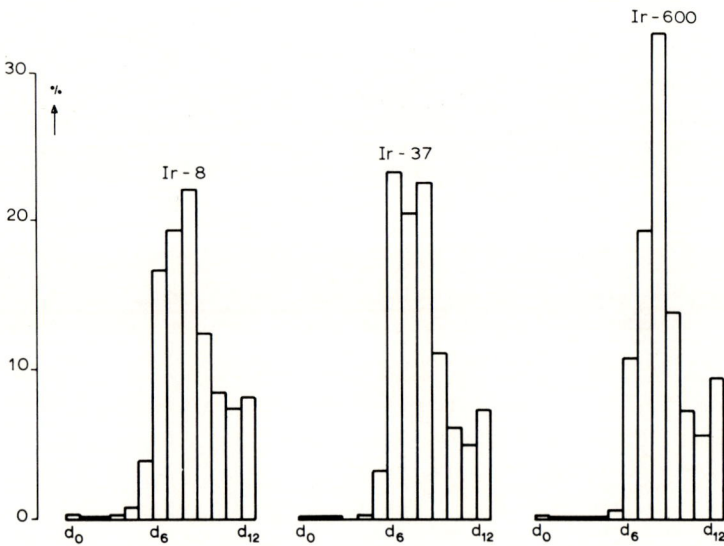

Fig. 15. Cyclohexane isotopic distributions for the iridium catalysts.

C. Reaction of Benzene with Deuterium on Iridium Catalysts

The results obtained with nickel raised the question whether the relation found between rate of exchange and particle size holds also for other metals of group VIII. We therefore carried out the benzene-D_2 reaction on some iridium catalysts widely differing in particle size. We chose iridium because we knew from earlier experiments that iridium black gives a very characteristic cyclohexane isotopic distribution pattern with a maximum for $C_6H_4D_8$, whereas the patterns of Ni, Ru, Pd, and Pt show a maximum for the d_6 compound.

Two of the three catalysts, viz, Ir-8 and Ir-45, have also been used in the infrared experiments. Ir-600 is an iridium-black catalyst. Figure 15 shows the cyclohexane isotopic distributions for the iridium catalysts.

The distribution found on Ir-600 displays a pronounced maximum for the d_8 compound and a smaller one for d_{12}. In this figure the heights of the column denote the mole fractions of the various deuterocyclohexanes in the cyclohexane fraction. A characteristic feature of the isotopic distribution on Ir-37 is that the amounts of cyclohexane-d_6 and -d_7 are larger than those in Ir-600, which gives rise to two maxima, one for d_6 and one for d_8. The amounts of cyclohexane-d_9,-d_{10},-d_{11}, and-d_{12} on the other hand, are smaller, but the proportion among them is approximately the same as in Ir-600. Considering the distribution found on Ir-600 to be characteristic of the sites at the crystal faces, and bearing in mind that the number of edge sites on particles of about 37 Å is relatively large, we arrive at the conclusion that the edge sites must be responsible for the additional amounts of cyclohexane d_6 and d_7. Moreover, as shown in Table VIII, R_H is greater in Ir-37 than in Ir-600.

Edge sites have thus a higher specific activity for deuteration than sites on the faces of the crystal, and their isotopic distribution pattern shows higher proportions of d_6 and d_7. The isotopic distribution pattern of the cyclohexane on the Ir-8 catalyst does not reflect the continuing trend

TABLE VIII

Rates of Exchange and Deuteration on Iridium

Catalyst	$\bar{d}_w$ (Å)	$\bar{d}_{vs}$ (Å)	R_E/R_H	$R_E \times 10^3$ (moles hr^{-1} m^{-2})	$R_H \times 10^3$ (moles hr^{-1} m^{-2})
Ir-8	≤10	8	0.16	0.38	2.4
Ir-37	45	37	0.36	6.9	19.0
Ir-600	215	590	0.70	0.99	1.4

toward higher deuteration rate and increased amounts of cyclohexane d_6 and d_7 with decreasing particle size. On the contrary, the deuteration rate is lower than on catalyst Ir-37, and there is less cyclohexane-d_6 and cyclohexane-d_7, though still more than in the pattern for catalyst Ir-600. Beyond d_8, the distribution differs distinctly from that for the other catalysts.

In view of the mean particle size of Ir-8 we have good reason to expect that with this catalyst the influence of corner sites becomes noticeable. Obviously, the specific rate of deuteration on these corner sites must be lower than that on the edge sites; in addition, the cyclohexane formed on them must contain a larger proportion of the more highly deuterated species than the deuteration product coming from either the edge sites or the sites at the faces.

A histogram of the benzene isotopic distribution pattern obtained on the three iridium catalysts is shown in Fig. 16.

The primary product of the exchange reaction on catalyst Ir-600 is

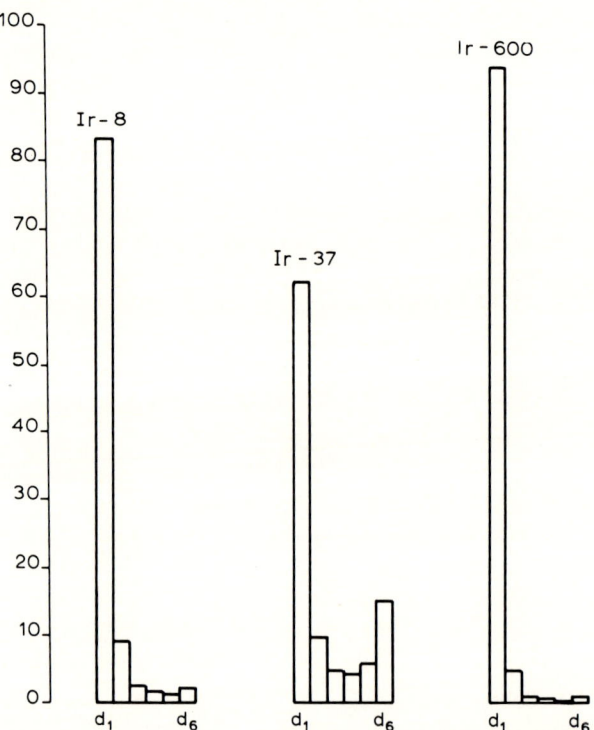

FIG. 16. Benzene isotopic distribution pattern obtained on the iridium catalysts.

almost exclusively benzene-d_1; in spite of the low conversion the main part of the benzene-d_2 is in this case due to further reaction of benzene-d_1. On catalyst Ir-37, unlike on Ir-600, there is a fair amount of multiple exchange, as is evidenced by the formation of considerable quantities of benzene-d_2 up to -d_6. This multiple exchange process will in all probability take place on edge sites. The specific rate of exchange, like that of deuteration, is higher on catalyst Ir-37 than on Ir-600. Looking now at the results obtained on Ir-8, we note that also in the case of the exchange reaction the trends reverse when the mean particle size is further decreased to 8 Å. There is less multiple exchange on catalyst Ir-37, and the specific rate of exchange is even lower than on Ir-600. This demonstrates that corner sites are less active for exchange than edge sites. A considerable part of the multiply exchanged benzene must come from the edge sites, which implies that predominantly single exchange takes place on the corner sites.

The outcome of the experiments on nickel and on iridium shows that there is a marked contrast between these metals as regards the influence of particle size upon the course of the reaction of benzene with deuterium. Clearly, further experiments should be done along the same lines with other metals of group VIII of the periodic system.

A single remark remains to be made about the mechanism of the reaction of benzene with deuterium though it has no direct bearing on the topic of this article. In principle, any of the following three reactions may be responsible for the total number of deuterium atoms in the deuterocyclohexane. The first reaction is, of course, the addition of deuterium to the benzene nucleus, which accounts for the presence of six deuterium atoms per molecule. The second is the benzene-D_2 exchange reaction, which causes the presence of deuterobenzene on the surface. When this benzene is deuterated while it is still on the surface, the resulting cyclohexane molecule will contain more than six deuterium atoms. A third possibility is that in the later stages of deuteration one or more exchange reactions are operative. Anderson and Kemball (43) conclude from their experiments on nickel, palladium, and platinum films that this third reaction does not occur. Our results show, however, that, at any rate on iridium, it does take place to a significant degree. From the isotopic distribution pattern of benzene, especially that obtained with catalyst Ir-600, it can be inferred that the benzene adsorbed at the surface consists predominantly of benzene-d_0, the remainder being C_6H_5D and insignificant amounts of benzene with two or more deuterium atoms. So, if the first two reactions were the only ones to take place, the deuterocyclohexane would consist primarily of $C_6H_6D_6$ with some $C_6H_5D_7$ and only very small amounts of the species with eight or more deuterium atoms. We may say therefore that, at least on iridium, additional exchange does take place during the deuteration process.

V. Conclusions

Whereas determination of chemisorption isotherms, e.g., of hydrogen on metals, is a means for calculating the size of the metallic surface area, our results clearly demonstrate that IR studies on the adsorption of nitrogen and carbon monoxide can give valuable information about the structure of the metal surface. The adsorption of nitrogen enables us to determine the number of B_5 sites per unit of metal surface area, not only on nickel, but also on palladium, platinum, and iridium. Once the number of B_5 sites is known, it is possible to look for other phenomena that require the presence of these sites. One has already been found, viz, the dissociative chemisorption of carbon dioxide on nickel.

Adsorption of carbon monoxide takes place all over the surface and there is distinct evidence that, at least on nickel, the CO stretching frequency depends upon the coordination number of the nickel atom to which it is attached. Hence, the adsorption of carbon monoxide yields information about the relative numbers of surface atoms with different coordination numbers. This information, however, is at best merely of a semiquantitative nature. Steric effects also play a role, as is evidenced by the fact that the subcarbonyl species can be formed only on nickel atoms with a low coordination number.

The most striking fact emerging from our experiments with benzene and deuterium is that there is a strong influence of particle size on activity and selectivity, both on nickel and iridium, but the nature of this influence is entirely different in these two cases. On nickel there is a dramatic increase in R_E/R_H with increasing particle size without there being any noticeable change in the isotopic product distributions. On iridium, variations in the mean metal particle size produce much smaller changes in R_E/R_H than were observed on nickel; however, the effect of these changes on the isotopic product distributions is very strong.

Moreover, there are indications that the edge sites are particularly active, both in the exchange and in the deuteration reaction. No such effect has been found on nickel.

VI. Preparation and Characterization of Catalysts

Catalyst A_f was prepared, shortly before use, by impregnating aerosil with a solution of $Ni(NO_3)_2$ according to the method of Eischens and Jacknow (8), and drying it at 120°C.

Catalyst A_0 was prepared in the same way as catalyst A_f, but was stored for about 1 year prior to use. During this period the $Ni(NO_3)_2$ recrystallized with the result that the number of small crystallites present after reduction was smaller than in sample A_f.

Catalyst B was made by pyrolizing catalyst A_f in air at 450°C for 16 hr.

Catalysts C, D, and E were prepared by the methods developed by Geus (*1, 2*). To this end, aerosil was suspended in a solution of $Ni(NO_3)_2$, after which the nickel was precipitated (as the hydroxide) under carefully controlled conditions.

Catalyst Y-G was prepared in conformity with the spray-drying method of Yates and Garland (*27*). $Ni(NO_3)_2$ was dissolved in a small amount of distilled water, whereupon acetone and the requisite amount of aerosil were successively added to it. The slurry was transferred to an atomizer and sprayed with continuous agitation. The spray was directed to a glass surface and the particles adhering to this surface were dried by leading a stream of air over them.

Catalyst F was made by the ion exchange method given by Hathaway and Lewis (*45*).

Aerosil was converted into the sodium form by treating it with a buffer solution (pH = 8.4) made of sodium hydroxide and sodium hydrogen carbonate solutions, after which it was filtered, washed free of alkali, and dried. This sodium-exchanged aerosil was then suspended in a solution of $Ni(en)_3(NO_3)_2$ prepared by adding the calculated amount of ethylenediamine to a solution of nickel nitrate. The suspension was agitated for about 30 min and filtered off. The catalyst was then washed and dried at 100°C.

Catalyst Pd-15 was prepared by the method of Zhmud et al. (*3*), i.e., by adsorption of $Pd(NH_3)_4^{++}$ on aerosil, filtration, careful washing and drying. After that, the catalyst was prereduced at 300°C with a hydrogen-nitrogen current containing 0.5 mole % of hydrogen.

Catalyst Pd-45 was made by boiling a suspension of alumina P 110 (Degussa) with a solution of Na_2PdCl_4. After 1 hr the pH of the slurry was brought to a value of 6–7 by addition of $NaHCO_3$. Boiling was continued for half an hour, after which the specimen was filtered off, washed and dried at 120°C. It was reduced in the infrared apparatus for 2 hr at 450°C.

Pd-105 was prepared by impregnation of aerosil with H_2PdCl_4. The catalyst was dried and prereduced with the N_2–0.5% H_2 mixture at 300°C.

The preparations of Pd-115 and Pd-190 were analogous to those of Pd-45 and Pd-105, respectively; the only difference was that the starting material was palladium acetate.

The preparation of catalyst Ir-8 was analogous to that of Pd-15, but for the use of the Ir ammine complex instead of palladium.

Catalyst Ir-37 was also prepared by adsorption of the ammine complex.

The catalyst was not washed, however, and the reduction was carried out with hydrogen instead of with the nitrogen–hydrogen mixture.

Catalyst Ir-100 was made by impregnation of aerosil with $(NH_4)_2IrCl_6$, drying, and reducing with the N_2–H_2 mixture at steadily increasing temperature (final temperature, 300°C).

Catalyst Ir-600 was made by reduction of IrO_2 (p.a.) with hydrogen at 200°C.

The specific metal surface area of our nickel samples was established by means of deuterium chemisorption, the amount of deuterium adsorbed being determined by exchange with a known quantity of hydrogen followed by mass spectrometric analysis. It was assumed in the calculation that 1 cm³ (NTP) of deuterium corresponds to 3.64 m² of nickel surface area.

The surface areas of the iridium and palladium catalysts were determined by chemisorption of hydrogen and carbon monoxide, respectively, the monolayer volume being determined from an adsorption isotherm taken at 20°C.

From the surface area S, the mean particle size $\bar{d}_{vs}$, is calculated as $\bar{d}_{vs} = 6/S\rho$ where ρ is the density of the metal. The weight mean diameter $\bar{d}_w = \sum_i n_i d_i^4 / \sum_i n_i d_i^3$ was determined from the degree of X-ray line broadening.

Photographs of the various preparations were taken by means of a Philips EM 300 electron microscope. To this end, the samples were embedded in polymethylmethacrylate and 200–400 Å thick sections cut from them with an ultramicrotome.

References

1. Geus, J. W., Dutch Patent Application 6,705,259 (1967); [*Chem. Abstr.* **72,** 36,325 K (1970)].
2. Geus, J. W., Dutch Patent Application 6,813,236 (1968).
3. Zhmud, E. S., Boronin, V. S., and Poltorak, O. M., *Russ. J. Phys. Chem.* **39,** 431 (1965).
4. Poltorak, O. M., and Boronin, V. S., *Russ. J. Phys. Chem.* **39,** 781 (1965).
5. Benesi, H. A., and Curtis, R. M., *J. Catal.* **10,** 328 (1968).
6. Schlosser, E. G., *Ber. Bunsenges. Phys. Chem.* **73,** 358 (1969).
7. Boudart, M., Aldag, A., Benson, J. E., Dougharty, N. A., and Girvin Harkin, C., *J. Catal.* **6,** 92 (1966).
8. Eischens, R. P., and Jacknow, J., *Proc. 3rd Int. Congr. Catal., Amsterdam, 1964* p. 627 (1965).
9. van Hardeveld, R., and van Montfoort, A., *Surface Sci.* **4,** 396 (1966).
10. van Hardeveld, R., and van Montfoort, A., *Surface Sci.* **17,** 90 (1969).
11. Nicholas, J. E., "An Atlas of Models of Crystal Surfaces." Gordon & Breach, New York, 1965.
12. Poltorak, O. M., and Boronin, V. S., *Russ. J. Phys. Chem.* **40,** 1436 (1966).
13. van Hardeveld, R., and Hartog, F., *Surface Sci.* **15,** 189 (1969).

14. Jaeger, H., and Sanders, J. V., *J. Res. Inst. Catal., Hokkaido Univ.* **16,** 287 (1968).
15. Alpress, J. G., and Sanders, J. V., *Aust. J. Phys.* **23,** 23 (1970).
16. Alpress, J. G., and Sanders, J. V., *Surface Sci.* **7,** 1 (1967).
17. Fukano, Y., and Wayman, C. M., *J. Appl. Phys.* **40,** 1656 (1969).
18. Drechsler, M., and Nicholas, J. F., *J. Phys. Chem. Solids* **28,** 2609 (1967).
19. Bond, G. C., *Proc. 4th Int. Congr. Catal., Moscow* Preprint 67 (1968).
20. Perry, A. J., and Brandon, D. G., *Surface Sci.* **8,** 422 (1967).
21. Peri, J. B., and Hannan, R. B., *J. Phys. Chem.* **64,** 1526 (1960).
22. Gruber, H. L., *J. Phys. Chem.* **66,** 48 (1962).
23. Spenadel, L., and Boudart, M., *J. Phys. Chem.* **64,** 204 (1960).
24. van Hardeveld, R., and Hartog, F., *Proc. 4th Int. Congr. Catal., Moscow* Preprint 70 (1968).
25. Eischens, R. P., Francis, S. A., and Pliskin, W. A., *J. Phys. Chem.* **60,** 194 (1956).
26. Garland, C. W., *J. Phys. Chem.* **63,** 1423 (1959).
27. Yates, J. T., and Garland, C. W., *J. Phys. Chem.* **65,** 617 (1961).
28. O'Neill, C. E., and Yates, D. J. C., *J. Phys. Chem.* **65,** 901 (1961).
29. Garland, C. W., Lord, R. C., and Troiano, P. F., *J. Phys. Chem.* **69,** 1195 (1965).
30. Bradshaw, A. M., and Pritchard, J., *Surface Sci.* **17,** 372 (1969).
31. Blyholder, G., *Proc. 3rd Int. Congr. Catal., Amsterdam, 1964* p. 657 (1965).
32. Blyholder, G., *J. Phys. Chem.* **68,** 2772 (1964).
33. Eischens, R. P., and Pliskin, W. A., *Advan. Catal. Relat. Subj.* **9,** 662 (1957).
34. Eischens, R. P., and Pliskin, W. A., *Advan. Catal. Relat. Subj.* **10,** 1 (1958).
35. Eischens, R. P., and Pliskin, W. A., *Actes 2nd Congr. Int. Catal., Paris, 1960* p. 789 (1961).
36. Schrage, K., and Burwell, R. L., *J. Amer. Chem. Soc.* **88,** 4549 (1966).
37. Hartog, F., Tebben, J. H., and Weterings, C. A. M., *Proc. 3rd Int. Congr. Catal., Amsterdam, 1964* p. 1210 (1965).
38. Bond, G. C., "Catalysis by Metals," p. 315. Academic Press, New York, 1962.
39. Ljubarskii, C. D., *Actes 2nd Congr. Int. Catal., Paris, 1960* p. 1559 (1961).
40. Nikolajenko, V., Bosáček, V., and Daneš, V., *J. Catal.* **2,** 127 (1963).
41. Nikolajenko, V., Daneš, V., and Krivanek, M., *Kinet. Katal.* **7,** 816 (1966).
42. Aben, P. C., Platteeuw, J. C., and Stouthamer, B., *Proc. 4th Int. Congr. Catal., Moscow* Preprint 31 (1968).
43. Anderson, J. R., and Kemball, C., *Advan. Catal. Relat. Subj.* **8,** 51 (1957).
44. Schrage, K., and Burwell, R. L., *J. Amer. Chem. Soc.* **88,** 4555 (1966).
45. Hathaway, B. J., and Lewis, C. E., *J. Chem. Soc. A* p. 6194 (1969).

Adsorption and Catalysis on Evaporated Alloy Films

R. L. MOSS AND L. WHALLEY

Warren Spring Laboratory
Stevenage, England

I. Introduction	115
II. Preparation of Alloy Films	117
A. Miscibility of Components	117
B. Successive Evaporation	120
C. Simultaneous Evaporation	125
III. Characterization of Alloy Films	134
A. Composition	134
B. Crystallite Size and Orientation	135
C. Surface Area	138
D. Bulk Structure	139
E. Surface Examination	143
IV. Adsorption and Catalysis on Alloy Films	147
A. Copper–Nickel Alloy Films	148
B. Palladium–Gold Alloy Films	158
C. Palladium–Silver Alloy Films	161
D. Palladium–Rhodium Alloy Films	172
E. Pt–Au, Pt–Ru, Ni–Au, and Au–Ag Alloy Films	178
V. Conclusions	184
References	185

I. Introduction

It is reasonable to ask two questions in relation to studies using evaporated alloy films, viz, why work with alloys and why prepare alloy catalysts in this particular form?

The activity of the transition metals, especially for the chemisorption of molecular hydrogen and in hydrogenation reactions has been correlated, in the past, with the existence of partially filled d bands. Many alloy studies were prompted by the expectation that catalytic activity would change abruptly once these vacancies were filled by alloying with a group IB metal. Examples of such behavior have been collected together for the Pd–Au system (1). It is to be expected also that various complications might superimpose on the simple activity patterns observed for primitive catalytic reactions, e.g., parahydrogen conversion, and these have been examined. An alternative approach supported by work on Cu–Ni alloy films (2) treats chemisorption essentially as a localized phenomenon where the bond strength between the adsorbing atom and the surface atom is modified by neighboring atoms. So, again, experiments with alloys are at the heart of efforts to describe the catalytic activity of metals.

There are a number of advantages in preparing alloys in the form of evaporated films, including those advantages which usually accrue from the use of metal films. Thin films of metals prepared by vacuum deposition which can provide surfaces of reproducible behavior and moderately large area have been used extensively in the field of gas adsorption and catalysis. Consequently, it has been possible to use various kinds of information, e.g., adsorptive capacities, adsorption heats, exchange studies, surface potential changes, in discussing the catalytic properties of a given metal. It is to be hoped that an equivalent literature will eventually exist for some of the more important alloy systems. Evaporated alloy films are also readily amenable to study by electron microscopy and valuable structural information can be gained about the alloy catalysts being compared. Some interesting alloy systems are subject to phase separation which may develop first at the catalyst surface and confuse the interpretation of results. It seems helpful to prepare such alloy catalysts as thin films because the rate of equilibration will be enhanced as a consequence of rapid surface diffusion.

There is also a very practical reason for studying the catalytic properties of alloys. Surprisingly few of the metals, which comprise the majority of the elements in the periodic table, are useful as catalysts, e.g., Pt, Pd, Ni, Ag, Cu. It would be an extreme coincidence if these pure metals had the optimum properties for the various processes in which they are used and, in fact, the addition of nonmetallic promoters can improve them. It seems more likely that some alloy involving these metals will have the electronic or other properties needed to provide the most active or selective catalyst for a particular reaction. There has been recent interest in the binary alloys of Ru, Rh, Pd, Pt, etc., and "synergistic" effects have been reported (3) for liquid-phase hydrogenations. Peaks in activity have been observed in a series of reduced (mixed) oxide catalysts (4, 5) and with supported Pt–Rh alloys (6, 7). Even the catalytic properties of a group IB metal may be improved by alloying with another of the same subgroup (8).

The pure metals are readily available to try out as catalysts, whereas the alloys are not, if a moderate surface area is required. Methods for preparing alloys as high-area powders, etc., raise questions about the unwanted introduction of promoters, e.g., chloride ions. Here, again, evaporated alloy films recommend themselves for the exploration of a whole new "territory" of alloy systems as catalysts in a variety of reactions.

II. Preparation of Alloy Films

There are a variety of options available for the preparation of evaporated alloy films and, indeed, suggested methods of preparation can be traced back over many years (9). Nevertheless, it is possible to distinguish some general methods for the preparation of binary alloys as follows:

(1) The two metals are deposited *successively* and then annealed in vacuum or in the presence of a gas, e.g., hydrogen.
(2) The two metals are evaporated *simultaneously*, either from separate sources containing the pure component metals or from an alloy source.

Method (2) can be further subdivided depending on whether the mixed metal vapor is quench-cooled or if a heated substrate is used to encourage atom mobility after condensation.

The first consideration is to decide if a series of solid solutions can be formed or not, bearing in mind that at normal temperatures for film preparation most pairs of metallic elements do not dissolve appreciably in one another to produce stable alloys. If a homogeneous alloy is expected e.g., Pd–Ag or Pd–Au alloys of any composition, then the objective seems relatively straightforward, viz, to find a method which will produce a homogeneous alloy with satisfactory surface area and surface cleanliness. The preparation of alloy films from metal pairs which are not completely miscible raises some questions about the best tactics to employ. In some cases where a miscibility gap exists at film preparation temperatures, and one component can diffuse readily, e.g., copper in Cu–Ni alloys, it can be arranged that the surface composition of the crystallites in the alloy film is the equilibrium composition of one of the phases, i.e., the Cu-rich phase, over much of the bulk composition range. As a result, only a very restricted number of significantly different surface compositions can be examined for adsorptive and catalytic behavior. The alternative approach might be to attempt to prepare a range of metastable solid solutions by the vapor-quenching method. These various aspects of alloy film preparation are discussed in subsequent sections.

A. Miscibility of Components

A convenient starting point is to consult the phase diagrams in Hansen's "Constitution of Binary Alloys" (10) and supplementary volumes (11, 12), but information may also be needed on miscibility at the temperatures normally employed for film preparation and catalytic reaction. Therefore, some consideration must be given to the thermodynamic properties of the

TABLE I

Thermodynamic Properties of Some Alloys

Alloy	Temp. (°K)	Function[a]	0.1	0.2	0.3	0.4	0.5	0.6	0.7	0.8	0.9	Ref.
Pd–Ag	1200	ΔH	−620	−1050	−1310	−1350	−1200	−1070	−780	−480	−250	15
		ΔS	0.480	0.731	0.840	0.899	0.939	0.848	0.809	0.715	0.479	15
Pd–Au	298	ΔH	−820	−1520	−1920	−1980	−1860	−1615	−1295	−900	−450	16
Pd–Cu	1350	ΔH	−569	−1994	−2834	−3350	−3169	−2432	−1875	−1395	−844	17
		ΔS	0.839	0.482	0.282	0.050	0.085	0.358	0.397	0.294	0.153	17
Pd–Rh	1575	ΔH	790	1520	2030	2290	2410	2310	2050	1880	1210	18
		ΔS	0.86	1.46	1.85	2.05	2.13	2.07	1.89	1.70	1.13	18
Pd–Co[c]	1273	ΔH	480	620	520	310	30	−260	−540	−590	−400	19
		ΔS	1.1	1.6	1.9	2.1	2.1	2.0	1.8	1.4	0.8	19
Pd–Ni[d]	1173	ΔH	133	141	66	−18	−121	−225	−309	−281	−189	20
		ΔS	0.847	1.243	1.531	1.719	1.784	1.746	1.554	1.296	0.983	20
Pt–Cu	1350	ΔH	−799	−898	−1331	−1852	−1972	−1679	−1364	−1034	−666	17
		ΔS	0.870	1.695	1.895	1.760	1.680	1.690	1.545	1.225	0.705	17
Pt–Ni	1625	ΔH	−235	215	357	438	958	1603	2231	1978	763	21
		ΔS	1.04	2.03	2.59	2.90	3.27	3.52	3.57	2.94	1.52	21
Ni–Au	1150	ΔH	550	970	1325	1590	1745	1780	1645	1215	635	13
		ΔS	0.87	1.38	1.75	1.98	2.07	2.04	1.85	1.38	0.77	13
Ni–Cu	973	ΔH	80	155	230	300	375	425	435	375	235	13
		ΔS	0.50	0.75	0.91	1.01	1.07	1.07	1.01	0.86	0.59	13
Au–Ag	800	ΔH	−430	−750	−970	−1090	−1110	−1050	−900	−670	−370	13
		ΔS	0.52	0.77	0.93	1.01	1.03	1.01	0.93	0.77	0.52	13

[a] ΔH, cal gm-atom^{-1}; ΔS, cal °K^{-1} gm-atom^{-1}.
[b] N_1, atomic fraction of the first element of each alloy listed in the first column.
[c] Estimated from graph.
[d] N_1 values rounded off.

system. Selected values of thermodynamic functions are available for a large number of binary alloys (*13, 14*). Table I summarizes values of ΔH and ΔS for some alloys of catalytic interest and, in particular, reproduces newer data from the literature.

Table I shows that the alloying of nickel with copper or gold and also the formation of palladium–rhodium alloys are endothermic processes. When such alloys are formed at high temperatures so that the free energy of formation is governed by the $T \Delta S$ term, then single-phase solid solutions are found. However, at lower alloy preparation temperatures, where ΔH largely determines the magnitude of ΔG, miscibility gaps are anticipated in these alloy systems. (In calculating ΔG for moderate preparation temperatures from the high-temperature data, it might be decided to assume that $\Delta C_p = 0$.) Alloys such as palladium–silver, which have large negative enthalpies of formation, can be expected to form a series of random solid solutions over the whole composition range. These alloys provide the simplest systems for studying the effects of electronic structure in chemisorption and catalysis. The role of the miscibility gap in the important copper–nickel system is fully discussed in Section IV.

In a two-component system a single-phase solution is thermodynamically stable when the free energy-composition diagram is concave upward [see Darken and Gurry (*22*) for a full discussion of free energy versus compo-

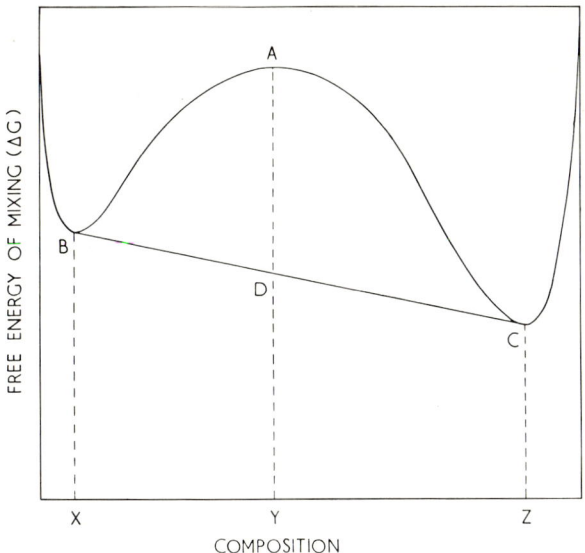

FIG. 1. Variation of free energy of mixing with composition for a binary alloy system possessing a miscibility gap in the range X–Z.

sition diagrams]. This condition holds, of course, for an ideal solution and it will also hold for a binary alloy system in which the heat of formation is exothermic. In the case of an endothermic alloy, however, the free energy versus composition curve at low temperatures may have regions in which it is convex upward. Consider, for example, the type of curve illustrated in Fig. 1. The common tangent drawn between points B and C represents the free energy of unmixed solutions of composition X and Z. At an intermediate composition, Y, in the part of the curve which is convex upward, the solution represented by point A will be unstable relative to point D on the tangent. The stable configuration corresponding to this composition is therefore two solutions of composition X and Z.

It is also of interest to note that starting from miscibility gap data, attempts have been made to derive thermodynamic properties, i.e., free energies of mixing, and Sundquist (*23*) summarizes information and discusses a number of binary alloy systems.

B. Successive Evaporation

When the vapors of two metals are deposited successively on a substrate, it would seem that some pairs of metals become alloyed in the process. Michel (*24*) reported that Ag–Zn and Ag–Sn alloyed but not Ag–Mg, Cu–Au, or Cu–Al, explaining the difference between the alloy systems in terms of calculated atom diffusion distances at 27°C. However, Fujiki (*25*) concluded that the temperature of the surface layer of the first deposit is raised by the deposition of the second metal vapor, on the basis of a study of the Au–Pb, Bi–Pb, Cu–Sn, Ag–Sn, and Cd–Sn systems by electron diffraction. This temperature rise, which promotes alloy formation, was thought to be mainly due to the latent heat of condensation of the metal vapor which deposits on the substrate layer [but there are calculations to the contrary (*26*)] rather than radiation heating by the evaporation source. Hence, metal pairs having large latent heats of condensation may form alloys more readily. When cadmium was deposited on tin, the β phase was formed which indicated that the temperature had risen during deposition from room temperature to 133°C at least. In support of these ideas about temperature increases at the surface during deposition, Belous and Wayman (*27*) measured temperatures some 400°–500° higher than the substrate temperature during the deposition of gold or silver.

The above work suggests that various factors might determine the extent of alloying and that the result would be specific to the experimental arrangements adopted, e.g., the amount of radiant heat will vary. Therefore, it seems unwise to rely on alloy formation during the deposition of the second layer and, in fact, Cu–Ni films prepared for surface studies by

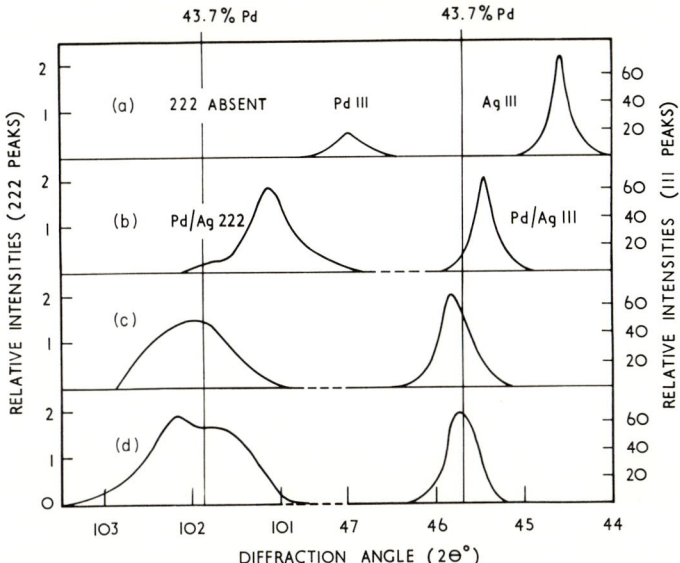

FIG. 2. X-ray diffractometer patterns from layers of Pd and Ag deposited at 0°C (a) and after heating at 530°C for 1 hr (b), 2 hr (c), and 3 hr (d). Vertical lines show expected peak positions (*30*).

successive deposition were either annealed in vacuum at 200°C (*28*) or heated in hydrogen at 500°C (*29*). So far, the alloy systems mentioned have not generally been expected to yield a range of random solid solutions when prepared as thin films, and it is of interest to examine the results of a systematic investigation of successive deposition for the Pd–Ag system (*30*).

Figure 2 shows the (111) and (222) X-ray diffractometer peaks from samples of Pd–Ag film, either immediately after successive deposition at 0°C (a) or after annealing for periods of 1, 2, or 3 hr at 530°C (b–d) in vacuum. Interdiffusion of the component metals can be described by X-ray equivalent concentration-penetration diagrams (*31*) and from these the progress of the dual layer toward homogeneity, following the different heat treatments, can be represented quantitatively in terms of the "degree of interdiffusion," m_t/m_∞ (*32*). This is defined as the ratio of the net number of atoms which have crossed the Matano interface to a total net number which will have crossed when homogeneity has been established. The film prepared by heating for 3 hr at 530°C [pattern (d)] was reasonably homogenized with $m_t/m_\infty = 0.90$, but clearly successive deposition is an "uphill" way of preparing such alloy films compared with simultaneous evaporation.

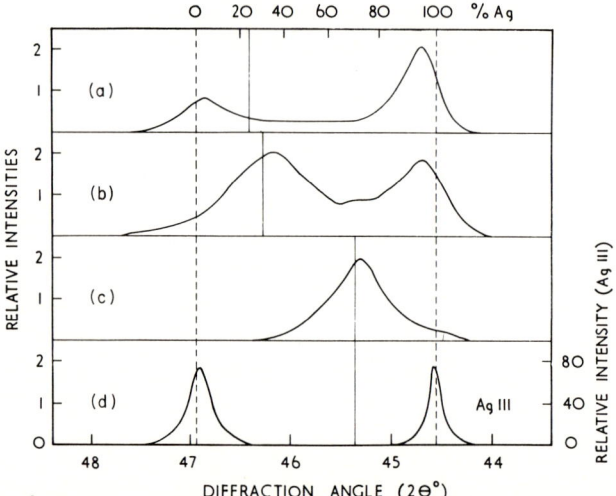

FIG. 3. X-ray diffractometer patterns from layers of Pd and Ag deposited at 0°C; Ag deposited first, (a)–(c), progressively lighter films; Pd deposited first (d) (*30*).

A certain amount of alloying was achieved by depositing palladium on top of silver (Fig. 3a–c) but very little using the reverse order (Fig. 3d). The evaporation of the more penetrative component, palladium from a high-temperature source (i.e., a tungsten heating loop) which also subjects the silver deposit to radiant heat, appears to be a major factor responsible for the importance of the order of deposition in successive evaporation. The effect was greatest when the silver layer (deposited first) was thinnest (Fig. 3c). Subsequent heat treatment of this film produced an apparently less well-homogenized alloy as highly degraded "unalloyed" material became included in the larger alloy crystallites composing the film. This shows again that the effect of annealing should be at least studied in developing a preparative method based on successive deposition.

It is particularly helpful that we can take the Cu–Ni system as an example of the use of successive deposition for preparing alloy films where a miscibility gap exists, and one component can diffuse readily, because this alloy system is also historically important in discussing catalysis by metals. The rate of migration of the copper atoms is much higher than that of the nickel atoms (there is a pronounced Kirkendall effect) and, with polycrystalline specimens, surface diffusion of copper over the nickel crystallites requires a lower activation energy than diffusion into the bulk of the crystallites. Hence, the following model was proposed for the location of the phases in Cu–Ni films (*33*), prepared by annealing successively deposited layers at 200°C in vacuum, which was consistent with the experimental data on the work function.

It was assumed that the nickel crystallites are rapidly enveloped in a skin of a copper-rich alloy, from which diffusion towards the center of each crystallite then takes place. If x_1 and x_2 are the atomic fractions of copper in the two equilibrium phases and x is the atomic fraction of copper in the alloy film under consideration, then the crystallites in the annealed film may have a variety of forms. Solid solutions occur at either end of the composition range but the values of x_1 and x_2 at 200°C are <0.1 and ~ 0.8. Hence, over much of the composition range (i.e., where x lies between x_1 and x_2), the Cu–Ni films should consist of crystallites with a kernel which is almost pure nickel (composition x_1) enveloped in a skin of a copper-rich alloy (composition x_2). Eventually, when x is only slightly larger than x_1, the alloy skin does not completely surround the nickel crystallites; small patches of alloy (x_2) and almost pure nickel (x_1) are both exposed.

Finally, with respect to successive evaporation, Pd–Rh films used for CO oxidation (34) are an example of preparing alloy films where a miscibility gap exists and interdiffusion rates are slow. These Pd–Rh films were prepared by depositing layers of palladium and rhodium at 0°C, followed by annealing in 50 Torr hydrogen at 400°C for 21 hr. The apparent surface compositions, evaluated from the CO oxidation rate as described in Section IV, and information on film structure obtained by X-ray diffraction (XRD) are recorded in Table II.

Film 1 appeared to be reasonably homogeneous when examined by X-ray diffraction but the apparent surface composition was only 62% Rh, whereas the overall composition was $\sim$90% Rh. It would seem that palladium has successfully penetrated the rhodium layer during the annealing period forming an alloy but that rhodium diffusion upwards is inadequate through the thin palladium layer so that the surface is Rh-deficient. When the order of deposition was reversed (film 3), but the overall composition was again $\sim$90% Rh, then the apparent surface composition was 95% Rh showing reasonable agreement in this case.

The overall compositions of films 2 and 4 was such that homogeneous alloys were not expected, and the X-ray diffraction peaks showed evidence of phase-separation with two maxima corresponding to the compositions recorded in column 8, Table II. When palladium was deposited on top (and the film was rather light) then the apparent surface composition after annealing was 10% Rh and X-ray diffraction indicated a phase also containing 10% Rh (film 2). This convenient result was not observed when the order of deposition was reversed (film 4). These differences between apparent surface composition and the overall composition of the homogeneous alloy (or one of the phases in the miscibility gap) are discussed in Moss and Gibbens (34), with further examples. The main point to be made here is the rather variable nature of the surface composition compared with that expected, due to the operation of a number of factors.

TABLE II

Pd–Rh *Films Prepared by Successive Evaporation*

Film no.	Order of deposition		Overall composition (at. % Rh)	Film weight (mg)	Film structure	Composition of XRD Sample[a] (at. % Rh)		Apparent surface composition (at. % Rh)
	Bottom layer	Top layer				XRF[b]	XRD[a]	
1	Rh	Pd	87.8	21.7	Homogeneous	88.3	91.4	62
2	Rh	Pd	57.4	8.1	Two-phase	52.3	10.2; 78.8	10
3	Pd	Rh	89.2	30+	Homogeneous	88.9	91.4	95
4	Pd	Rh	41.4	15.6	Two-phase	38.3	4.0; 81.6	28

[a] XRD, X-ray diffraction.
[b] XRF, composition analysis by X-ray fluorescence.

C. Simultaneous Evaporation

A number of aspects of this method are discussed below, with reference to both the principles involved and the experimental techniques adopted.

1. *Evaporation Sources*

The use of separate sources containing the pure metals avoids difficulties arising from the preferential volatilization of one component in an alloy source but raises some experimental problems. If the surface composition of the alloy cannot be determined (e.g., by the methods discussed in Section III), then it seems that a film of uniform composition *in depth* should be attempted and so the evaporation rates of the individual components need to be held constant. Usually, where alloy films were prepared for adsorption and catalytic studies, evaporation rates were controlled by manual adjustment of the source temperature. For example, the sources might be hairpins or spirals of the pure metals or beads of the pure metal attached to tungsten heating loops, through which a controlled electrical current is passed. An apparatus for depositing films simultaneously with a known ratio of the two components has been described (*35*) consisting of a servomechanism which controls the rate of vaporization of one of the substances with respect to the other. In another automatic control device, sensing with a crystal oscillator (*36*), the long-term variation in chemical composition was less than 0.5%. A different method for the controlled deposition of alloy films from two sources employed an electron beam which oscillated between the two sources (*37, 38*); the dwell-time of the beam on each source determined the relative rates of evaporation. There is, of course, the problem of adapting elaborate pieces of equipment so that clean films can be prepared and unwanted catalyzing surfaces avoided.

The geometry and positioning of separate sources is important if alloy films of uniform composition *over the area* of the adsorption/catalytic reaction vessel are required. The uniformity of composition can be tested as described in Section III, but some trial and error can be avoided by attention to the principles involved. The distribution of thin films condensed on surfaces from point, strip, wire, and ring sources has been reviewed (*39*) with regard to both the theoretical considerations involved and their practical application. A bead of metal (on a tungsten heating loop) positioned at the center of a spherical vessel, or a straight metal wire down the central axis of a cylindrical vessel, should give a uniformly thick film of that metal. However, the second metal source has to be designed and positioned with some care if a uniform *composition* is to be achieved, because small displacements from the exact center of the evaporation vessel may produce an unacceptable result. Short concentric spirals of the pure metals (in wire form), carefully positioned in a spherical vessel have proved effective for preparing binary alloys of uniform composition over the area of the vessel (40).

Non-uniformity of composition sometimes can be exploited to advantage, e.g., in observing work function changes at a series of compositions, a sliding cathode photo tube (41), Fig. 4, was used. A curved substrate, either cylindrical or spherical, can also be used in preparing specimens with a broad range of concentrations (42).

Evaporation from an alloy source largely avoids the problems of source geometry which may be encountered in simultaneous evaporation from separate pure metal sources. The success of the method depends on the relative vapor pressures of the component metals in the alloy source because one component metal with a substantially higher vapor pressure will evaporate preferentially. The alloy film will then be rich in the more volatile component compared with the composition of the source and may also be stratified into layers of different composition. Evaporation from alloy sources is discussed in detail by Dushman (43) and Holland (44) on the basis of Raoult's law, and it is possible to predict which metal pairs

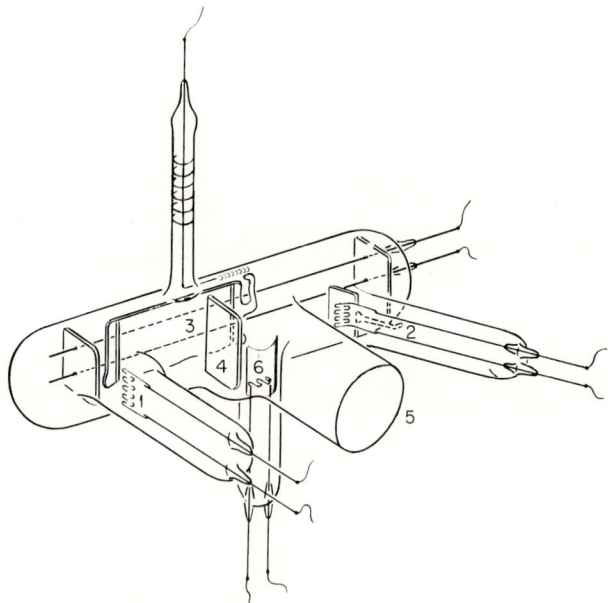

Fig. 4. Sliding cathode phototube (41) showing evaporation sources (1, 2) for depositing the alloy film on the photocathode (3), shielded partly as required by the glass screen (4). A quartz window (5) faces the cathode and connection is made to a gold film anode evaporated from (6).

TABLE III

Data Relating to Evaporation from Alloy Sources

Metal	Vapor pressure P at 1200°C (Torr)	$P/M^{1/2}$
Copper	3.39×10^{-3}	4.25×10^{-4}
Silver	2.14×10^{-1}	2.06×10^{-2}
Gold	1.82×10^{-4}	1.30×10^{-5}
Iron	9.55×10^{-5}	1.28×10^{-6}
Cobalt	4.07×10^{-5}	5.30×10^{-6}
Nickel	3.02×10^{-5}	3.93×10^{-6}
Palladium	1.45×10^{-4}	1.41×10^{-5}
Platinum	9.33×10^{-10}	6.66×10^{-11}
Rhodium	1.10×10^{-9}	1.08×10^{-10}

are likely to provide alloys which will evaporate without fractionating; such alloys are known as "constant evaporation rate alloys." The ratio of the evaporation rates of metals A and B (E_A/E_B) will be equal to their concentrations by weight in the alloy source (W_A/W_B) when

$$P_A(M_B)^{1/2}/P_B(M_A)^{1/2}$$

is unity, where P and M are the vapor pressure and molecular weight, because it can be shown that

$$E_A/E_B = (W_A P_A/W_B P_B)(M_B/M_A)^{1/2}.$$

Values of $P/M^{1/2}$ are tabulated by Holland (44), but we provide in Table III further values [calculated from data in Smithells (45)] which enable us to discuss some recent observations on the preparation of alloy films, most of which were used in catalytic experiments.

According to the data in Table III, the value of the ratio $P/M^{1/2}$ is approximately the same for the metals Au, Fe, Co, Ni, and Pd. Binary alloys formed from any pair of these metals can therefore be expected to evaporate without substantial fractionation. On the other hand, films evaporated from Ag–Pd and Cu–Ni alloys can be expected to be enriched in Ag and Cu, respectively. These predictions are largely confirmed by experiment. For example, the composition of Pd–Au films was found to be the same as the wires which were evaporated (46), but in the case of Pd–Ag, evaporation of a 30% Ag–Pd alloy wire yielded a 50% Ag–Pd alloy film (47). Alexander and Russell evaporated a number of alloys from pellets in the reaction vessel as shown in Fig. 5 (48). The alloy pellet was placed in a small quartz cup with its surface equidistant from the hemispherical top of the reaction vessel. The pellet was evaporated by

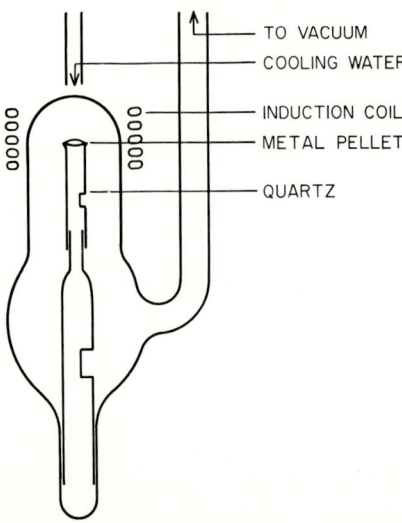

Fig. 5. Reaction vessel, in the dome of which, alloy films are deposited by evaporation from an alloy pellet subjected to induction heating (*48*).

induction heating and then the resulting alloy film was annealed. The results showed that there was very little fractionation in the case of Fe–Ni alloys, but Pd–Ni films were richer in Pd, and Cu–Ni films were highly enriched in Cu.

If the proper techniques are used to homogenize the alloy (and to determine its composition), then a moderate degree of fractionation during evaporation is not really a disadvantage. Otherwise fractionation may result in some degree of stratification in the film. Michel found that Al–Cu films were rich in Al at the beginning of the evaporation, but progressively richer in Cu as the evaporation proceeded (*24*). Similarly, the initial layers of Ag–Pd films deposited at room temperature were richer in Ag (*47*), though deposition at 400°C followed by annealing yielded homogeneous films. In the case of Ni–Fe films, stratification was attributed to a variation in sticking coefficient as the nature of the substrate surface became modified by the deposit (*49*). It has been shown that fractionation can be avoided by "flash-evaporation," in which the alloy is very rapidly evaporated by dropping fine particles onto a refractory strip maintained at a high temperature (*50*). In this method, considerations such as diffusion rates in the source also apply.

2. *Heated Substrates and Vacuum Effects*

During the formation of an alloy film, regions rich in one component may develop (e.g., due to preferential nucleation of one metal), and it would seem that heating the substrate could ensure that the condensing atoms are able to diffuse a sufficient distance through the bulk to produce a random solution (assuming this is the equilibrium form). The mean square displacement Δx^2 is related to the diffusion coefficient D (cm^2 sec^{-1}) and the time t (sec) by the expression:

$$\Delta x^2 = 2\, Dt.$$

The diffusion coefficients of palladium in a Pd–Ag alloy and silver in a range of Pd–Ag alloys are known, and the diffusion of palladium and silver atoms in a 20% Pd–Ag alloy was calculated (*30*) for $t = 3600$ sec representing the film preparation time. At temperatures of 100°, 200°, 300°, and 400°C, silver atoms would diffuse in this time distances of 3×10^{-4}, 0.15, 9, and 150 Å, respectively whereas at the corresponding temperatures, palladium atoms would diffuse 26, 460, 3000, and 11,000 Å. Palladium atoms can thus penetrate the alloy lattice at moderate temperatures, whereas silver atoms have a probability of diffusing distances equivalent to a few unit cells only when the substrate temperature is greater than 300°–400°C.

When Pd–Ag alloy films were prepared by simultaneous evaporation in conventional vacuum, then indeed a substrate temperature of 400°C was necessary to form a homogeneous alloy (*30*). Now, surface diffusion should be substantially faster than bulk diffusion [the activation energy for surface diffusion is 55–60% that for bulk diffusion in Cu–Au and Ni–Au (*51*)], but the above finding suggests that the latter process played an important part in this instance. The literature contains data (*52*) usually for alloys in massive form which would enable a similar calculation to be carried out for other binary alloys; diffusion data for Au–Pd alloys in thin film form have been measured (*53*). At 400°C, the loss of surface area by sintering can be severe and also surface contamination is increased. Therefore, it seemed that further consideration should be given to preparing Pd–Ag films which would be very much less sintered while still exhibiting as good a homogeneity (*54*).

It might be expected that the structure of an evaporated alloy film would be dependent on the partial pressure of the residual gases during evaporation by analogy with pure metal films. For example, the vacuum conditions have a marked effect on the crystallite size in evaporated silver films deposited at 0°C on glass (*55*). Now, the activation energy for surface diffusion in a number of systems is altered by adsorbed gases and higher *surface* diffusion rates, if a consequence of depositing alloy films in ultrahigh vacuum (UHV) should lead to satisfactory homogeneity at moderate substrate temperatures, i.e., well below the 400°C, reported above. Hence, reasonably unsintered alloy films might be prepared with a useful surface area for adsorption work.

TABLE IV

Homogeneity of Pd–Ag Alloy Films

Group	Composition (at. % Pd)	Vacuum (Torr)	Weight[b] (mg)	Lattice constant deviation, (Å)[c] Δa_0	X-ray profile shape[d]
1	9	8×10^{-6}	—	0.008	Sym
	32	2×10^{-6}	16.5	0.013	Sym
	54	5×10^{-6}	14.0	0.017	$\sim$ Sym $\rightarrow$ low θ
	63	8×10^{-6}	14.3	0.005	Very asym $\rightarrow$ low θ
	65	2×10^{-6}	19.0	0.002	$\sim$ Sym $\rightarrow$ high θ
	80	5×10^{-6}	17.9	0.015	$\sim$Sym
				Av. 0.010	
2	21	8×10^{-10}	—	0.004	Sym
	35	8×10^{-10}	—	0.004	Sym
	42	8×10^{-10}	—	0.001	Sym
	68	5×10^{-9}	17.6	0.001	Sym
				Av. 0.0025	

[a] The substrate was maintained at a temperature of 0°C.

[b] Total weight on inside of spherical glass vessel, area 220 cm².

[c] Deviation of a_0 from expected value based on data in Coles (77).

[d] Symmetrical or asymmetrical; arrow means a tail toward angle indicated.

Table IV shows X-ray data (55) on the homogeneity of Pd–Ag films prepared by simultaneous evaporation from separate sources, either in conventional vacuum or in UHV, with the substrate maintained at 0°C. The second group of films was prepared using a stainless steel system incorporating a large (100 l/sec) getter-ion pump, sorption trap, etc., but deposited inside a glass vessel. By the tests of homogeneity adopted, alloy films evaporated in conventional vacuum were not satisfactory, i.e., the lattice constants were generally outside the limits of the experimental error, ± 0.004 Å, and the X-ray line profiles were not always symmetrical. In contrast, alloy films evaporated in UHV were satisfactorily homogeneous. Further, electron micrographs showed that these latter films were reasonably unsintered and thus, this method provides clean Pd–Ag alloy films with the required characteristics for surface studies.

In this context, it is interesting to note that it has been claimed (56) that single-crystal Fe–Ni alloy films can be prepared by deposition on heated rock salt substrates in vacua of 10^{-3}–10^{-4} Torr. Other workers (57) have found that the use of UHV permits single-crystal films of Fe–Ni to be formed (at deposition rates of 14 Å/min) without the annealing necessary after deposition at 10^{-5} Torr. Single-crystal Au–Pd films have also been prepared (58) and after quenching from 500°C gave an electron diffraction pattern interpreted as showing a superlattice of the CuAu I type.

The nucleation and growth of metal films is a subject of some complexity and this is increased when two metals which do not form a full range of solid solutions are substituted for a single pure metal. Nevertheless, the effect of using a heated substrate to prepare alloy films of such metal pairs requires brief consideration of the processes involved, as an insight into the disposition of the phases within the crystallites formed. A process of continuous nucleation is envisaged on a substrate which initially is the glass wall of the reaction vessel but subsequently, and for most of the time, is the film material itself. The crystallites formed sinter together during the film deposition period at a substrate temperature of, say, 400°C. Most observations relating to heterogeneous nucleation are consistent with the general theory which assumes metastable equilibrium between single adatoms and polyatomic aggregates on the substrate surface to which macroscopic thermodynamic properties are ascribed (59), although perhaps not entirely justified for small aggregates.

There is a point at which these aggregates reach a critical size of minimum stability r^* and the free energy of formation ΔG^* is a maximum. Further addition of material to the critical nucleus decreases the free energy and produces a stable growing nucleus. The nucleation rate is the product of the concentration of critical nuclei N^* given by

$$N^* = N_{ad} \exp(-\Delta G^*/kT),$$

and the rate at which adsorbed atoms migrate and become incorporated into the nuclei, where N_{ad} is the adatom concentration on the substrate. The free energy of formation of an aggregate is expressed as the sum of terms including surface free energies and ΔG_v, the free energy difference between supersaturated vapor of pressure p and condensate of equilibrium vapor pressure p_e and atomic volume Ω where

$$\Delta G_v = -(kT/\Omega) \ln(p/p_e).$$

The nucleation rate is strongly dependent on the ΔG_v term in ΔG^*, and for any one deposition rate, the supersaturation ratio at the substrate temperature T is given principally by the bulk heat of sublimation. The formation of Pd–Rh films (60) may be cited here, because catalysis over these films is discussed in Section IV. The respective heats of sublimation for palladium and rhodium are 89 and 133 kcal/gm, and therefore rhodium should form more nuclei than palladium on a glass substrate which has a weak affinity for the film material. The interfacial (aggregate-substrate) free energy and substrate surface energy terms in ΔG^* are more important for nucleation on a metal substrate. After the initial deposition on glass, condensation occurs on the film material itself, and there might not be an effective nucleation barrier for either metal ($r^* < 2$ atoms) during much of the film formation process if heated substrates are avoided.

Hence, the decision to use a heated substrate with simultaneous evaporation of the component metals as an aid to homogenization requires consideration of whether or not it might have an adverse effect, i.e., causing preferential nucleation of one component, which interdiffusion may not be able to remedy. It was believed (60) that in preparing Pd–Rh alloys by simultaneous deposition on a substrate at 400°C, rhodium nucleated preferentially and that crystallites grew by the addition of palladium (and rhodium) atoms. The diffusion of palladium atoms into this kernel formed a phase with $88 \pm 5\%$ Rh (phase II). The outer shell of the crystallite, phase I, was in effect a solid solution deficient in rhodium compared with the overall film composition, and the Rh content of phase I therefore increased as the Rh flux was increased.

3. *Vapor Quenching*

In this method the component metals are evaporated simultaneously onto a substrate which eliminates appreciable atom mobility, because it is very cold and also a good enough heat sink to absorb the heat of condensation, thus preventing a significant rise in surface temperature. Since the vapor phase has no miscibility restrictions, the condensing atoms are bound to mix randomly and homogeneously. It is therefore possible, in principle, to prepare a homogeneous alloy in systems in which, at equilibrium, there is essentially no solubility in the solid state. Vapor quenching has been used to prepare solid solutions of a number of combinations of metals which are metastable at and above room temperature. Systems which have been studied include Co–Cu (61), Fe–Cu (62), Cu–Ag and Co–Au (63), Cu–Mg (64), and Gd–Ag (65). The rapid quenching of liquid droplets, known as "splat-cooling," has also been successful (66), although it has been shown to be less effective than vapor quenching in the Co–Cu and Cu–Ag systems (67).

The structure of a vapor-quenched alloy may be either crystalline, in which the periodicity of the unit cell is repeated within the crystallites, or amorphous, in which there is no translational periodicity even over a distance of several lattice spacings. Mader (64) has given the following criteria for the formation of an amorphous structure: the equilibrium diagram must show limited terminal solubilities of the two components, and a size difference of greater than 10% should exist between the component atoms. A ball model simulation experiment has been used to illustrate the effects of size difference and rate of deposition on the structure of quench-cooled alloy films (68). Concentrated alloys of Cu–Ag (35–65%

Ag) and Co–Au (25–65% Au) were amorphous when deposited on an amorphous substrate held at 80°K (*63*). On warming, the films were found to anneal in two stages. In the first stage a transformation to a metastable, single-phase crystalline solid solution occurred in a narrow temperature range. The second stage involved the transition to the equilibrium two-phase state. These transitions are illustrated (*26*) by the changes which occur in the resistivity of Cu–50% Ag films in Fig. 6. The effect of depositing the film at different substrate temperatures is shown by the dashed curve. For a number of different alloy systems (*64*) it was found that crystallization occurred at about 0.3 T_m, where T_m is the average melting point of the component metals. Decomposition of the crystalline state to the equilibrium two-phase state occurred in the range 0.45–0.5 T_m. When crystalline alloy films were formed on deposition, the crystallite size was very small, e.g., about 100 Å for a Co–65% Cu film deposited on cleaved rock salt at room temperature (*64*).

Vapor quenching provides a method of bridging the miscibility gap which exists in many alloy systems, and makes a range of novel alloys available for study. Such films, of course, would not be ideal for catalytic studies. They could not be used at high temperatures, and indeed the heat of reaction might be sufficient to induce a transformation to a more stable structure. In addition, characterization by X-ray diffraction would be difficult, even for the crystalline films, because of line broadening by the small crystallites. Nevertheless, alloy films which are metastable above room temperature can be prepared, and their high surface area would

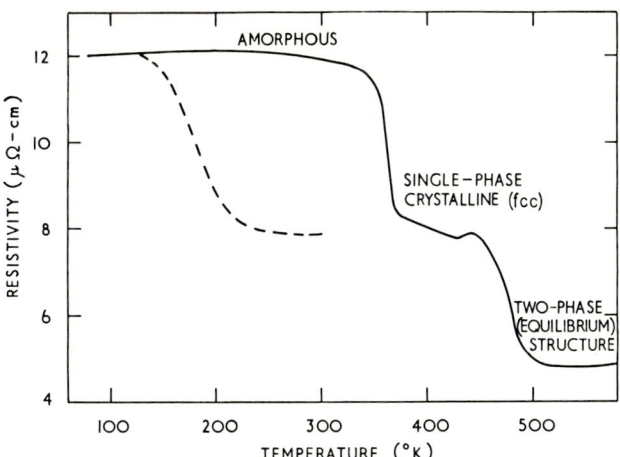

Fig. 6. Variation of resistivity with temperature for 50% Cu–Ag films deposited at 80°K and then warmed up progressively (solid line) compared with films (broken line) deposited at different temperatures above 80°K (*26*).

enable them to be used for adsorption studies and for reactions which proceed at moderate temperatures.

III. Characterization of Alloy Films

In discussing the principles involved in alloy film formation, reference had to be made to alloy systems which are uncommon or unused in studies of adsorption and catalysis. This section is specifically concerned with the characterization of alloy films prepared for such purposes. However, the various aspects of alloy film structure mentioned in Section II have to be kept in mind when discussing results of catalytic experiments using evaporated alloy films.

The characterization of evaporated alloy films can be carried out at widely different levels of sophistication. At the very least, it is necessary to determine the bulk composition, probably after the film has been used for an adsorption or catalytic experiment. Then various techniques can be applied, e.g., X-ray diffraction, electron diffraction, and electron microscopy, to investigate the homogeneity or morphology of the film. The measurement of surface area by chemisorption presents special problems compared with the pure metals. Finally, there is the question of the surface composition (as distinct from the bulk or overall composition), and a brief account is given of techniques such as Auger electron spectroscopy which might be applied to alloy films.

A. Composition

With due regard to the lateral variations in composition which can arise as a consequence of source geometry and positioning (discussed in Section II), it is wise to analyze the alloy film at a number of representative points. For example, if a catalytic reaction was carried out over an alloy film deposited inside a spherical vessel maintained at a constant temperature over its entire area, then the mean alloy composition (and the uniformity of composition) is required. A convenient procedure is to cut glass reaction vessels carefully into pieces at the end of the experiment and to determine the composition by X-ray fluorescence analysis of a number of representative pieces. Compositions of Pd–Ag alloy films (40) determined at 12 representative parts of a spherical vessel from the intensities of the $AgK_{\alpha_{12}}$ and $PdK_{\alpha_{12}}$ fluorescent X-ray emissions are shown in Table V; mean compositions are listed in the first column. (The Pd and Ag sources were separate short concentric spirals.) In other applications of evaporated alloy films to adsorption and catalytic studies, as good or better uniformity of composition was achieved. Analyses of five sections of a cylindrical

TABLE V

Uniformity of Pd–Ag *Alloy Film Composition*

Mean composition (% Pd)	% Pd analyzed at representative points on vessel surface										
12	14	10	13	12	11	12	11	11	13	13	13	12
51	50	54	53	55	50	51	51	54	53	46	50	46
84	88	86	88	86	83	84	78	83	84	83	81	84

vessel (*47*) showed Pd–Ag alloy compositions which varied by ±1 or ±2% from the mean. Also, in the account of this work, a detailed description of the analysis, again by X-ray fluorescence, is given. Other workers (*48*) have used spectrophotometric methods with Cu–Ni, Fe–Ni, and Pd–Ni alloy films and found excellent uniformity of composition in films prepared by induction-heating alloy pellets.

Some other points worth noting in connection with alloy film composition are: The loss in weight from separate sources is a *guide* to mean composition but not an exact measure because the sources become themselves alloyed. It is often important to determine the composition of the actual specimen on which other characterizing measurements have been made. If there is confidence that the films are reasonably homogeneous, lattice constants determined by X-ray diffraction can be used to examine the uniformity of composition (*69*), but the change of lattice constant with composition may be inconveniently small.

B. Crystallite Size and Orientation

Electron microscopy and electron diffraction provide information on the state of crystallite aggregation, orientation, shape, etc. It is of some interest to know if these features are varying with composition as well as the adsorptive or catalytic properties under investigation, so that one source of misinterpretation can be avoided. Electron micrographs showing that the crystallites are little affected by changes in composition would usefully supplement surface area measurements. Specimens are readily obtained by stripping the film from fragments of the reaction vessel.

Figure 7(a, b, d, and e) shows transmission electron micrographs from Pd–Ag films of comparable weight, prepared and annealed at 400°C, and used once to catalyze the oxidation of ethylene at 240°C (*40*). The structure of this series of alloy films varied consistently with composition. Silver-rich films (e.g., Fig. 7a, ~13% Pd) showed extensive coalescence of the crystallites, while at the other end of the composition range (e.g., Fig. 7e,

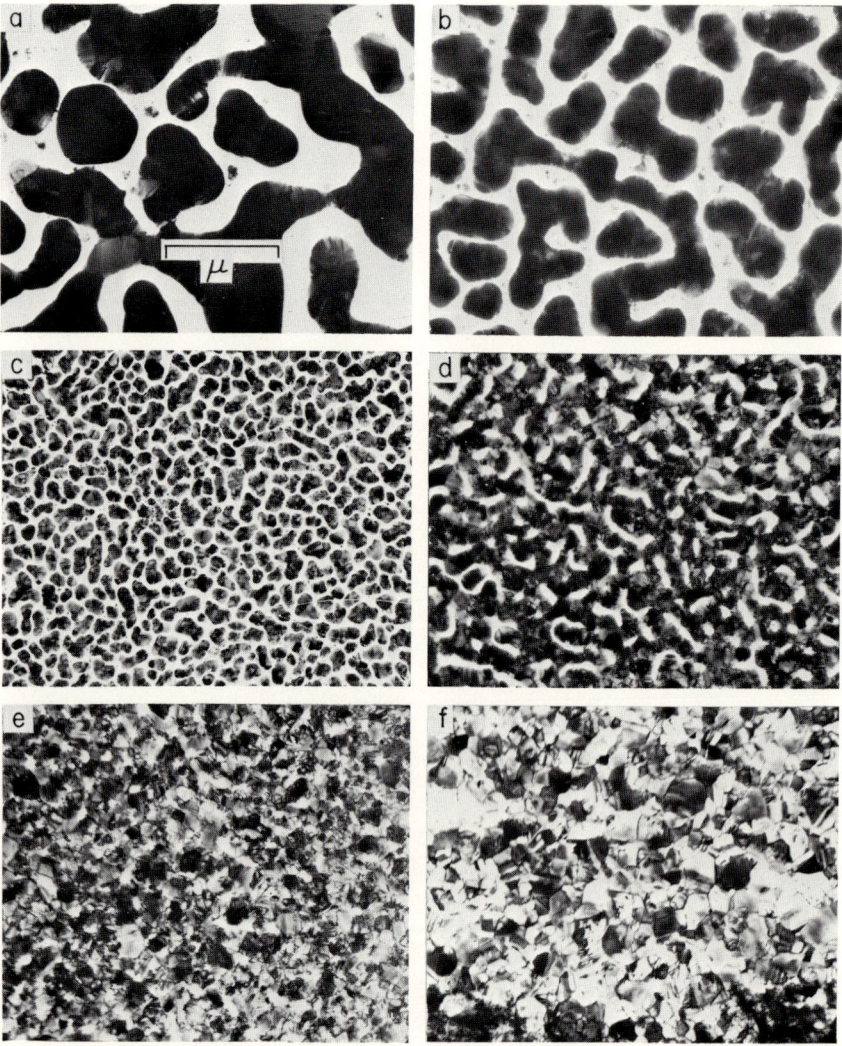

Fig. 7. Transmission electron micrographs of Pd–Ag films used for C_2H_4 oxidation after annealing at 400°C [(a), (b), (d), (e)] and containing 12.6, 36.7, 51.0, and 94.0% Pd, respectively. Film (c), 51% Pd, was similarly prepared but unused and film (f), 19.4% Pd, was not annealed but used (40).

$\sim$94% Pd), the films were composed of a compact mass of small crystallites. These differences in texture are explained in terms of the inverse correlation between melting point and sinterability which exists when the forces between condensed atoms and the substrate are weak.

Figure 7c, 51% Pd, fits neatly into the series, although the film was not used in catalytic reaction showing that the structure had been determined at the preparation stage. Figure 7f shows an electron micrograph of a film also deposited at 400°C but not annealed before catalyzing the oxidation of ethylene. Although it possessed a low Pd content, ~19%, the film was unlike the corresponding annealed films and had suffered much less aggregation. Thus, low melting-point alloy films would be better not annealed if surface area must be preserved, but annealing appears to be a useful addition to simultaneous deposition in conventional vacuum (see Section II). In catalytic experiments compensation for the smaller area can be made by working at somewhat higher temperatures.

Electron micrographs of Pd–Au films used for formic acid decomposition (69) showed two main differences due to change in composition:

(1) The grain size was greatest (about 4000 Å) for pure Pd and pure Au and least (700 Å) for a 39% Pd film.

(2) All compositions were subject to preferred orientation. Pure Pd had a very strong and 74% Pd a less strong (111) orientation, whereas 39% Pd, 15% Pd, and pure Au films had a (110) orientation of increasing strength. It was inferred that these changes indicated differences in the relative extents of exposed crystal face, and this information was valuable in discussing the catalytic results (see Section IV).

When both the component metals have reasonably high melting points, as in Pd–Rh alloys, then the variation of crystallite size with composition (60) is limited, despite the use of high substrate temperatures (400°C).

TABLE VI

Crystallite Orientation in Pd–Rh Films

Composition (at. % Rh)	Film wt. (mg)	Corrected X-ray intensities				
		(111)	(200)	(220)	(311)	(222)
1.09	26.85	100	0.2	<0.1	0.1	9
4.85	27.20	100	5	1	2	6
12.9	30.50	100	0.6	<0.1	0.4	11
24.6	25.30	100	3	0.3	2	11
54.8	27.65	100	13	3	9	10
76.1	26.00	100	8	1	3	4
97.2	29.25	100	39	25	25	6
Pd powder (ASTM)		100	42	25	24	8
Rh powder (ASTM)		100	50	26	33	11

Crystallite orientation was studied by measuring the areas of the X-ray diffraction peaks obtained with a counter-diffractometer (Table VI). The areas were corrected for changes in the X-irradiated area by dividing each by $\sin \theta_{111}/\sin \theta_{hkl}$ and are directly related to the volume of alloy in the hkl orientation, but only reflecting planes parallel to the glass substrate contribute to the X-ray intensity. Data from the ASTM Index for random powder samples of Pd and Rh are also included in Table VI. The occurrence of preferred orientation in most of the alloy films is apparent, i.e., a strong (111) texture has developed. Only as the Rh content increases does this preference weaken so that an almost pure Rh film shows nearly random orientation.

C. Surface Area

Alloy films are commonly sintered during preparation by deposition on substrates heated to, say, 400°C or by subsequent annealing at such temperatures, and, consequently, rather small surface areas have to be measured, perhaps in vessels of substantial volume. Krypton adsorption at liquid nitrogen temperature was used with induction-evaporated Cu–Ni, Fe–Ni, and Pd–Ni films, and BET surface areas of 1000–2000 cm² were recorded (48), after correction for bare glass. The total area of Cu–Ni films was measured by the physical adsorption of xenon at −196°C (70); in addition, the chemisorption of hydrogen on the same samples enabled the quantity α to be determined where

$$\alpha = \frac{\text{No. of H atoms adsorbed at saturation}}{\text{No. of Xe atoms adsorbed at saturation}}.$$

(The observed ratio $\alpha_{\text{alloy}}/\alpha_{\text{Ni}}$ agreed well with the Ni content predicted.) As discussed later, it was believed that hydrogen chemisorption was proportional to the surface nickel concentration (see Section IV). It is clear, however, that chemisorption as a method of surface area measurement must be used with discretion in the case of alloy films.

In earlier work with pure metals, it was generally accepted that the area of films deposited at, say, 0°C was proportional to their weight (with the exception of group IB and low melting-point metals). Information was available on the surface areas of films of Ni, Pt, Pd, Rh, etc. (71), and hence absolute reaction rates could be calculated. It would be a considerable undertaking to establish similar data for alloy systems, bearing in mind that various compositions would have to be examined and also a method for preparing exact compositions would be required. However, for sintered alloy films, approximate methods can be proposed.

It has been found for sintered nickel films (72) that the surface area a

is related to the film weight w by the expression:

$$a = B + Dw,$$

but the constant D for films sintered at 400°C is only 3.3 cm²/mg ($B = 490$ cm²); the relationship does not hold for films weighing ~5 mg. Therefore, it could be assumed for, say, Pd–Ni alloy films, prepared at 400°C and weighing 20–30 mg, that the surface area is approximately constant (with due regard to the accuracy of rate measurements) *at each composition*. Then the surface areas of the pure metal films prepared in the same way could be determined; on the basis of the results obtained it could be decided if areas for a few alloy compositions were also needed to construct a diagram of surface area versus composition.

Alternatively, it may be possible to demonstrate for the pure metals that the *catalytic activity* is independent of film weight in a certain weight range. For example, rates of ethylene oxidation were constant over pure palladium films, deposited and annealed at 400°C and weighing between ~4 and ~40 mg (*73*). Then, if electron micrographs show that the crystallite size is relatively independent of composition, a satisfactory *comparison* of catalytic activity can be made at the various alloy compositions. Finally, surface area measurements are less urgently needed when activity varies by orders of magnitude, or where the main interest lies outside the determination of absolute reaction rates.

D. Bulk Structure

It is probably not essential to have an alloy film which is homogeneous throughout, but some means of determining the surface composition would then be required. Usually, the experimenter will want to prepare homogeneous alloy films (where this is possible) because it is easier to justify the assumption that the surface and bulk compositions are equal. It would seem unlikely that an alloy film exhibiting poor bulk homogeneity, when a random solid solution was required, would be satisfactory as some inhomogeneity would also exist at the surface. In the case of alloys where phase separation is expected under the conditions of preparation and use, examination of the bulk film structure is again of considerable interest.

1. *Homogeneous Alloys Expected*

X-ray diffraction with facilities for determining the intensity profile of each diffraction "line" is particularly informative. Criteria which have been adopted as evidence of good bulk homogeneity (*54*) are (1) a correct lattice constant a_0, and (2) a symmetrical X-ray diffraction profile. A method has also been described (*30*) for determining if a range of lattice parameters exists due to incomplete alloying, which might be used as the basis of a third criterion.

(1) Sintered alloy films of reasonable thickness, e.g., opaque, mirror-like films, can provide an adequate number of diffraction peaks for the determination of a lattice constant of adequate accuracy for present purposes. Thus, the apparent lattice constants calculated from the centroids of individual diffraction peaks, observed with a counter-diffractometer, may be extrapolated to $\theta = 90°$, using the Nelson–Riley function to give a value of a_0. There has been some discussion about differences in lattice constants for thin films compared with bulk metals; values of a_0 for pure silver films ($\sim$1000 Å nominal thickness) were found (74) to be consistently small compared with bulk silver but only by 0.05%. For alloy films a similar deviation would correspond to a variation of 1% in the composition of the alloy. [Larger deviations have been reported for very thin films, e.g., -0.2% in copper films of 100 Å nominal thickness (75).]

The lattice constants observed for a series of alloy films are then compared with the lattice constants observed in bulk samples, e.g., values

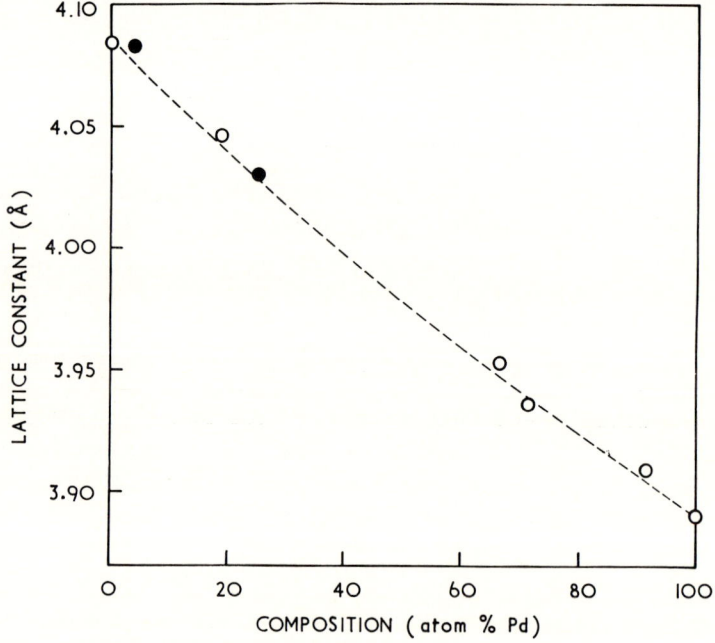

Fig. 8. Lattice constants in Pd–Ag alloy films (30) prepared by simultaneous deposition onto glass at 400°C from separate sources (○) and from an alloy source (●); lattice constants for Pd–Ag wires (77), broken line.

recorded in Pearson's "A Handbook of Lattice Spacings and Structures of Metals and Alloys" (76). Figure 8 shows lattice constants for Pd–Ag films prepared by simultaneous evaporation onto glass at 400°C (30) and for a series of Pd–Ag alloy wires (77), (broken line). The homogeneity of these films, which showed only small random deviations in the lattice parameter compared with the bulk values, therefore appears to be reasonably satisfactory on the first of the three criteria mentioned above.

(2) Apart from providing evidence for the existence of only those diffraction peaks expected, at about the correct position, visual observation of the diffractometer chart record, can be informative about alloy homogeneity. However, due regard has to be paid to possible artefacts such as the existence of the $K\alpha$ doublet; the $K\alpha$ doublet can be resolved where necessary using the graphical method described by Rachinger (78). These considerations apart, the symmetry of the diffraction profile appears to be a sensitive indicator of the state of alloy homogeneity.

(3) A closer analysis of the X-ray data can be made for alloy films which appear to have satisfactory bulk homogeneity (30). The X-ray diffraction peaks may be broadened compared with the breadth observed for bulk specimens of pure metals due to the small size of the crystallites composing the films. This can mask any broadening due to the occurrence of a range of lattice parameters arising from some degree of incomplete alloying. The broadening B_c due to crystallite size is given by the Scherrer expression

$$B_c = K\lambda/L \cos \theta,$$

where K is the Scherrer constant, λ is the radiation wavelength, L is the crystallite size normal to the substrate, and θ is the Bragg angle. The broadening B_a due to any incomplete alloying is given by

$$B_a = 2 \tan \theta (da/a),$$

where da is the range of lattice parameter and a is the measured lattice parameter. By measuring the breadth of (111) and (222) diffraction peaks from Pd–Ag (30, 40), Cu–Ni (79), and Pd–Au (69) alloy films, it was possible to separate the two effects and to obtain values of da/a. For greatest accuracy, Fourier analysis should be used but a simpler treatment seems sufficient for the present purposes; if a triangular distribution of lattice parameter is assumed, the expression yielding da/a is

$$\left(\frac{da}{a}\right)^2 = \frac{3}{\pi^2} \frac{B_2^2 \cos^2 \theta_2 - B_1^2 \cos^2 \theta_1}{\sin^2 \theta_2 - \sin^2 \theta_1},$$

where the subscripts 1 and 2 denote the (111) and (222) peaks having an observed breadth, B. The analysis gives nonzero values for pure metal films, and the approach is to make comparisons between these values of da/a and those determined for alloy films.

2. *Phase-Separation Expected*

Extensive use has also been made of both X-ray diffraction and electron diffraction to examine thin alloy films where solid solutions were not formed. Where it is intended to study the interdiffusion of metals, then both the thin film technique and diffraction methods can be employed, e.g., in studying the interdiffusion of gold and nickel (*80, 81*). The use of vapor quenching as a preparative method (Section II) also raises questions about whether the alloy film is microcrystalline or truly amorphous, which can be investigated by analyzing the diffuse diffraction patterns in terms of the radial distribution function (*82*). The following examples of the simple use of X-ray diffraction are given because the alloy films examined had been used in adsorption/catalytic experiments. X-ray reflection photographs of Cu–Ni films deposited and subsequently annealed at 200°C provided clear evidence of two phases, each belonging to the fcc lattice type. The photometric curve of the X-ray diffraction pattern, Fig. 5 in Sachtler and Dorgelo (*28*) shows that the (111) and (200) diffraction "lines" have split up into pairs.

Figure 9 shows a selection of (111) diffraction profiles from Pd–Rh alloy films (deposited and annealed at 400°C) which were used to catalyze ethylene oxidation (*60*) at ~150°–200°C. The profile for the film with 24.6% Rh is symmetrical, and inspection of the (222) profile (not illustrated) after resolution of the α_1–α_2 doublet showed no evidence of phase

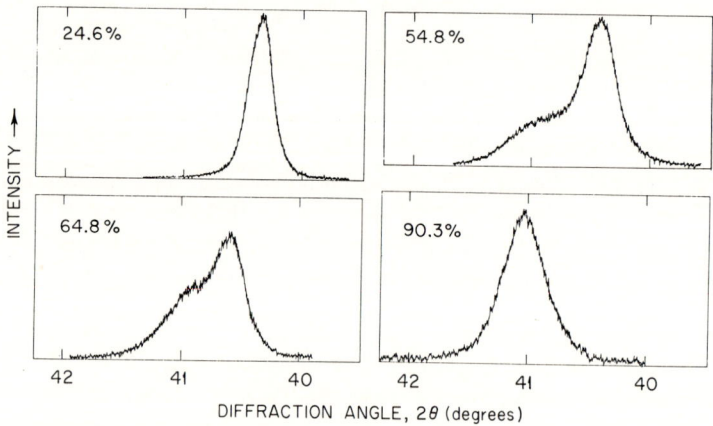

Fig. 9. X-ray diffraction profiles from (111) planes of Pd–Rh films, prepared at 400°C and used in catalytic reaction (*60*).

separation. At 54.8% Rh, the profile is very asymmetrical and two peaks can be recognized at 64.8% Rh, but at still higher Rh contents (90.3%) the profiles are again symmetrical. Thus, the pattern of solubility at either end of the composition range with a tendency to phase separation at intermediate compositions, is observed, as anticipated from the phase diagram of the Pd–Rh system.

E. Surface Examination

Alloys which are homogeneous when prepared at high temperatures may be subject to phase-separation when used as catalysts at lower temperatures. Even if phase-separation is not expected, the surface will tend to acquire the composition which has a minimum surface energy. Examples are given in Section IV where enrichment of the surface by one component occurred either during alloy catalyst preparation or use. These effects can give rise to a situation in which the composition of the catalyzing surface is very different from the bulk composition of the alloy. Unfortunately, very few attempts have been made to determine the surface composition of alloys in any form used in catalytic studies even though such knowledge is highly desirable for the proper interpretation of results.

1. *Work Function Measurements*

Work function measurements have been used to follow the changes which occurred when freshly evaporated films of Cu–Ni (*28*) and Pt–Au (*41*) alloys were annealed. After annealing, all the alloys in each system had nearly the same work function, about 4.61 eV in the case of Cu–Ni alloys, and 5.34–5.37 eV in the case of Pt–Au alloys. These values were close to, but lower than, the work functions of pure Cu (4.67 eV) and pure Au (5.38 eV), respectively. The results were interpreted as a verification of the prediction that equilibrated films of these alloys should have a constant surface composition within the limits of the miscibility gap, and chemisorption and catalytic studies using the alloy films appeared to confirm this view (see Section IV). It has been suggested, however, on statistical grounds, that the work function of an alloy should be close to that of the component which has the lower work function and rather insensitive to large changes in the atomic concentrations of the components (*83*). If this latter suggestion is correct, it is not possible to use work function measurements as a means of indicating the surface composition of an alloy. It should be possible to decide which of these viewpoints is correct by measuring the work functions of an alloy system such as Ag–Pd, where the surface composition is expected to change with the bulk composition.

It has indeed been found (*83a*) that the work function of Ag–Pd alloy films equilibrated at 300°C only changes from 4.38 eV at pure silver to 4.50 eV at 86% Pd (increasing to 5.22 at pure Pd). It was proposed (*83a*) that this work function pattern is a consequence of surface enrichment by silver and so confirmation of the theory awaits further work function measurements on alloy surfaces for which compositions have been determined by, say, Auger electron spectroscopy.

2. *Electron-Probe Microanalysis*

There are two reports (*84, 85*) of the use of this technique in the characterization of alloy catalysts. In the microanalyzer the sample is subjected to a beam of electrons and the secondary X-radiation is collected and analyzed to yield information on the nature and distribution of elements in the sample (*86*). Commonly, a probe of diameter 0.2–1 μ is used so that the instrument is very suitable for checking the constancy of composition of different parts of an alloy sample. It has been used in this way in a study of catalysis on evaporated Cu–Ni films (*84*). Quantitative analysis has also been carried out on Cu–Al, Fe–Al, Fe–Cu, and Al–Ag films (*87*), and the diffusion of gold into a copper film has been detected qualitatively (*88*). However, the usefulness of conventional electron-probe microanalysis for the measurement of the *surface* composition of an alloy (*85*) is open to question. The depth of penetration depends upon the experimental conditions, and although a surface layer of 100 Å might be detected the beam would normally penetrate to a much greater depth, e.g., 10,000 Å (*86*). Some unpublished work by the present authors illustrates this for Cu–Ni films. A 400-Å layer of nickel was evaporated on to a 1200-Å copper film at 0°C. X-ray diffraction analysis of the layered film showed no evidence of alloying, but electron-probe analysis using the minimum possible beam potential (6 kV; Cu and Ni $L\alpha$ radiation) detected both copper and nickel. Thus, the electron beam had completely penetrated the nickel film and the analysis obtained was certainly not representative of the surface. Indeed, silicon X-radiation was also detected, showing that the beam had also penetrated through the copper film to the glass substrate.

Although conventional electron-probe microanalysis appears to be unsuitable for analysis of the exposed surface layer of atoms in an alloy catalyst, recent developments have shown that X-ray emission analysis can still be used for this purpose (*89, 90*). By bombarding the surface with high energy electrons at grazing incidence, characteristic $K\alpha$ radiation from monolayer quantities of both carbon and oxygen on an iron surface was observed. Simultaneously, information about the structure of the surface layer was obtained from the electron diffraction pattern.

3. Auger Electron Spectroscopy

Auger emission spectroscopy is a most promising technique for the chemical analysis of solid surfaces. In this technique the surface is bombarded with a beam of electrons having typical energies of from 100 to 3000 volts, and the secondary electron spectrum is observed. It was shown by Harris (91) that high sensitivity could be obtained by electronic differentiation of the energy distribution of the secondary electrons. A standard three-grid low-energy electron diffraction (LEED) apparatus can easily be modified for the measurement of Auger peaks (92), and most investigations of Auger analysis of surfaces have been carried out in association with LEED studies (93, 94). This does not mean that Auger spectroscopy is limited to single crystal surfaces, and there is no reason, in principle, why the technique cannot be applied to the study of alloy films.

The transitions involved in Auger emission are illustrated in Fig. 10. The primary process is the ionization of an inner shell by bombardment with electrons. The vacancy is then filled by an electron from an outer shell, and the energy released can either appear as an X-ray quantum, or

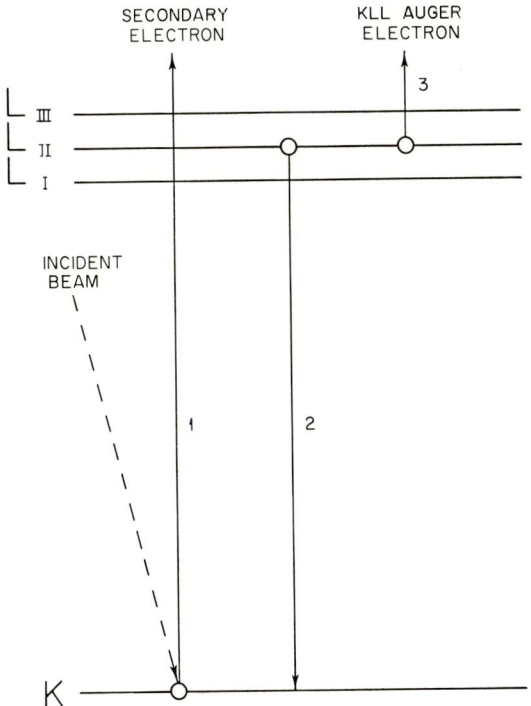

FIG. 10. Transitions involved in Auger emission.

a radiationless process can occur in which the energy is transferred to another electron in an outer shell. In the latter case an Auger electron is ejected. The process illustrated in Fig. 10 is referred to as a *KLL* transition, i.e., the filling of a *K*-shell vacancy by an *L*-shell electron followed by ejection of an *L*-shell Auger electron. *KLL* spectra are the simplest and easiest to interpret, but for the range of exciting energies normally used, they can only be observed for the lightest elements in the periodic table. The Auger spectra obtained from the heavier elements (*LMM*, *MNN*, etc.) are more complex, and overlapping peaks from different elements can give rise to difficulties in calibration (*93*). An accurate calibration can be obtained by depositing known amounts of foreign atoms on a clean surface in UHV, though only a few quantitative studies have been reported (*92*, *95*). One such study has shown that concentrations of as little as 2% of a monolayer of cesium could be easily detected on a silicon (100) surface (*95*).

The usefulness of Auger electron spectroscopy for *surface* analysis depends upon the very small mean free paths of the low energy Auger electrons (<1000 eV). Palmberg and Rhodin (*94*) studied the Auger spectra of silver and gold when a silver film was evaporated on to a (100) gold surface. They were able to show that the mean escape depth through silver was about two monolayers or 4 Å at 72 eV (the energy of the Au peak) and four monolayers or 8 Å at 362 eV (the energy of the Ag peak). Thus an analysis which is truly representative of layers of atoms close to the surface is achieved. They also showed how LEED can be used to calibrate the Auger spectrum. This was done by noting the time of deposition required to convert the Au (100) 1 × 5 structure to a Au (100) 1 × 1–Ag structure. This time was assumed to correspond to the formation of a silver monolayer. By adapting this type of technique it should be possible to study the effect of annealing on the surface structure and composition of the individual crystal planes of an alloy surface, and to study chemisorption and catalysis on such a surface.

Spectra for a series of Cu–Ni alloys have been obtained (*91*) and these are reproduced in Fig. 11. Because of overlapping of peaks from the component metals, separate indications of each element are only obtained from the 925 eV Cu peak and the 718 eV Ni peak. The results have only qualitative significance because the quoted nickel concentrations are bulk values. Nevertheless, they do suggest that for these particular samples of Cu–Ni alloys, the surface composition varies smoothly from pure copper to pure nickel. Auger spectroscopy has subsequently shown that the surface composition of the (110) face of a 55% Cu–Ni crystal was identical with the bulk composition (*95a*). Ono et al. (*95b*) have used the technique to study cleaning procedures; argon ion bombardment caused nickel enrichment of

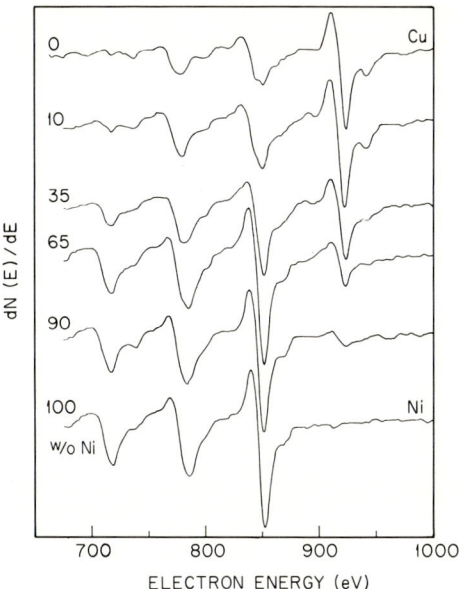

FIG. 11. Auger spectra for Cu–Ni alloys (*91*).

a 60% Cu–Ni alloy while oxidation–reduction had the opposite effect. None of the published studies refers to alloys prepared entirely as evaporated films but we are aware of a combined Auger, LEED, and RHEED examination (*95c*) of nickel grown epitaxially on (111)—orientated copper films. Copper was detected at the surface even after the deposition of a nickel layer, several hundred Angstroms thick, when films prepared at 70°C were annealed at 200°C and these results have relevance to the discussion in Section II, B.

IV. Adsorption and Catalysis on Alloy Films

Much of the work described in this section has inevitably to be reviewed against the background of earlier work using alloy catalysts where it was hoped to correlate activity or adsorptive properties with the electronic structure of the alloy (cf. Introduction). Therefore, it seems useful to summarize some current ideas about d-band structure with particular reference to the Pd–Ag system which has been extensively studied. However, it has been stated (*96*) that the differences between the Cu–Ni and the Pd–Ag systems, with respect to electronic structure, may be more impressive than their similarities and this must be kept in mind, i.e., ideas about, say, Pd–Ag cannot necessarily be used to discuss catalysis over group VIII–IB metal alloys in general.

The original conclusions from magnetic susceptibility measurements on alloys are not in accordance with modern measurements of the number of vacancies per atom in the d-band of palladium. For some time it has been realized that a simple deduction about the number of d-band vacancies (n_d) cannot be made from the composition at which paramagnetism disappears (*97*). With increasing concentration of the IB metal, some electrons enter the 5s-band of palladium and, in conjunction with new specific heat values, the rigid band analysis gives a value of $n_d = 0.48$ (*96*). Much more direct evidence on the value of n_d is available from de Haas–van Alphen measurements on pure palladium which yield a value of 0.36 (*98*) and suggests the rejection of the rigid band model, although it successfully explains the transport properties of Pd–Ag alloys (*99*) in terms of the density of states derived from specific heat measurements. The position can be maintained if it is assumed (*100*) that the s-and d-bands remain unchanged in shape by alloying, but the s-band shifts its position linearly in energy relative to the d-band so that n_d varies from 0.36 for pure Pd to zero for the 60% Ag alloy. However, a recent X-ray isochromat study (*101*) of the Pd–Ag system showed that 38 ± 5 at. % Ag was sufficient to fill the d-band, and the results could be described within a rigid band model. In summary, the number of d-band vacancies per atom of palladium may be nearer to 0.4 than the widely accepted value of 0.6, yet the latter value correlates rather better with the catalytic properties of the Pd–Au and Pd–Ag systems.

It has also to be remembered that the band model is a theory of the bulk properties of the metal (magnetism, electrical conductivity, specific heat, etc.), whereas chemisorption and catalysis depend upon the formation of bonds between *surface* metal atoms and the adsorbed species. Hence, modern theories of chemisorption have tended to concentrate on the formation of bonds with localized orbitals on surface metal atoms. Recently, the directional properties of the orbitals emerging at the surface, as discussed by Dowden (*102*) and Bond (*103*) on the basis of the Goodenough model, have been used to interpret the chemisorption behavior of different crystal faces (*104, 105*). A more elaborate theoretical treatment of the chemisorption process by Grimley (*106*) envisages the formation of a surface compound with localized metal orbitals, and in this case a weak interaction is allowed with the electrons in the metal.

A. Copper–Nickel Alloy Films

Early catalytic studies on Cu–Ni alloys were prompted by the suggestion of Dowden and Reynolds (*107, 108*) that d-band vacancies are essential for the alloys to be active as catalysts. This suggestion, based upon the observed complete loss of activity for styrene hydrogenation when nickel was alloyed with 38% copper, stimulated considerable interest and further catalytic studies on Cu–Ni alloys quickly followed. Soon, however, conflicting results obtained using alloys prepared by different methods gave rise to some confusion. Whereas Dowden and Reynolds had observed maximum activity with pure nickel on a series of foil catalysts, Best and Russell (109), using alloy powders, measured ethylene hydrogenation rates on catalysts containing 90 and 63% Cu that were higher than on pure nickel by factors of 10–100, respectively. Hall and Emmett (110) confirmed that addition of copper enhanced the activity of nickel, and also found that hydrogen adsorbed on Cu–Ni alloy catalysts acted as a promoter, whereas on pure nickel it was an inhibitor for the hydrogenation of ethylene. These alloy powder catalysts had been prepared by reduction of the oxides, and it has been shown (111) that the uptake of many monolayers of hydrogen by such catalysts can be associated with residual oxides. A good review of catalysis on Cu–Ni alloy powders is given by Emmett (112).

In order to avoid contamination of the alloys with oxygen, Gharpurey and Emmett (113) decided to attempt to prepare Cu–Ni catalysts by a film technique. They prepared thin films of copper on nickel or nickel on copper by successive evaporation and heated them overnight at 300°C in

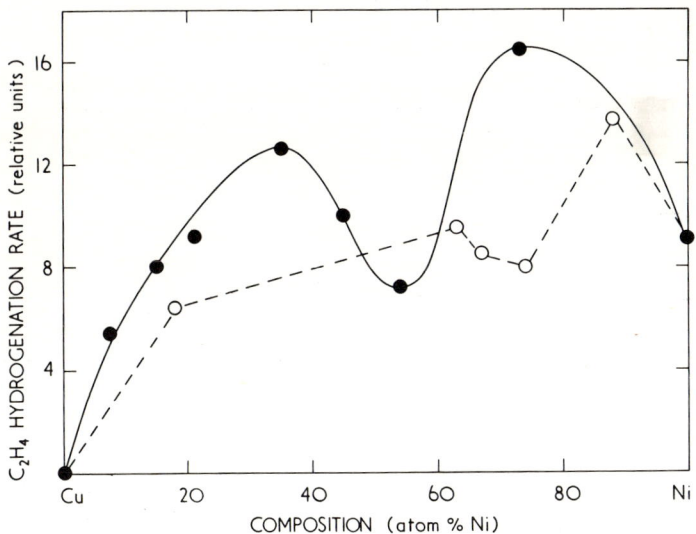

FIG. 12. Variation of C_2H_4 hydrogenation rate with composition (113) for Cu–Ni films (○) compared with earlier results (110) using powder catalysts (●).

hydrogen. This produced apparently homogeneous films which had the same activity for ethylene hydrogenation regardless of the order in which the metals had been deposited. The variation in specific activity for ethylene hydrogenation with alloy composition is shown in Fig. 12, where it is compared with the data of Hall and Emmett (110). These results do not appear to support the d-band theory, but the films were not characterized by X-ray diffraction and it is not known to what extent they were homogenized.

Further progress in the study of the Cu–Ni system awaited the preparation and careful characterization of alloy films of known bulk and surface composition. The essential step was taken by Sachtler and his co-workers (28, 33, 114) who prepared Cu–Ni alloy films by successive evaporation of the component metals in UHV. After evaporation the films were homogenized by heating in vacuum at 200°C. The bulk composition of the alloys was derived from X-ray diffraction, and the photoelectric work function of the films was also measured. A thermodynamic analysis, summarized by Fig. 13, indicated that alloy films sintered at 200°C should consist, at equilibrium, of two phases, viz., phase I containing 80% Cu and phase II containing ~2% Cu. Evidence was presented that alloys within the

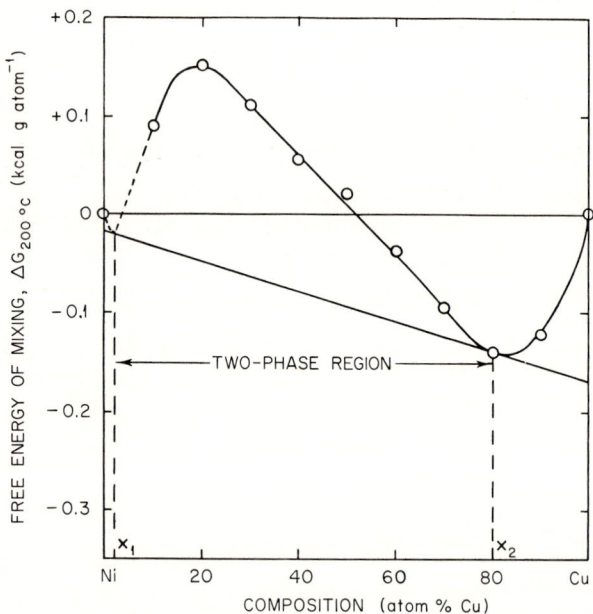

Fig. 13. Free energy of mixing at 200°C for Cu–Ni alloys with miscibility gap indicated (33).

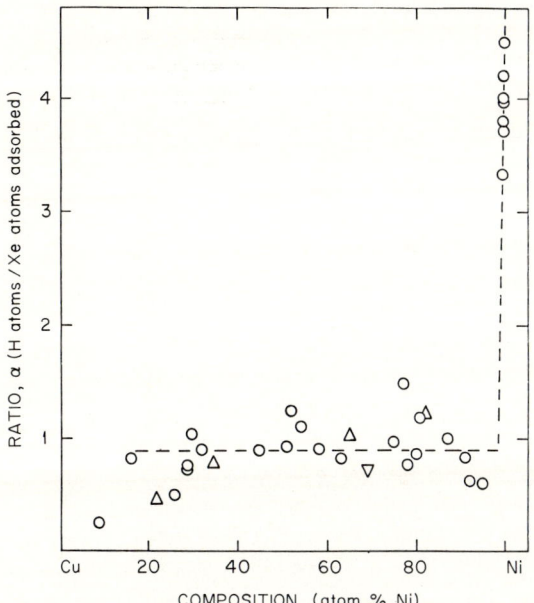

Fig. 14. Adsorption ratio α as a function of composition for Cu–Ni alloy films; Ni deposited on Cu (○), Cu deposited on Ni (△), both sintered at 200°C; Cu deposited on Ni and sintered at 300°C (▽) (70).

miscibility gap had a constant surface composition, 80% Cu–20% Ni, independent of the nominal bulk composition (Section II).

The chemisorption of hydrogen, the catalytic hydrogenation of benzene, and its rate of exchange with deuterium on these films were subsequently studied (2, 70, 115). The ratio α of the number of hydrogen atoms chemisorbed to the number of xenon atoms physically adsorbed, and the catalytic activity for benzene hydrogenation are shown as functions of alloy composition in Figs. 14 and 15. Clearly, these results support the idea that for most of the composition range, the Cu–Ni alloy films had a constant surface composition. The activity associated with a surface composition of 80% Cu–20% Ni in hydrogen chemisorption and benzene hydrogenation was appreciably higher than that of pure copper, but lower than the activity of pure nickel.

It was claimed that this model helps to explain earlier catalytic results using Cu–Ni alloys, but comparisons with alloys in granular, or other massive form, are difficult. The available catalytic results on Cu–Ni alloys show that the method of preparation of the catalyst can have a profound influence upon the observed activity pattern. The promoting effect on the catalytic activity, caused by cooling in hydrogen rather than in vacuum

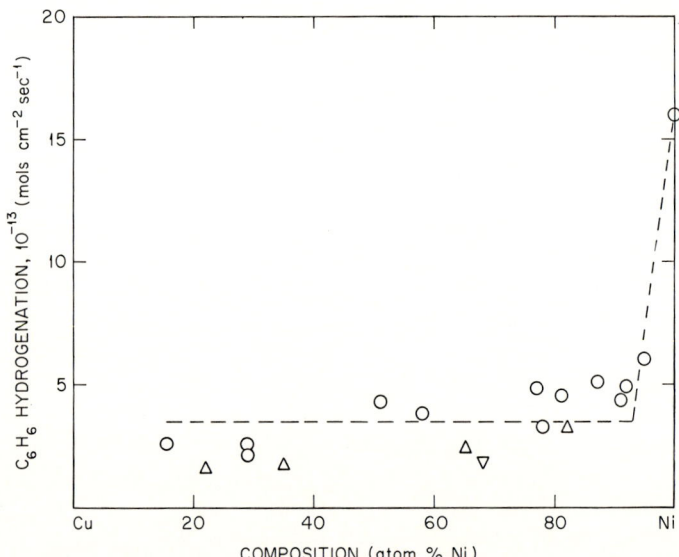

FIG. 15. Benzene hydrogenation at 150°C over Cu–Ni alloy films; symbols are as defined in Fig. 14 (70).

(110, 116) is probably the main reason for the somewhat different results obtained on powders, prepared by reduction of the oxides, compared with films. Although this promotion may be due in part to a reaction with residual oxides in powder catalysts (111), it has been pointed out that the surface composition of an equilibrated alloy should depend on the ambient atmosphere (33). Thus, for alloys equilibrated in hydrogen, one might expect surface enrichment by nickel because hydrogen is strongly chemisorbed by nickel but only negligibly adsorbed on copper. Indeed, Clarke and Byrne (117) offered evidence based on butene-1 hydrogenation rates (Fig. 16) in support of this idea; their activity pattern is consistent with the presence of the 2% Cu–98% Ni phase at the surface for Cu–Ni films annealed in hydrogen. Subsequently, however, in discussing H_2–D_2 equilibration rates, they remarked (118) that atomic displacement of nickel caused by chemisorption was an unnecessary supposition. Activity persisted down to 3% Ni, at which composition, one complication, i.e., phase separation, would at least be absent.

Copper–nickel alloy films similarly deposited at high substrate temperatures and annealed in either hydrogen or deuterium were used to study the hydrogenation of buta-1,3-diene (119) and the exchange of cyclopentane with deuterium (120). Rates of buta-1,3-diene hydrogenation as a function of alloy composition resemble the pattern for butene-1 hydrogenation except that pure nickel was two orders of magnitude more active than the alloys. The ratio but-1-ene:but-2-ene tended to increase with increasing Cu content showing that π-diolefin attachment had become more difficult. The ratio of *cis*- to *trans*-but-2-enes for the alloy catalysts compared with the pure metals suggested that the alloys were acting as mixtures of active nickel centers in copper. Activity for cyclopentane exchange with deuterium persisted down to 4% Ni, although activation energies changed below 24% Ni and an even-number distribution pattern developed. The latter feature was ascribed to intermediates, responsible for severe self-poisoning of pure nickel, which desorb more readily from Cu-rich alloys. Similarly, severe self-poisoning was found during propane exchange with deuterium (*121*) at 50°C over nickel and Cu–Ni alloy films, leading to the result that some alloys were more active than nickel itself. Copper–nickel films sintered in hydrogen were also used to study ethane hydrogenolysis (*121a*) which required a sufficiently high reaction temperature to encourage the view that the single-phased alloy films prepared, remained in this condition. A sharp break in activity at median compositions was not found, instead the rate decreased moderately from 100% to 30% Ni, and then more rapidly to 4% Ni, behavior at variance with the simple d-band filling interpretation.

The use of hydrogen annealing to ensure homogenization and surface cleanliness is an attractive procedure in alloy film preparation but for the question of surface enrichment. The possibility of surface enrichment in

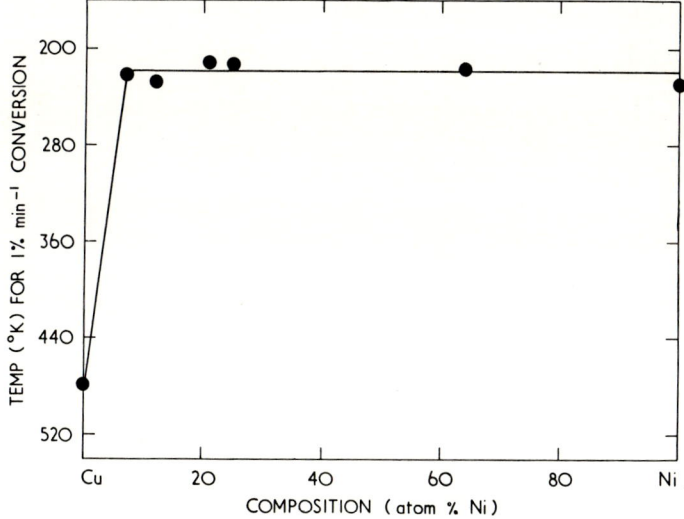

FIG. 16. Butene-1 hydrogenation over Cu–Ni alloy films, annealed in hydrogen at 530°C (*117*).

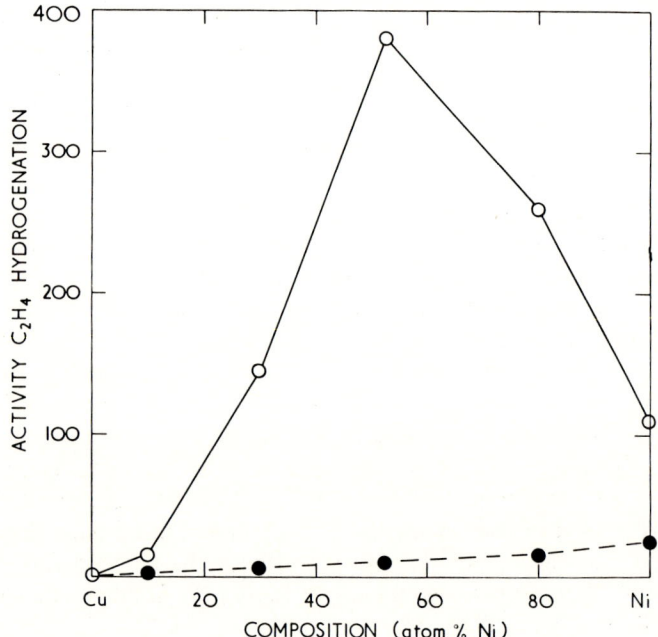

FIG. 17. Ethylene hydrogenation over Cu–Ni alloy films at 30°C; films deposited at −183°C and annealed at 30°C (○) or 250°C (●) (123).

the presence of a chemisorbed gas is supported by evidence from CO adsorption on Pt–Au alloy films (41), Section IV.E, and also FIM atom probe studies (122) which showed that some surface tungsten atoms are mobile in the presence of CO and nitrogen. However, results obtained using Cu–Ni films prepared without hydrogen annealing are rather variable, probably mainly as a result of the different substrate temperatures adopted by different workers in attempts to bring about alloying. A number of recent studies serve to illustrate this point.

Figure 17 shows the completely different activity patterns observed for ethylene hydrogenation on Cu–Ni films deposited successively at −183°C and annealed at either 30° or 250°C (123). Campbell and Emmett (29) stated that Cu–Ni films deposited at 0°C showed no evidence of alloy formation when heated at 300° or 500°C in the absence of hydrogen. (In contrast, simultaneous deposition followed by annealing in hydrogen at 500°C yielded the 38% Ni–Cu alloy and excess copper; an alloy of this composition agrees with some thermodynamic data.) Sachtler et al., (28), using films deposited successively in UHV, found that 200°C was a sufficient annealing temperature to equilibrate the films in two phases. Zhavoronkova et al. (84) deposited their Cu–Ni films at 300°C in 10^{-8} Torr and sintered them at 400°C but did not detect phase separation. A more detailed study of Cu–Ni alloy film preparation has now been made by Clarke and Spooner (*124*); one of their conclusions is that deposition at 300°C, followed by annealing for 1 hr, both in UHV, gave complete mixing.

Although differences in the vacuum conditions may partly account for some of these divergent results, the effect of temperature is likely to be more important. It has been shown in Section II that metastable single-phase alloys can be obtained by formation at sufficiently high temperatures where there is no miscibility gap. Figure 18 reproduces the phase diagram for the Cu–Ni system, as calculated (*70*) from two available sets of thermodynamic data. If the data yielding the solid curve are correct then no phase segregation should be found in Cu–Ni alloys deposited and annealed above about 400°C. In fact, the average temperature of films during evaporation can be substantially higher than the apparent substrate temperature (*27*), Section II.B, perhaps by as much as 400°–500°C at the condensing surface. Not only would a single-phase alloy form, but larger crystallites would be produced. Subsequently, during the annealing stage at, say, 400°C (now the true temperature), the rate of equilibration may be very slow and a metastable state may be maintained.

We have seen that the early hope that catalytic studies on Cu–Ni alloys would provide clear confirmation of the simple d-band theory has

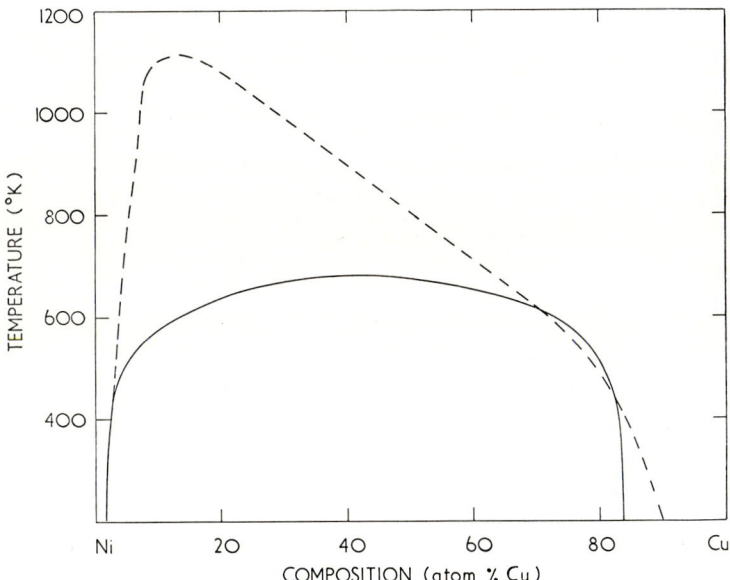

Fig. 18. Phase diagram for the Cu–Ni system calculated from two available sets of thermodynamic data (*70*).

not been fulfilled. Of course, there are results which, to a greater or lesser extent, provide a measure of support for the theory, in that abrupt changes in activity occur at a certain concentration of Cu in the alloys. The simple theory predicts that the concentration at which such effects should occur is about 60% Cu (although, as pointed out at the start of this section, modern data on the analogous Ag–Pd system indicate that the d-band is filled at about 40% Ag). Catalytic results show apparent d-band effects (*125*) at 40–60% Cu for benzene hydrogenation on powder catalysts and at about 80% Cu for ethylene hydrogenation on foils. The latter result has been interpreted as being due to the creation of additional d-band holes by thermal excitation at the high temperatures (400°–600°C) at which the reactions were studied. The activation energy for formic acid decomposition on foils was also observed to increase sharply at 80% Cu, though the activity pattern between 0 and 60% was the same as that observed by Dowden and Reynolds (*108*). Recently, a study of the recombination of hydrogen atoms on Cu–Ni foils (*85*) showed a fall in activity at about 60% Cu which was associated with the filling of d-band vacancies. High activity relative to pure Cu persisted up to 90% Cu, however, and this was rationalized by the authors, who quoted evidence for the existence of d-band vacancies even at small nickel concentrations. Thus we can see that even when apparent d-band effects occur they often require a special explanation.

On the basis of results obtained using Cu–Ni alloy films, an alternative description of the processes occurring in chemisorption and catalysis has been advanced by Sachtler and co-workers. The constant value of α for all the alloys within the miscibility gap corresponded with one hydrogen atom adsorbed per surface nickel atom, i.e., for alloys with 80% Cu at the surface. Pure Cu did not chemisorb appreciable quantities of hydrogen, and because the d-bands of the alloy and pure Cu are equally filled it was proposed (*2*) that the chemisorption of hydrogen was independent of the bulk electronic structure of the alloy. Since the hydrogen chemisorption is proportional to the nickel concentration in the adsorbing surface, the nickel atoms in the surface of Cu–Ni alloys can be "titrated" by chemisorbing hydrogen. The results for benzene hydrogenation were similar (Fig. 15), suggesting that the catalytic activity for this reaction was also not controlled by d-band vacancies. According to Sachtler and van der Plank these results indicate that chemisorption and catalysis are not collective phenomena. Instead, they prefer a model in which chemisorption is envisaged as an interaction with localized d-orbitals on individual surface atoms.

The conclusions of Sachtler and co-workers are entirely reasonable for the Cu–Ni films which they studied. Nevertheless, they were effectively restricted to a single surface composition, and a generalization of these conclusions, even to the whole Cu–Ni system, has its dangers. Therefore, it is of considerable interest to examine the results obtained by Zhavoronkova *et al.* (*84*) for Cu–Ni films which were also prepared without annealing in hydrogen apparently and showed no evidence of phase separation. As discussed above, a metastable solid solution may have been formed at each composition, so that we could have results in this case for H_2–D_2 exchange at various nickel contents. In Fig. 19, values of specific activity (K_{sp}) *divided by nickel content* (n), and expressed as $\log(K_{sp}/n)$, for a reaction temperature of $-40°C$, are plotted as a function of composition. The activity of pure copper was extremely low compared with nickel or the alloys. Between $\sim 20\%$ Cu and almost pure copper, the specific activity appears to be directly proportional to the Ni content of the alloy, in agreement with the conclusions of Sachtler *et al.* There is, however, a break in the plot at 20% Cu of the type (if not at the composition) that would be expected if the specific activity was dependent upon a change in the collective properties of the alloy. It appears that while the individual surface atom approach provides a valid description of the behavior of the Cu-rich alloys, d-band vacancies might have some influence on the catalytic activity of the Ni-rich alloys. A similar effect has been noted by the present

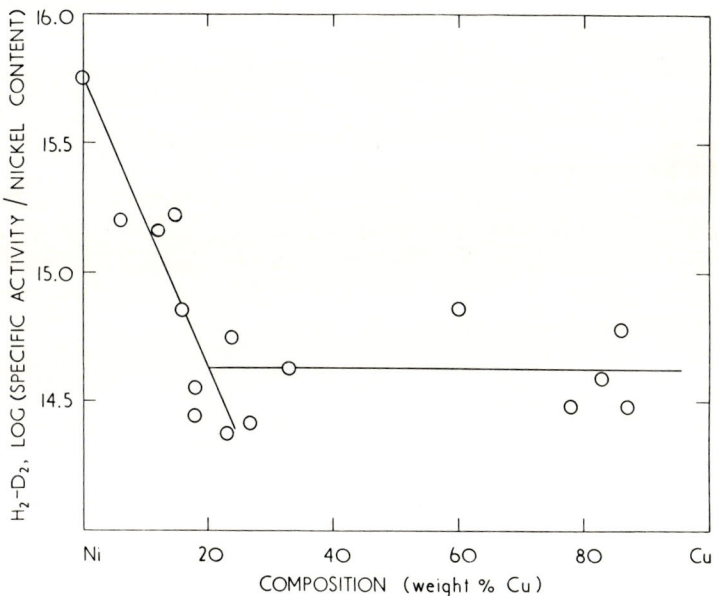

FIG. 19. Rate of H_2–D_2 exchange (at $-40°C$) *divided by nickel content* as a function of composition for Cu–Ni alloy films deposited at 300°C and sintered at 400°C (*84*).

authors in the chemisorption of CO on Pd–Ag alloy films (*126*). These latter results, and their implications for chemisorption and catalysis on alloys, are discussed further in Section IV.C. However, as a final illustration of the confusing behavior of Cu–Ni alloys, it must be recorded that Zhavoronkova *et al.* (*84*) could alter the pattern shown in Fig. 19 by adsorbing hydrogen at 40°–60°C. After this treatment, the specific activity was reduced and further (K_{sp}/n) was constant from pure nickel to 96% Cu, ascribed by the authors to catalytic activity arising from nickel atoms in a copper diluent.

B. Palladium–Gold Alloy Films

Gold forms a continuous series of solid solutions with palladium, and there is no evidence for the existence of a miscibility gap. Also, the catalytic properties of the component metals are very different, and for these reasons the Pd–Au alloys have been popular in studies of the "electronic factor" in catalysis. The well-known paper by Couper and Eley (*127*) remains the most clearly defined example of a correlation between catalytic activity and the filling of d-band vacancies. The apparent activation energy for the ortho-parahydrogen conversion over Pd–Au wires was constant on Pd and the Pd-rich alloys, but increased abruptly at 60% Au, at which composition d-band vacancies were considered to be just filled. Subsequently, Eley, with various collaborators, has studied a number of other reactions over the same alloy wires, e.g., formic acid decomposition (*128*), CO oxidation (*129*), and N_2O decomposition (*130*). These results, and the extent to which they support the d-band theory, have been reviewed by Eley (*1*). We shall confine our attention here to the chemisorption of oxygen and the decomposition of formic acid, which have been studied on Pd–Au alloy films.

1. *Oxygen Chemisorption*

In one of the earliest reports of the use of clean evaporated alloy films in surface studies, Stephens described the preparation and characterization of Pd–Au films and presented some results for the adsorption of oxygen on them (*46*). Films of pure Pd and 60% Au were evaporated directly from wires, while films of 80% Au and pure Au were evaporated from a pre-outgassed tungsten support wire. The films were evaporated in a UHV system and the pressure was kept below 10^{-8} Torr during evaporation. After evaporation, the films were stabilized by cycling between $-195°$ and $30°C$ four times. They were characterized by X-ray diffraction and chemical analysis; surface areas were measured by the BET method using krypton adsorption.

TABLE VII

Oxygen Adsorption on Pd–Au *Alloy Films*

Film composition	Fast uptake (molecules/100 mg)	Total uptake	Fast/total uptake
Pd	3.10×10^{18}	3.70×10^{18}	0.84
60% Au–Pd	0.76×10^{18}	1.66×10^{18}	0.46
80% Au–Pd	0.27×10^{18}	0.46×10^{18}	0.59

[a] From Stephens (46).

The results for oxygen chemisorption are summarized in Table VII. Palladium and the alloys behaved similarly. There was a rapid initial uptake at 10^{-3} Torr and 0°C, followed by a slow adsorption over 16 hr. There was no oxygen chemisorption on pure gold films at 0°C. The ratios of the total oxygen uptakes for palladium and the two alloys were approximately equal to the ratios of the surface areas of the films. This was interpreted as implying that the same number of surface atoms had reacted in each case. In other words, the presence of palladium had made it possible for oxygen to react with gold atoms in the surface, although oxygen would not adsorb on pure gold films. This discussion of the results in terms of the properties of individual surface atoms rather than holes in the d-band was unusual at the time, though, as we have seen, the individual surface atom approach has been used to explain high activity on Cu-rich copper–nickel alloys in which the d-band is expected to be full.

2. *Formic Acid Decomposition*

The decomposition of formic acid over evaporated Pd–Au alloy films has been studied by Clarke and Rafter (69); the same reaction on Pd–Au alloy wires was studied by Eley and Luetic (128). The alloy films were prepared in a conventional high vacuum system by simultaneous evaporation of the component metals from tungsten hairpins. The alloy films were characterized by X-ray diffraction and electron microscopy. The X-ray diffractometer peaks were analyzed by a method first used by Moss and Thomas (30). It was found that alloys deposited at a substrate temperature of 450°C followed by annealing for one hour at the same temperature were substantially homogeneous. Electron microscopy revealed that all compositions were subject to preferred orientation (Section III).

The variation of the apparent activation energy with alloy composition is shown in Fig. 20. These results show very close agreement with those of Eley and Luetic in the Au-rich region, i.e., there is a fall in apparent acti-

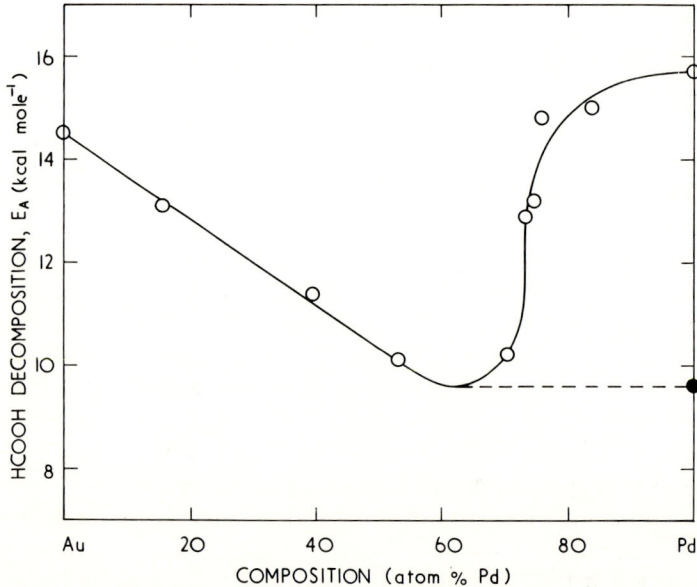

Fig. 20. Apparent activation energy, E_A, for formic acid decomposition over Pd–Au alloy films deposited and annealed at 450°C (○); pure Pd film deposited at −196°C and annealed at 200°C (●) (69).

vation energy as gold is alloyed with palladium until the composition is about 60% Pd. At this point there is a marked deviation in the two sets of results. Eley and Luetic obtained a constant activation energy of 8 kcal/mole on alloy wires containing more than 70% Pd and on pure Pd. Clarke and Rafter's results on alloy films showed an increase in activation energy on the Pd-rich alloys and the activation energy on pure Pd was higher than on pure Au. The Pd-rich films, however, had shown a strong (111) texture, whereas the Au-rich films had a (110) texture. It was concluded that this variation in the relative exposure of different crystal faces was not present in the catalysts used by Eley and Luetic, and that the change in crystal face predominance becomes kinetically significant at the Pd-rich end. This conclusion was supported by the result shown as a dashed line in Fig. 20. This pure Pd film was evaporated with the substrate at −198°, and then sintered in order to give a randomly orientated sample. The activation energy on this film was 9.6 kcal/mole, so that when the effect of preferred orientation in Pd–Au alloy films was eliminated the results agreed very closely with those obtained on wires.

These results show that, even in an alloy system where no phase segregation is expected, a variation in preferred orientation in alloy films of different composition may impose a substantial effect upon the observed activity pattern. This is, of course, a limitation of the evaporated alloy film technique, but it is one which can be overcome by proper characterization of alloy films using the methods described in Section III.

When the effects of texture are eliminated from the Pd–Au alloy film results, very close agreement is obtained with the results of Eley and Luetic for Pd–Au wires. The latter authors pointed out that the total change in activation energy from pure Pd to pure Au was about 6 kcal/mole, which compares quite closely with the change observed for the parahydrogen conversion (127). There was also an abrupt rise in the apparent activation energy in both reactions, but whereas it occurred at 60% Pd for the formic acid reaction, the rise was observed at 40% Pd for the parahydrogen conversion. The latter result, of course, is often quoted for its excellent correlation with the value of 0.6 d-band holes per atom in palladium obtained from magnetic susceptibility measurements. In the case of the formic acid decomposition, it was argued that the reaction needs a greater concentration of d-band holes because it is to be expected that any individual palladium atom will either have two holes with parallel spin or none at all. The reaction is probably a two-site reaction and, therefore, in order to have two adjacent sites associated with holes, a higher overall concentration of holes is necessary.

We have, then, another example of an alloy and reaction in which the simple d-band theory has to be modified in a rather speculative way in order to explain experimental results. Actually, this is unnecessary for the formic acid reaction if we take the more recent value of about 0.4 for the number of d-band holes per palladium atom. This is not a satisfactory solution, because it is then difficult to explain the low activation energy for the parahydrogen conversion on Pd–Au alloys containing between 40 and 60% Pd.

C. Palladium–Silver Alloy Films

Palladium and silver form a continuous series of solid solutions which have been selected for various investigations into the effect of electronic structure on magnetic and electrical properties. The thermodynamic properties of the system have been studied thoroughly over a wide temperature range, 330°–1200°K, showing large negative enthalpies of formation. Surface studies relevant to the present discussion include heats of oxygen chemisorption on coprecipitated Pd–Ag (131) and the oxidation of CO over a 5% Pd–Ag alloy prepared by smelting the component metals in argon (132). Ribbons of Pd–Ag have been used to investigate the effect of electronic structure on the rates of hydrogen absorption and its reaction with oxygen or ethylene (*133*). Bulk Pd–Ag alloys were also used in studying the decomposition of formic acid (*134*). Evaporated alloy films have been used in studies of CO oxidation (*30*), ethylene oxidation (*40, 135*), parahydrogen conversion (*47*), and surface potential changes accompanying CO chemisorption (*83a, 126*).

1. CO *Oxidation*

Alloy films were prepared by simultaneous deposition of the component metals onto a glass substrate at 400°C and then annealed at the same temperature (*30*). The uniformity of composition was checked by X-ray fluorescence analysis and the structure examined by X-ray diffraction. Each film was used for a number of experiments to determine the activation energy, requiring the development of a reactivation treatment for use between rate determinations, e.g., the activity of Ag-rich films was largely restored by heating in hydrogen at 250°C. Evacuation at 250°C between experiments was sufficient to maintain the activity of Pd and Pd-rich films, but this treatment doubled the activity of Ag-rich films. Activity was also enhanced if Pd-rich films were cooled in hydrogen to 0°C, after hydrogen treatment at 250°C, but the rate observed initially over these films was restored if the hydrogen was removed at 250°C. [The effect of dissolved hydrogen on the activity of palladium has long been established (*127*).]

Table VIII records the Arrhenius parameters and the activity of four alloy films and the two pure metals; the results are insufficient to provide a neat correlation with bulk electronic structure such as observed for CO oxidation over Pd–Au wires (*129*), but the familiar pattern is discernible. The rate of CO oxidation is approximately constant for Ag and Ag-rich films but decreases by a factor of 10^4 over pure Pd and a Pd-rich film.

TABLE VIII

E, log A, and k for CO *Oxidation on* Pd–Ag

Film composition	Activation energy (kcal/mole)	Log_{10} frequency factor, A (Torr CO_2/ min cm^2 apparent area)	Activity, 150°C (molecules/sec cm^2)
Ag	22	9.46	1.5×10^{17}
11% Pd	18	7.56	2.2×10^{17}
17% Pd	11	3.68	1.2×10^{17}
20% Pd	10	2.29	1.6×10^{17}
54.5% Pd	29	9.16	1.9×10^{13}
Pd	36	12.92	2.6×10^{13}

Daglish and Eley (*129*) found that with increasing Pd content in gold alloys the apparent activation energy increased slowly (from 0 to 5.8 kcal/mole between 0 and 40% Pd) and then sharply to 30.2 kcal/mole at 55% Pd, remaining at high values to 100% Pd. With Pd–Ag films the apparent activation energy decreased from 22 to 10 kcal/mole with additions of up to 20% Pd in Ag, but higher values are again observed with a further increase in Pd content, e.g., 29 kcal/mole at ~55% Pd. A small addition of Pd to Ag foil (*132*) was similarly found to cause a decrease in the apparent activation energy compared with pure Ag. The kinetics of CO oxidation over Pd–Ag alloy films also change markedly with composition. Thus, while there was a first-order dependence in oxygen at all compositions studied, the CO dependence represented by:

$$\text{rate} = k p_{O_2} p_{CO}^y$$

had values of y varying from -1 (Pd-rich) to $+1$ (Ag-rich). The apparent activation energies measured contain terms for the heats of adsorption of the reactants and so reflect these differences in kinetics. However, all the information presently available for CO oxidation over Pd–Ag films points to the same general conclusion, viz, abrupt rather than smooth changes in catalytic properties across the composition range.

2. *Ethylene Oxidation*

The oxidation of ethylene appears to be an attractive choice for studies on the effect of geometric and electronic structure on *selectivity* in catalytic oxidation reactions, observing the relative rates of ethylene oxide and CO_2 formation over silver-based catalysts. Background information on the behavior of silver catalysts is summarized by Voge and Adams (*136*) in a series of tables dealing with various aspects including process studies, specific activities, moderators, catalyst configuration, and kinetics. However, some basic choices have to be made in setting up such studies, e.g., the selectivity of pure silver, used with an excess of air, is about 45% compared with a selectivity of perhaps 75% over a properly moderated catalyst (moderators may be chlorine or sulfur compounds). Twigg (*137, 138*) considered that the mechanism involved the chemisorption of oxygen as atoms followed by the reaction of gaseous or weakly adsorbed ethylene, either with one oxygen atom to form ethylene oxide or with two oxygen atoms to form species which are oxidized further to CO_2 and water. In addition, the ethylene oxide may isomerize to acetaldehyde which is oxidized rapidly to CO_2 and water. It might be thought desirable to arrange conditions so that this secondary oxidation is minimized. Finally, it has to be recognized that the nearly constant selectivity over moderated silver suggests (*136*) that we are dealing with coupled reactions. In a recently proposed mechanism (*139*) incorporating this concept, it was suggested that ethylene oxide is formed by reaction of ethylene with a molecular oxygen–silver complex. Reaction of ethylene with the atomic oxygen–silver complex which forms gives a chemisorbed ethylene oxide which can rapidly isomerize to acetaldehyde and, subsequently, oxidize to CO_2 and water (Scheme I).

In Scheme I, flexibility in the prediction of the selectivity is provided by allowing the competition of ethylene oxide desorption with the isomerization to acetaldehyde and by allowing atomic oxygen complexes to recombine to form Ag_2O_2 complexes.

On the basis of the mechanisms outlined above, it is not unreasonable to expect that geometrical factors could affect the activity/selectivity in ethylene oxidation, e.g., the spatial separation of the chemisorbed oxygen atoms could be important. Kummer (*140*) studied the reaction over the (211), (110), and (111) faces of silver single crystals but found that the rate and selectivity were almost unaffected. Similarly, no marked differences in activity or selectivity were found for silver films (*141*) in which the crystallites were initially either randomly orientated or had the (110) plane parallel to the glass substrate, probably because under reaction conditions the orientated films recrystallized. In contrast, when silver–gold alloy powders were used (*8*), the activity for overall ethylene oxidation could be correlated with the lattice spacing. Also it was proposed that the addition of small amounts of gold increased the number of stronger sites

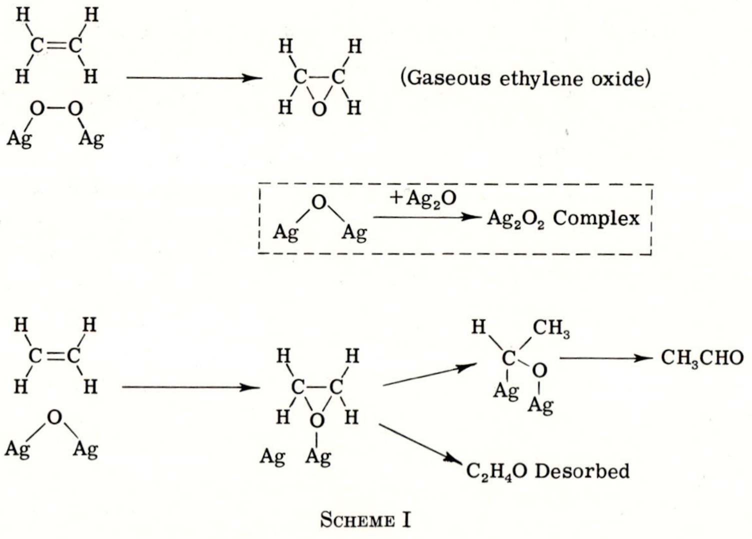

SCHEME I

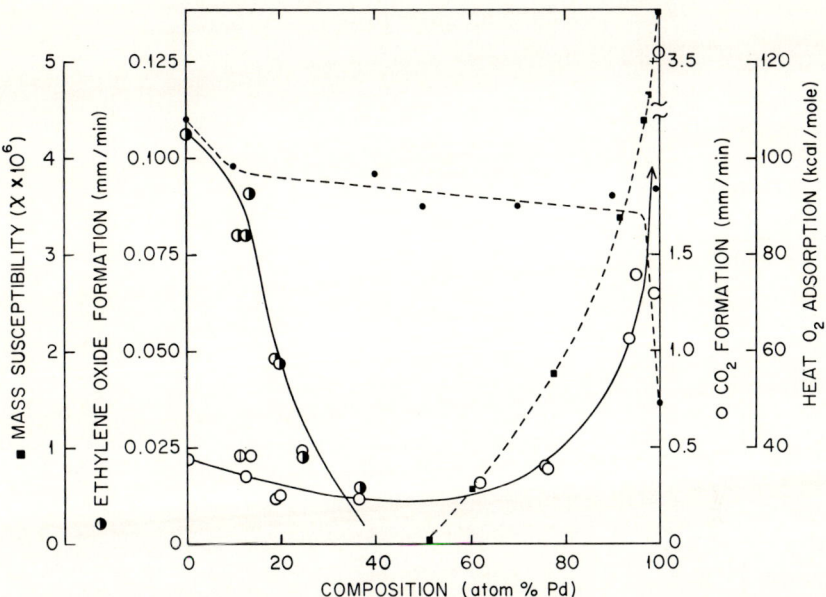

Fig. 21. Ethylene oxidation to ethylene oxide (◐) and CO_2 (○) over Pd–Ag alloy films. Magnetic susceptibility data (■) (*142*) and heats of oxygen chemisorption (●) (*131*) are also included (*135*).

for oxygen adsorption while decreasing the total number, thereby promoting oxygen dissociation and leading to the observed increase in selectivity at the expense of conversion.

Palladium–silver alloy films (*40, 135*) prepared by simultaneous deposition on to glass at 400°C and then annealed at the same temperature had sufficiently large surface areas to permit ethylene oxidation to be studied at reaction temperatures which would be typical for technical catalysts. Figure 21 shows rates of ethylene oxide and CO_2 formation at 240°C as a function of alloy composition; the figure also includes information on heats of oxygen chemisorption on alloy powders (*131*) and magnetic susceptibility data (*142*) taken from earlier literature. The yield of ethylene oxide was greatest over pure silver, decreasing with increasing Pd content to zero over alloy films with more than 40% Pd. The rate of CO_2 production was almost constant in the range 0–80% Pd, before building up sharply to a substantial maximum over pure palladium. The variation in activity did not arise from a simple variation in activation energy or frequency factor. Table IX shows these parameters for CO_2 formation; E for ethylene oxide formation could only be measured satisfactorily over pure silver and was 11.7 kcal/mole. The sharp decrease in the rate of

CO_2 formation between pure Pd and alloys with a low Ag content corresponded with a large increase in E, 13.6 and 30.2 kcal/mole over pure palladium and the 62% Pd alloy, respectively, somewhat compensated by an increase in frequency factor. As the concentration of d-band vacancies falls to zero, E again decreased (13.9 kcal/mole at 43.5% Pd) and remained at low values.

It is a feature of the application of alloy films in catalytic studies that we are readily reminded of the complications which can superimpose on any pattern of activity related to the electronic properties of the bulk. The intervention of phase separation in the Cu–Ni system has been discussed (Section IV.A), and two complications which occur when ethylene is oxidized over Pd–Ag alloy films must now be mentioned before discussing the observed rates in terms of apparent electronic properties. Thus hydrogen may dissolve in at least the surface layers of the alloy, filling d-band holes expected to exist from the bulk composition. After use as catalysts for ethylene oxidation, Pd-rich alloys, i.e., alloys with more than 60% Pd up to 99% Pd (*but not pure* Pd) exhibited phenomena associated with Pd–Ag–H alloys. Figure 22 shows the (111) diffraction profiles observed in two alloy films with 99% Pd. Specimen (a) was protected from the reaction mixture and the peak is at the correct 2θ position (vertical line). After reaction (in a closed system) (b) which removed 75% of the oxygen, double peaks appear ascribed to the coexistence of the α and β phases of Pd–Ag–H, whereas reaction to complete oxygen removal left the film (c) so charged up with hydrogen that only the expanded lattice of the β phase was observed. The dissolved hydrogen could subsequently be removed by heating in vacuum (d). (The extent of the deviation in the lattice constant correlated with the utilization of oxygen from the reaction mixture for a series of films with ~78% Pd.) It seems clear that hydrogen coming from

TABLE IX

E and log A for CO_2 Formation in C_2H_4 Oxidation

Film composition	Activation energy (kcal/mole)	Log_{10} frequency factor, A (Torr CO_2/min cm^2 apparent area)
Ag	12.5	2.64
11% Pd	16.4	4.19
26% Pd	18.5	5.29
43.5% Pd	13.9	3.21
62% Pd	30.2	10.01
Pd	13.6	4.03

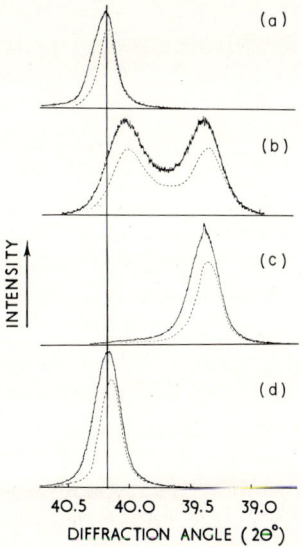

Fig. 22. X-ray diffraction profiles (111 planes) from 99% Pd–Ag films treated as described in text; unresolved trace, solid line; Kα_1 component, broken line (40).

adsorbed ethylene can dissolve in Pd-rich films, during or toward the end of the reaction.

A second possible complication is the enrichment of the surface by silver under reaction conditions. The adsorption free energy of oxygen on silver is higher than on palladium and may provide enough driving force to establish the diffusion of silver atoms from the bulk to the surface of Pd–Ag alloys at reaction temperatures (above 200°C). The extreme case of "dealloying" was investigated, viz, the effect of oxygen alone on Pd-rich films, and evidence was found for a range of solid solutions (40). This experiment shows the possibility that during ethylene oxidation over Pd–Ag films, the surface of the alloy may become enriched by silver, confusing the correlation with the properties expected from the bulk composition.

It is tempting to associate directly the absence of ethylene oxide over catalysts with more than 40% Pd with the appearance of holes in the d-band. It could be assumed that ethylene is chemisorbed directly on Pd-rich alloys and rapidly decomposed, whereas on Ag-rich alloys ethylene is only adsorbed on top of an oxygen-covered surface leading to selective oxidation. However, the general conclusion from earlier kinetic studies (143) is that the rate-determining step over pure palladium also involves the latter mode of ethylene chemisorption.

The activity of pure silver and palladium and the Pd–Ag alloys for the complete oxidation to CO_2 is examined next (Fig. 21). Observed rates for the complete combustion of various olefins, including ethylene, have been correlated (144) with the strength of oxygen chemisorption, i.e., the highest activity was found over platinum and palladium, the metals which adsorb oxygen least strongly of those studied (silver was not included). The mean integral heat of oxygen chemisorption on palladium was 67 kcal/mole (145); other calorimetric results for porous silver (146) gave an initial heat of 120 kcal/mole, remaining high to beyond half-coverage (80–90 kcal/mole at $\theta = 0.5$). From the dissociation equilibria of water vapor over powders, Parravano and co-workers (131, 147) found 49 kcal/mole for oxygen on palladium and 108 kcal/mole for oxygen on silver, both values at very low coverages. The high activity of palladium for complete oxidation is therefore seen in terms of its moderate heat of oxygen chemisorption, whereas silver with a higher heat is less active.

Figure 21 also shows heats of oxygen chemisorption for Pd–Ag alloy powders (131). A large increase in the heat occurred when palladium was alloyed with only 1% Ag, but thereafter the heat increased rather slowly with increasing silver content. The satisfactory inverse correlation is again apparent between activity for complete oxidation and the heat of oxygen chemisorption, with the activity decreasing sharply on addition of a few percent of silver to palladium, then remaining relatively constant as the Ag content increased. A rapid change in adsorptive and catalytic properties by the addition of a small concentration of a group IB metal to a transition metal is not unusual, e.g., Daglish and Eley (129) observed a 50-times decrease in the initial rate of CO oxidation on addition of 10% gold to palladium. However, in the Pd–Ag work cited above, only 1% Ag effected marked changes, and also, while pure Pd films were not charged up with hydrogen after reaction, alloy films with only 1% Ag showed clear evidence of dissolved hydrogen. Silver-enrichment of the surface must be suspected of causing a more rapid filling of d-band vacancies than bulk composition would indicate. The combination of adsorbed hydrogen (dissociated from ethylene) with oxygen is normally rapid on pure palladium (143), but Kowaka (133) found by using ribbons that the H_2–O_2 combination activity falls off rapidly between pure palladium and 20% Ag/Pd. So, as a further consequence of Ag-enrichment the bulk solution of hydrogen can compete with the desorption of hydrogen as water.

3. *Parahydrogen Conversion*

The parahydrogen conversion has been studied on Pd–Ag films (47), wires (148), and foils (149). The films were prepared by evaporation from alloy wires onto a cylindrical Pyrex reaction vessel held at 400°C. After annealing for 1 hr at 400°C, parahydrogen conversion studies were commenced. The films were characterized by the usual combination of X-ray fluorescence and X-ray diffraction analysis (Section III).

The variation of activation energy as a function of alloy composition is shown in Fig. 23. There is a fairly sharp rise in activation energy from about 0.4 kcal/mole at 50% Ag to 5.9 kcal/mole on pure Ag, but the increase observed with Pd–Au wires (127) was more abrupt. Pd–Ag alloy wires (148) showed a gradual increase from about 2 kcal/mole for pure Pd to 4 kcal/mole at 60% Ag, followed by a more rapid, smooth increase from 4.9 kcal/mole at 80% Ag to 11.5 kcal/mole on pure Ag. The results on films were also used to derive heats of adsorption at equilibrium coverage. Values increased from 1.29 kcal/mole on the 68% Ag alloy to 2.89 kcal/mole on pure Ag.

The rise in activation energy on films occurred at a Ag content slightly higher than the 60% commonly expected from d-band theory. Various reasons why an exact correspondence should not be expected were discussed, e.g., the possibility of d–s promotion of electrons in Ag and absorption of hydrogen in the Pd-rich alloys. In the case of the Pd–Ag wires, most of the increase in activation energy occurred beyond 80% Ag. The authors of the latter work demonstrated a correlation between the experi-

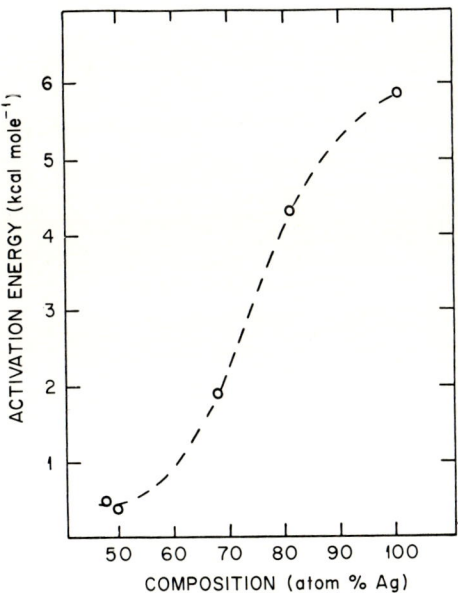

FIG. 23. Activation energy for parahydrogen conversion over Pd–Ag alloy films (47).

mental activation energy and the relative heights of the Fermi surfaces for the various alloys (148). They suggested that d-band vacancies were created at the surface by excitation of electrons from the necessary levels up to the Fermi surface, allowing a vacancy mechanism to proceed even though d-band vacancies do not exist in the bulk alloy. This point will be considered further in the next part of this section.

4. CO Surface Potentials

There have been few studies of chemisorption on alloys, particularly where the surface has been prepared under very clean conditions, although such studies should help us to understand the catalytic process. It has been shown that homogeneous Pd–Ag alloy films can be prepared without extensive sintering (54) by simultaneous evaporation in UHV with the substrate cooled to 0°C, and so it is now possible to study chemisorption on Pd–Ag alloy films. Surface potentials accompanying CO chemisorption were measured (126) by the diode method. Comparisons of surface potentials were made after exposure to CO at 0°C at a pressure of 2×10^{-6} Torr for 10 min.

The variation of the observed surface potential with palladium content is shown in Fig. 24. No chemisorption was detected on pure silver but CO adsorbed rather slowly on the Ag-rich alloys and the measured surface potential increased with increasing Pd content. The surface potential in-

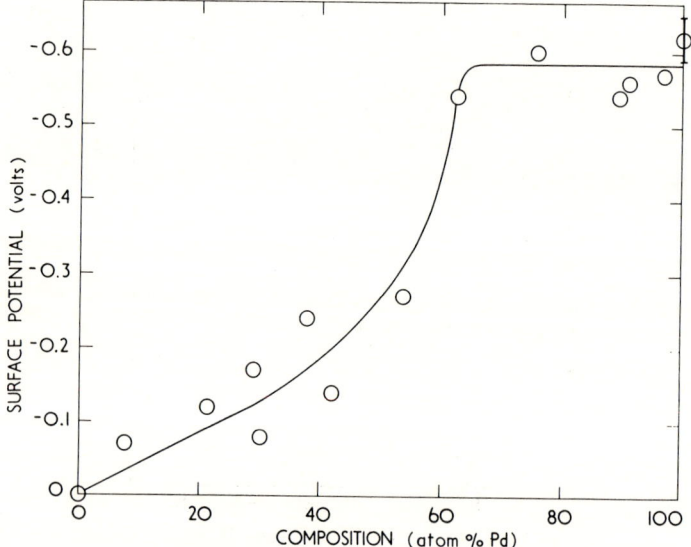

FIG. 24. CO surface potentials on Pd–Ag alloy films (126).

creased sharply at ~60% Pd which corresponds to the modern value of 0.4 for the number of d-band vacancies. The Pd-rich alloys all behaved exactly like pure palladium with a rapidly established surface potential of ~0.6 eV. Results are also available (*83a*) for Pd–Ag alloy films deposited simultaneously at −195°C and equilibrated at 300°C before admission of CO at 10^{-4} Torr. From pure silver to ~55% Pd, the (photoelectric) work function change, $\Delta\phi$, immediately after CO admission was in agreement with our data (Fig. 24). In contrast, values of $\Delta\phi$ for alloys with 72% and 86% Pd were 0.32 and 0.34 eV, respectively; for pure palladium, the Fowler plot had two linear portions and the steeper slope yielded $\Delta\phi = 0.82$ eV. It was proposed by Bouwman (*83a*) that these Pd–Ag films equilibrated at 300°C had surfaces enriched with silver. The following comments apply to the results in Fig. 24 where it was assumed that the *unannealed* films had surface compositions representative of the bulk.

It is apparent that the Pd-rich alloys behave in a way which is consistent with the idea that d-band vacancies are important for CO chemisorption, while the persistence of some capacity to adsorb CO (although more slowly) to almost pure Ag supports the idea that palladium atoms in the surface are acting as individual chemisorption centers with silver atoms acting merely as a diluent. It is interesting to note that a similar pattern was obtained on Cu–Ni alloy films prepared in such a way that no phase separation was detected (*84*), Section IV.A. Two methods of resolving this dichotomy have been suggested. It has been argued that d-band vacancies created by the excitation of electrons up to the Fermi surface exist even at small concentrations of the transition metal (*148*). There is evidence that such electron promotion may be energetically feasible (*100*), and thus it is possible to explain results on alloys throughout the composition range on the basis of d-band vacancies. On the other hand, the present authors have made a suggestion (*126*) which might have some merit in explaining why chemisorption and catalysis may sometimes, but not always, be correlated with the bulk electronic structure of transition metal alloys.

The existence of a band of surface states caused by the termination of the periodic potential can be predicted by quantum mechanics. This band of surface states will overlap the normal crystal band if $|z|$ is not too large; z is given by

$$z = (\alpha - \alpha')/\beta,$$

where α is the Coulomb integral and α' is the Coulomb integral on the end atom in a chain of atoms (*150*). In a system of alloys, $|z|$ will depend mainly on β, the resonance integral, as the composition is varied. If the band of surface states overlaps the normal crystal band, the bulk prop-

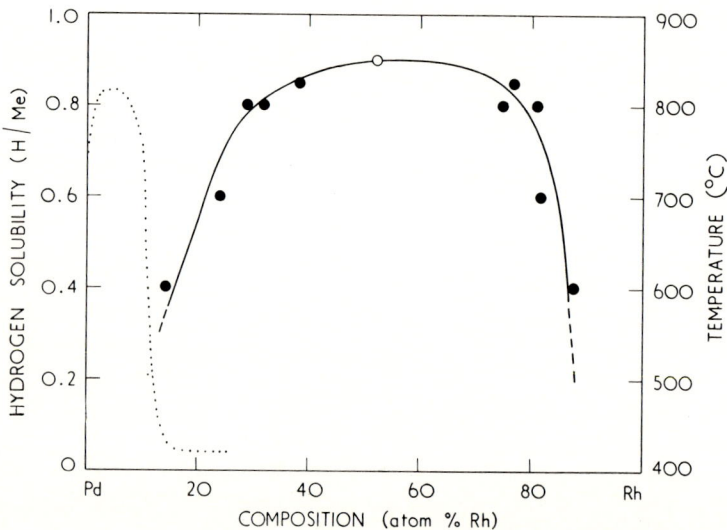

FIG. 25. Limits of the miscibility gap in (○) homogeneous, (●) two-phase Pd–Rh alloys calculated from lattice constants (151), solid line; hydrogen solubilities at 760 Torr (152), dotted line (60).

erties of the metal will relate to the surface properties. It is suggested that increasing the Ag content, and hence increasing the separation of Pd atoms, would decrease the magnitude of β. Hence, $|z|$ might become sufficiently large for the band of surface states to separate from the bulk d-band, and Ag-rich alloys would behave like a two-dimensional atomic array. Unlike the Pd-rich alloys, the surface properties of these Ag-rich alloys need not relate to the electronic properties of the corresponding bulk alloy. The viewpoint from which the surface properties of an alloy system such as Pd–Ag are discussed may therefore depend upon the concentration of the group IB metal additive. Where the existence of a miscibility gap limits the range of surface compositions available for study, it may not be valid to generalize conclusions made from such results to the whole composition range.

D. Palladium–Rhodium Alloy Films

Lattice constants for Pd–Rh alloys quenched from 1300°C vary smoothly with composition, showing only a small positive deviation from Vegard's law, but prolonged vacuum annealing below 850°C revealed the existence of a wide miscibility gap (151). Figure 25 shows the limits of the miscibility gap calculated from the lattice constants of the two-phase system between 825° and 600°C. Recent thermodynamic data (Table I) confirm the tendency to phase separation. The enthalpies of formation are endothermic and only temperatures above ~860°C produce a sufficiently large $T \Delta S$ term to yield negative values of ΔG (assuming the temperature independence of ΔH and ΔS) at compositions between 10 and 90% Rh. Hence, it is to be expected that Pd–Rh alloy films, deposited and annealed at 400°C will not be solid solutions at intermediate compositions and some of the X-ray diffraction evidence (60) supporting this view has already been presented (Section III). Figure 25 also shows the isobaric solubilities at 760 Torr of hydrogen as a function of Rh content (152). At compositions beyond ~20% Rh, the hydrogen solubility is extremely limited, but alloys with less than 10% Rh dissolve more hydrogen than palladium itself. Other data (153, 154) suggest that the hydrogen solubility limit occurs at about 30% Rh. When ethylene was oxidized over Pd–Rh films (73), the variation of activity with bulk composition was complex, but an explanation was possible on the basis that these two complicating factors, i.e., hydrogen solubility and phase separation, were involved.

When small amounts of rhodium were added to palladium, the activation energy E, increased from initial values of 12–14 kcal/mole to a maximum beyond 10% Rh of more than 20 kcal/mole. Beyond 60% Rh, the activation energy again increased rapidly from a minimum to ~25 kcal/mole at 97.7% Rh but was ~5 kcal/mole *less* when the reaction was carried out over pure rhodium. Values of the pre-exponential term, A, in the Arrhenius

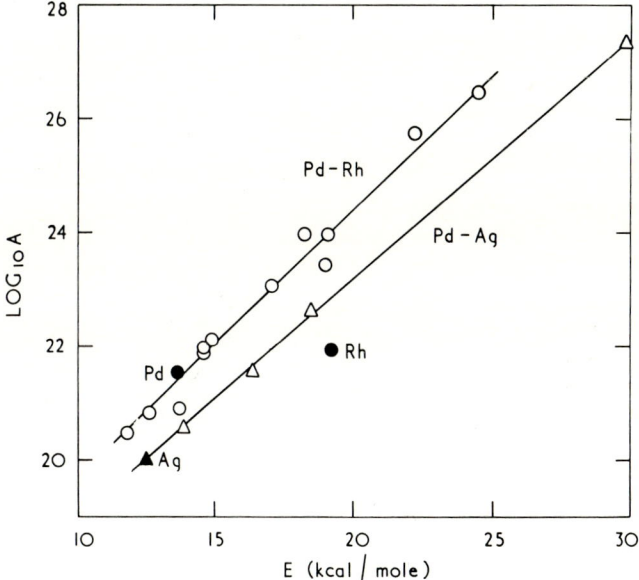

FIG. 26. Compensation effect plots for C_2H_4 oxidation over Pd–Rh alloy films (○) and Pd–Ag alloy films (△); pure metals are indicated by solid symbols (73).

equation, calculated in terms of the number of CO_2 molecules formed per second per square centimeter of apparent surface, are shown plotted as $\log_{10} A$ versus E in Fig. 26 (circles). For comparison, a similar plot is given for ethylene oxidation over Pd–Ag alloy films converting $\log_{10} A$ values appropriately (triangles).

The points for Ag and Pd–Ag alloys lie on the same straight line, a "compensation effect," but the pure Pd point lies above the Pd–Ag line. In fact, the point for pure Pd lies on the line for Pd–Rh alloys, whereas the other pure metal in this series, i.e., rhodium is anomalous, falling well below the Pd–Rh line. Examination of the many compensation effect plots given in Bond's *Catalysis by Metals* (155) shows that often one or other of the pure metals in a series of catalysts consisting of two metals and their alloys falls off the plot. Examples include CO oxidation and formic acid decomposition over Pd–Au catalysts, parahydrogen conversion (Pt–Cu) and the hydrogenation of acetylene (Cu–Ni, Co–Ni), ethylene (Pt–Cu), and benzene (Cu–Ni). In some cases, where alloy catalysts containing only a small addition of the second component have been studied, then such catalysts are also found to be anomalous, like the pure metal which they approximate in composition.

The variation of activity (for ethylene oxidation) across the range of Pd–Rh alloy films (Fig. 27) followed in general the variation in apparent activation energy, although changes in activity were a consequence of

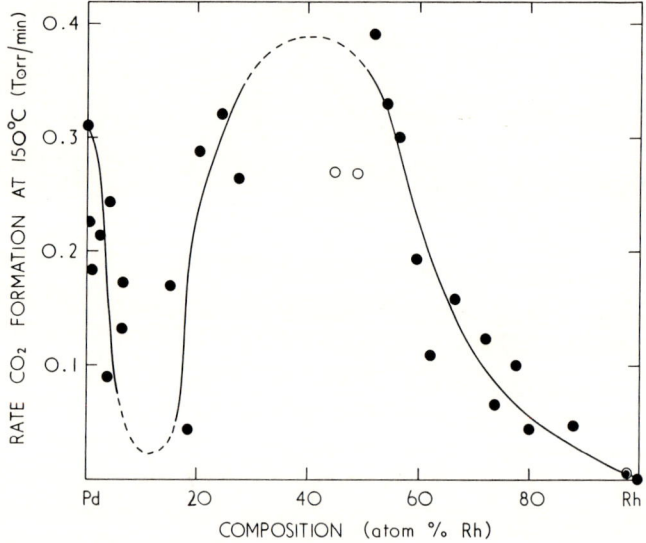

FIG. 27. C_2H_4 oxidation on Pd–Rh alloy films at 150°C as a function of bulk composition (73).

variation in both E and A. Thus the minimum in activity at $\sim$10–15% Rh reflects the maximum in E at these Rh contents, and the steady fall in activity beyond 50% Rh follows the rise in E up to $\sim$98% Rh. As indicated above, pure Rh is the exception because it was the least active catalyst, although E was 5 kcal/mole lower compared with the 98% Rh alloy. It is convenient to analyze this complex activity pattern in terms of three composition ranges.

Composition range 0–30% Rh. In this composition range the alloys appear to be reasonably homogeneous but after reaction show evidence of dissolved hydrogen at compositions up to $\sim$10% Rh. There is an obvious association of the minimum in activity with the filling of d-band vacancies by electrons from dissolved hydrogen. As a parallel with the Pd–Ag system, there is again the curious absence of evidence for dissolved hydrogen in pure palladium itself, but at least its activity is appropriately high compared with those Pd–Rh alloys which show evidence of dissolved hydrogen. Further support for the view that dissolved hydrogen reduces the activity is the simpler activity pattern observed for CO oxidation over similarly prepared Pd–Rh films with this composition range (see subsequent text).

Composition range 30–80% Rh. In this composition range phase separation occurs, and the structure of such Pd–Rh alloy films has been reviewed (Section II). "Phase" I varied in composition and "phase" II contained $88 \pm 5\%$ Rh. It was proposed that these results could be explained by the preferential nucleation of rhodium so that the crystallites consisted of a phase II kernel surrounded by an outer shell (phase I), the Rh content of which increased with an overall increase in the Rh content of the alloy film. Note the essential difference to the Cu–Ni films (*28, 33*) discussed in Section IV.A where complete separation into two phases of fixed equilibrium composition is envisaged, and over a wide composition range the crystallite surfaces have the same composition.

Composition range 80–100% Rh. At this end of the composition range, reasonably homogeneous alloys were again formed, and for present purposes it is unnecessary to reiterate the comments made on the activity of these alloys and about pure rhodium (*73*).

It is clear from the change in activity over this composition range that the surface composition of the Pd–Rh crystallites varied and, as expected, the direction of change indicates an increasing Rh surface content (Fig. 27). There is a maximum in activity at 52% Rh, but at slightly smaller Rh content the rate is distinctly less; these anomalous results (open circles in Fig. 27) are explained by the following treatment of the data. The *surface* composition was equated with the composition corresponding to the observed lattice constant of phase I. In Fig. 28 the catalytic activity

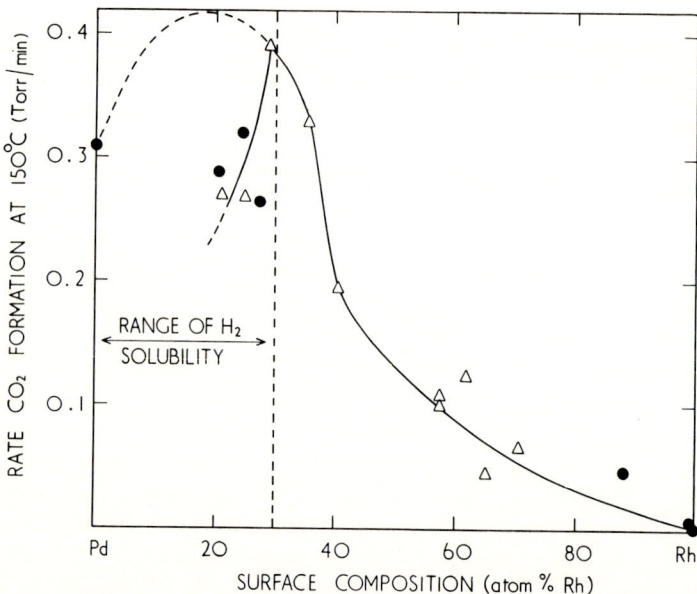

Fig. 28. Results in Fig. 27 replotted using apparent surface compositions derived from X-ray data where phase separation was detected (△) or assumed for pure metals and homogeneous alloys, (●) (*73*).

of alloys in the 30–80% Rh range is plotted against the apparent surface composition thus derived (open triangles). For comparison, results for pure palladium and rhodium and alloys with 20–30% Rh or >80% Rh are included (filled circles), where the equivalence of surface and bulk composition was assumed. Figure 28 shows that starting from pure rhodium, the activity increased until an apparent surface Rh content of ∼30% was reached; at this point the surface composition is entering the range where the effects of hydrogen solubility become important. Now the apparently anomalous points (open circles in Fig. 27) fall into place with results for alloys with ∼20–30% Rh (filled circles) which are at the tail end of the range where hydrogen solubility occurs and the activity is depressed.

It is a matter of speculation as to whether or not the activity would pass through a significant maximum at a surface composition between 0 and 30% Rh. It is interesting to note in this connection that the magnetic susceptibility (*156, 157*) and the electronic specific heat coefficient (*156*) increase from low values at ∼60% Ag–Pd through pure palladium and reach a maximum at ∼5% Rh–Pd, thereafter decreasing smoothly to pure rhodium. Activity maxima have also been reported for reduced mixed oxides and supported alloys of group VIII metal pairs. For example, in the liquid-phase hydrogenation of nitrobenzene (5), maximum activity occurs at 25% Ru–Pt, 25% Ru–Pd, 15% Ir–Pt, and 25% Rh–Pd. Yoshida (7) gives a table which compares maxima in activity for a variety of reactions over binary alloys of group VIII metals with magnetic susceptibility maxima to support his conclusion for Pt–Rh alloys that activity (for hydrogenating aromatic compounds) is closely related to the number of unpaired d-electrons present. For some of the alloy systems mentioned, the possibility of the onset of phase separation or the limit of hydrogen solubility cannot be ignored as factors responsible for the observed enhanced activities or for the shape of the activity versus composition plots. Nevertheless, interesting variations of activity with composition can occur. Rates of CH_4–D_2 exchange were measured by McKee and Norton (158–160) (using alloys in powder form) and expressed in satisfactory units because surface areas were also determined in this work. These authors note that it is surprising that this exchange reaction, which is well understood (161), should not have been used in studying alloy systems. These considerations might be extended to other exchange reactions which could advantageously be used in conjunction with evaporated alloy films as catalysts.

When Pd–Rh alloy films (prepared by simultaneous deposition at 400°C)

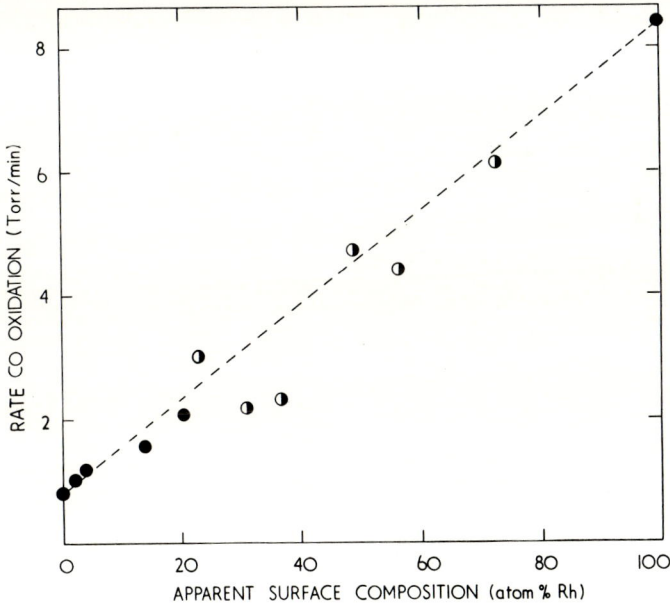

FIG. 29. CO oxidation at 240°C over Pd–Rh alloy films as a function of apparent surface composition; homogeneous alloys and pure metals (●); alloys showing phase separation (◐) (34).

were used in CO oxidation (*34*), then a simpler variation of activity with composition was observed than that found for ethylene oxidation. The effect of equating the surface composition with the composition of the "phase I" derived from X-ray diffraction was again tried. Treated in this way, the CO oxidation rates show a linear relationship with Rh content (Fig. 29). Hence it was proposed that CO oxidation rates can be used to estimate the surface composition in the following way: it would be assumed that the rate of CO oxidation, r_{obs}, is related to the fraction of the alloy surface, θ, due to Rh atoms by

$$r_{obs} = r_1(1 - \theta) + r_2\theta,$$

where r_1 and r_2 are the rates of CO oxidation on pure palladium and pure rhodium, respectively, for a given preparative method. If a Pd–Rh alloy film had been formed at, say, a low sintering temperature and perhaps with a somewhat larger surface area, then experiments with the pure metals at the same sintering temperature would be required to derive new values of r_1 and r_2. Thus, for the alloy film in question, the apparent surface composition would be obtained from the observed rate of CO oxidation. This approach was used to measure the apparent surface composition of Pd–Rh alloy films prepared by annealing in hydrogen separately deposited layers of palladium and rhodium. It was shown (*34*) that the order of deposition, film weight, and composition were all important variables in determining the surface composition in this alloy system where a miscibility gap occurs and interdiffusion rates are slow (see Section II). In these systems an alternative approach to preparation can be adopted, viz, to attempt to prepare the alloy as a metastable solid solution by the vapor-quenching method (see Section II) which would be assisted by the low surface atom mobilities. It follows that the catalytic reactions studied over such alloy films (and any stabilizing/sintering treatment) would have to be carried out at modest temperatures to prevent phase separation. Table X shows that Pd–Rh films deposited at 0°C (and annealed in hydrogen at 400°C) have indeed the correct lattice constants, and apparently phase separation is still absent. However, the apparent surface compositions derived from CO oxidation rates (last column, Table 10) show considerable deviations in two out of three cases when compared with the overall film composition (first column). This is yet another example of desegregation at an alloy surface being in advance of changes in the bulk.

E. Pt–Au, Pt–Ru, Ni–Au, and Au–Ag Alloy Films

The Pt–Au system (*162*) is analogous to the Cu–Ni system in that it exhibits a wide miscibility gap (below 1258°C), and one metal (gold)

TABLE X

Annealed Films Deposited Simultaneously at 0°C

Overall composition (at % Rh)	Film weight (mg)	Composition of X-ray sample (at % Rh)	Lattice constant (Å) Expected	Lattice constant (Å) Observed	CO oxidation rate at 240°C (Torr/min)	Apparent surface composition (at % Rh)
43.0	20.5	41.3	3.859	3.855	2.64	27.4
52.3	34.3	52.3	3.849	3.847	4.02	48.0
—	32.9	68.7	3.835	3.835	—	—
79.0	26.7	79.1	3.825	3.822	7.2	95.4

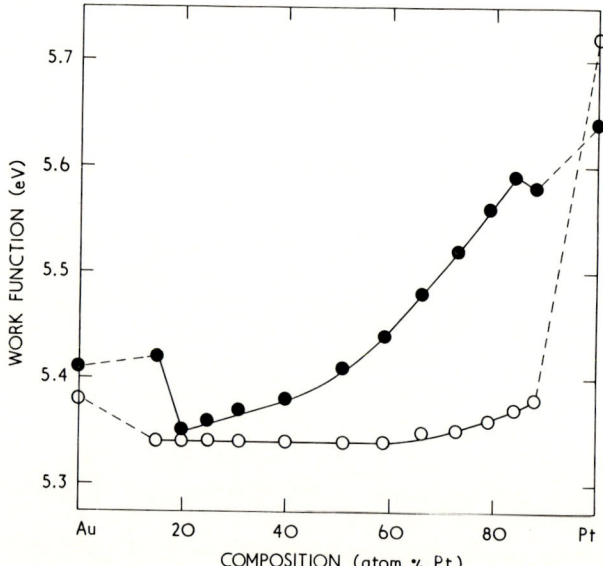

FIG. 30. Work function of Pt–Au alloy films prepared by simultaneous deposition at −196°C; measured at room temperature (●) and after sintering at 300°C (○) (41).

diffuses considerably faster than the other. Conclusions about the surface composition of Cu–Ni films (Section IV.A) should therefore apply to Pt–Au films prepared in a similar way, i.e., the equilibrated alloy films should consist of crystallites with a Pt-rich kernel and a Au-rich skin. In order to verify these conclusions, the photoelectric work functions of Pt–Au evaporated films were measured by Bouwman and Sachtler (41) after evaporation, after sintering at 200°C, and then after chemisorption of CO.

Two designs of phototube were used, so that both successive and simultaneous evaporation of the two metals could be studied. The tube used for successive evaporation was similar to that used with Cu–Ni films (Fig. 4). The films were evaporated in UHV; pretreating the gold by heating it above its melting point in 5–10 Torr of hydrogen before bakeout enabled a pressure in the 10^{-10} Torr range to be maintained during evaporation. Bulk compositions of the alloys were measured by neutron activation analysis, X-ray diffraction, and also by calculation from the weight lost by the filaments during evaporation and the geometry of the system.

The interdiffusion of gold and platinum is much slower than that of copper and nickel, and films formed by successive evaporation of the component metals were not brought to equilibrium by sintering at 200°C. Films formed by simultaneous evaporation were equilibrated by sintering at 300°C. The work function data for the latter films are shown in Fig. 30. The freshly deposited films contained crystallites of platinum and gold in random distribution, and the work function gradually increased from the Au-rich alloys to the Pt-rich alloys. After sintering, however, the work function was approximately constant (5.34–5.38 eV) for alloys of bulk compositions ranging from 15% Pt to 88% Pt. It was concluded that all these alloys had a constant surface composition of about 15% Pt, the stable composition on the left-hand side of the phase diagram in Bouwman and Sachtler (*41*).

The Pt–Au films were not used in catalysis, but the chemisorption of CO was studied. The work function of Pt was only raised by 0.03 eV and there was no change with the alloys after short exposures to CO. It was therefore not possible to titrate the Pt content of the surface with CO in the same way as hydrogen was used with Cu–Ni alloy films (*2*). Long-term exposure of the films to 10^{-5}–10^{-4} Torr CO at 20°C for periods up to four days caused the work function of the alloys to increase slowly Fig. 31. After 16 hr this increase was more evident in the Pt-rich region, but the effect was observed on the Au-rich regions after longer exposures. The effect was accelerated if the films were maintained at 100°C. These results were cited as *direct* evidence for the enrichment of the surface with platinum

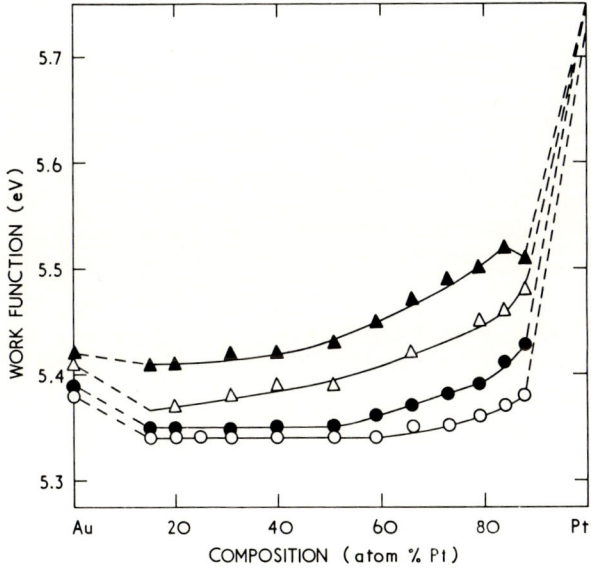

FIG. 31. Work function of Pt–Au alloy films (see Fig. 30) measured after sintering at 300°C (○); then after exposure to 10^{-4} Torr CO at 20°C for 16 hr, (●), for 90 hr (△), and for 16 hr at 100°C (▲) (*41*).

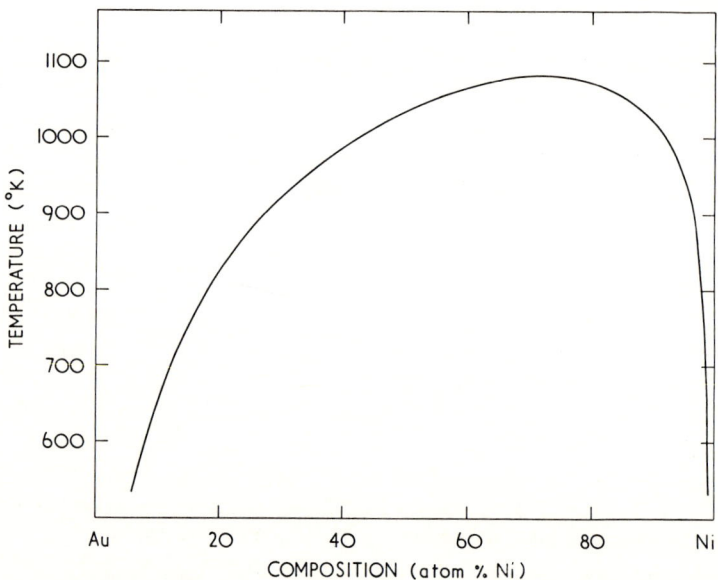

FIG. 32. Phase diagram for Ni–Au system from data in Hultgren et al. (13).

in the presence of CO, a gas which is chemisorbed on only one of the component metals. We have already seen the possible enrichment of Cu–Ni alloys with Ni in the presence of hydrogen (117), and surface enrichment of Pd–Ag alloys with Ag in the presence of oxygen was also suspected (135). After heating to 300°C and pumping, the original work function was regained, as would be expected, when the CO was desorbed. Oddly enough, long-term exposure to CO also appeared to have a small effect on the work function of pure gold films, but this was not commented upon.

Platinum–ruthenium alloy films provide a further opportunity to examine the consequences of CO chemisorption on surface composition (83a). It was known that the binding energies between CO and ruthenium or platinum are roughly comparable and therefore there should be, in this case, insufficient net driving force to enrich the surface by one of the metals. It was found that the CO surface potentials changed gradually with composition from 0.60 eV for pure Ru to 0.04 eV for pure Pt. Prolonged exposure to CO at 100° and 200°C did not significantly alter values for the pure metals or for the Pt–Ru alloy films, and this result confirms the prediction made.

A few results were reported by Campbell and Emmett (29) for the hydrogenation of ethylene on Ni–Au alloys. Two alloy films containing 45 and 84% Ni were prepared by simultaneous evaporation followed by annealing at 500°C in hydrogen. The rates of reaction were equal on the two alloys, and about one tenth as fast as on pure nickel, while pure gold was catalytically inactive. The Ni–Au system exhibits a miscibility gap (163); reference to the phase diagram (Fig. 32) shows that at 500°C the stable phases contain ca. 17 and 97% Ni, respectively. It is not possible to predict which phase is likely to be present at the surface because the diffusion rates of the two metals are not very different and there is no pronounced Kirkendall effect (164). The catalytic results (Fig. 33) however, indicate that the gold-rich phase is probably present at the surface even though the alloys were annealed in hydrogen. If this is so, then there is appreciable catalytic activity at 17% Ni, although at this composition d-band vacancies might be expected to be filled.

The use of Au–Ag alloy powders (8) as catalysts for ethylene oxidation was mentioned in Section IV.C, and results are now available for isotopic oxygen exchange over Au–Ag alloy films (165). The maximum rate of homomolecular exchange was measured on alloys with 50–60% Au and was about 5 times greater than the exchange rate on silver. Initial rates of exchange between adsorbed and gaseous oxygen were approximately equal to the rates of homomolecular exchange. The retained oxygen (after pretreatment in 0.1 Torr oxygen at 400°C) decreased almost linearly with decrease in Ag content, but the "reaction capability" of oxygen on the alloys was higher than on pure silver.

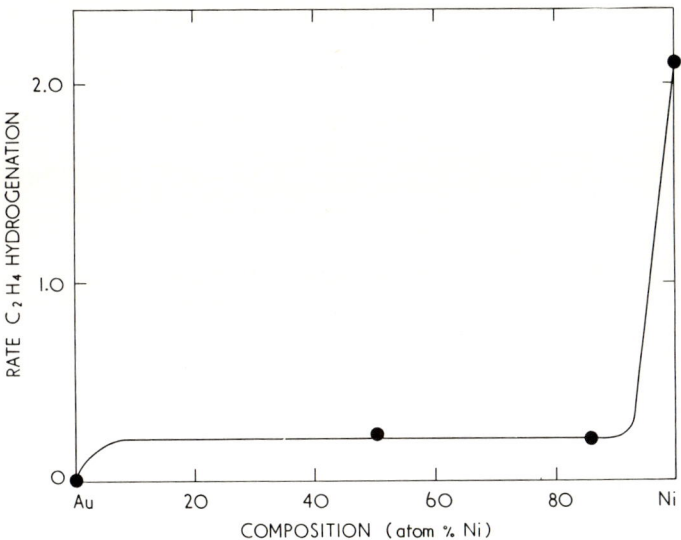

Fig. 33. Initial rate (Torr/min 1000 cm²) at 21°C of C_2H_4 hydrogenation over Ni–Au alloy films (29).

V. Conclusions

There is now available a substantial amount of information on the principles and techniques involved in preparing evaporated alloy films suitable for adsorption or catalytic work, although some preparative methods, e.g., vapor quenching, used in other research fields have not yet been adopted. Alloy films have been characterized with respect to bulk properties, e.g., uniformity of composition, phase separation, crystallite orientation, and surface areas have been measured. Direct quantitative measurements of surface composition have not been made on alloy films prepared for catalytic studies, but techniques, e.g., Auger electron spectroscopy, are available.

Studies of adsorption and catalysis over alloy films are most numerous for the Cu–Ni system where phase separation is expected at normal temperatures (<400°C) for film preparation and use. Hence, the results obtained illustrate various complications which can arise and require elucidation before any fundamental statement can be made about the nature of catalytic activity. Where phase separation was detected and one of the phases exists at the surface over much of the bulk composition range, then only a limited number of significantly different alloys are available on which to carry out adsorption or catalytic experiments. In other studies, metastable Cu–Ni alloys may have been formed, subject to unknown changes at the surface during pretreatment or use. Annealing in hydrogen before catalytic experiments may induce surface enrichment by nickel atoms. Alloy systems containing palladium have been involved in most of the remaining studies, e.g., Pd–Au, Pd–Ag, or Pd–Rh. Although only the Pd–Rh system is subject to phase separation, hydrogen solubility from dissociatively chemisorbed hydrocarbons (used as reactant) is another complication over parts of the composition range. Structure-sensitive reactions will raise additional problems in systems where there is a change of crystallite orientation with composition.

Once, however, these complications are recognized, it is then possible to begin to evaluate the general relevance of the role of d-band vacancies or the individual surface atom approach. Throughout the review, we have given for each reaction and alloy system, the point of view expressed by the original authors in addition to our own comments. Undoubtedly, studies using evaporated alloy films provide further examples of sharp variations in catalytic properties when group VIII and IB metals are alloyed, e.g., formic acid decomposition on Pd–Au films and parahydrogen conversion on Pd–Ag films. Some results using Cu–Ni films require a special explanation if they are to be discussed in terms of collective electronic properties. Other Cu–Ni film results, e.g., hydrogen chemisorption and benzene hydrogenation, have been interpreted in terms of the individual surface atom approach. It would seem that further experimental results obtained using well-characterized alloy films, including surface analysis, could contribute to improving our understanding of catalysis by metals.

REFERENCES

1. Eley, D. D., *J. Res. Inst. Catal., Hokkaido Univ.* **16**, 101 (1968).
2. Sachtler, W. M. H., and van der Plank, P., *Surface Sci.* **18**, 62 (1969).
3. Rylander, P. N., and Cohn, G., *Actes 2nd Congr. Int. Catal., Paris, 1960* **1**, 977 (1961).
4. Bond, G. C., and Webster, D. E., *Platinum Metals Rev.* **9**, 12 (1965).
5. Bond, G. C., and Webster, D. E., *Platinum Metals Rev.* **10**, 10 (1966).
6. Yoshida, K., *Nippon Kagaku Zasshi* **88**, 125, 222 (1967).
7. Yoshida, K., *Shokubai* **10**, 2 (1968).
8. Flank, W. H., and Beachell, H. C., *J. Catal.* **8**, 316 (1967).
9. Belser, R. B., *J. Appl. Phys.* **31**, 562 (1960).
10. Hansen, M., "Constitution of Binary Alloys," 2nd Ed. McGraw-Hill, New York, 1958.
11. Elliott, R. P., "Constitution of Binary Alloys," 1st Suppl. McGraw-Hill, New York, 1965.
12. Shunk, F. A., "Constitution of Binary Alloys," 2nd Suppl. McGraw-Hill, New York, 1969.
13. Hultgren, R., Orr, R. L., Anderson, P. D., and Kelley, K. K., "Selected Values of Thermodynamic Properties of Metals and Alloys." Wiley, New York, 1963.
14. Kubaschewski, O., Evans, E. Ll., and Alcock, C. B., "Metallurgical Thermochemistry," 4th Ed. Pergamon, Oxford, 1967.
15. Chan, J. P., and Hultgren, R., *J. Chem. Thermodyn.* **1**, 45 (1969).
16. Darby, J. B., *Acta Met.* **14**, 265 (1966).
17. Myles, K. M., and Darby, J. B., *Acta Met.* **16**, 485 (1968).
18. Myles, K. M., *Trans. AIME* **242**, 1523 (1968).
19. Bidwell, L. R., Rizzo, F. E., and Smith, J. V., *Acta Met.* **18**, 1013 (1970).
20. Bidwell, L. R., and Speiser, R., *Acta Met.* **13**, 61 (1965).
21. Walker, R. A., and Darby, J. B., *Acta Met.* **18**, 1261 (1970).
22. Darken, L. S., and Gurry, R. W., "Physical Chemistry of Metals." McGraw-Hill, New York, 1953.
23. Sundquist, B. E., *Trans. AIME* **236**, 1111 (1966).
24. Michel, P., *Ann. Phys. (Paris)* **1**, 719 (1956).
25. Fujiki, Y., *J. Phys. Soc. Jap.* **14**, 913 (1959).
26. Nowick, A. S., and Mader, S., *in* "Basic Problems in Thin Film Physics" (R. Niedermayer and H. Mayer, eds.), p. 212. Vandenhoeck & Ruprecht, Göttingen, 1966.
27. Belous, M. V., and Wayman, C. M., *J. Appl. Phys.* **38**, 5119 (1967).
28. Sachtler, W. M. H., and Dorgelo, G. J. H., *J. Catal.* **4**, 654 (1965).
29. Campbell, J. S., and Emmett, P. H., *J. Catal.* **7**, 252 (1967).
30. Moss, R. L., and Thomas, D. H., *Trans. Faraday Soc.* **60**, 1110 (1964).
31. Rudman, P. S., *Acta Crystallogr.* **13**, 905 (1960).
32. Fisher, B., and Rudman, P. S., *J. Appl. Phys.* **32**, 1604 (1961).

33. Sachtler, W. M. H., and Jongepier, R., *J. Catal.* **4,** 665 (1965).
34. Moss, R. L., and Gibbens, H. R., *J. Catal.* (in press).
35. Gerber, R., *Rev. Sci. Instrum.* **38,** 77 (1967).
36. Behrndt, K. H., and Love, R. W., *Vacuum* **12,** 1 (1962).
37. Oron, M., and Adams, C. M., *J. Sci. Instrum.* **2,** 183 (1969).
38. Oron, M., and Adams, C. M., *J. Mater. Sci.* **4,** 252 (1969).
39. Holland, L., and Steckelmacher, W., *Vacuum* **2,** 346 (1952).
40. Moss, R. L., and Thomas, D. H., *J. Catal.* **8,** 151 (1967).
41. Bouwman, R., and Sachtler, W. M. H., *J. Catal.* **19,** 127 (1970).
42. Palatnik, L. S., Ignat'yev, O. M., and Ignat'yeva, L. K., *Fiz. Metal. Metalloved.* **21,** 700 (1966).
43. Dushman, S., "Scientific Foundations of Vacuum Technique," 2nd Ed. Wiley, New York, 1962.
44. Holland, L., "Vacuum Deposition of Thin Films." Chapman & Hall, London, 1960.
45. Smithells, C. J., (Ed.) "Metals Reference Book," 4th Ed., Vol. 1. Butterworth, London, 1967.
46. Stephens, S. J., *Nat. Symp. Vac. Technol., Trans.* **4,** 34 (1958).
47. Rossington, D. R., and Runk, R. B., *J. Catal.* **7,** 365 (1967).
48. Alexander, E. G., and Russell, W. W., *J. Catal.* **4,** 184 (1965).
49. Oldfield, R. C., *J. Phys. D: Appl. Phys.* **3,** 1495 (1970).
50. Harris, L., and Siegel, B. M., *J. Appl. Phys.* **19,** 739 (1948).
51. Pines, B. Y., Grebennik, I. P., and Gribko, V. F., *Ukr. Fiz. Zh. (Ukr. Ed.)* **13,** 280 (1968); [*Phys. Abstr.* No. 35486 (1968)].
52. Smithells, C. J., (Ed.) "Metals Reference Book," 4th Ed., Vol. 2. Butterworth, London, 1967.
53. Boiko, B. T., Palatnik, L. S., and Lebedeva, M. V., *Fiz. Metal. Metalloved.* **25,** 845 (1968).
54. Moss, R. L., Thomas, D. H., and Whalley, L., *Thin Solid Films* **5,** R19 (1970).
55. Moss, R. L., and Thomas, D. H., *Brit. J. Appl. Phys.* **15,** 673 (1964).
56. Kirenskii, L. V., Sukhanova, R. V., Pyn'ko, V. G., and Edel'man, I. S., *Bull. Acad. Sci. USSR, Phys. Ser.* **30,** 53 (1966).
57. Baltz, A., *J. Appl. Phys.* **34,** 1575 (1963).
58. Nagasawa, A., *J. Phys. Soc. Jap.* **19,** 2344 (1964).
59. Neugebauer, C. A., *in* "Physics of Thin Films" (G. Haas and R. E. Thun, eds.), Vol. 2, p. 1. Academic Press, New York, 1964.
60. Moss, R. L., Gibbens, H. R., and Thomas, D. H., *J. Catal.* **16,** 117 (1970).
61. Kneller, E., *J. Appl. Phys.* **33,** 1355 (1962).
62. Kneller, E. F., *J. Appl. Phys.* **35,** 2210 (1964).
63. Mader, S., Nowick, A. S., and Widmer, H., *Acta Met.* **15,** 203 (1967).
64. Mader, S., *J. Vac. Sci. Technol.* **2,** 35 (1965).
65. Bates, P. A., Popplewell, J., and Charles, S. W., *J. Phys. D: Appl. Phys.* **3,** L15 (1970).
66. Duwez, P., and Willens, R. H., *Trans. AIME* **227,** 362 (1963).
67. Mader, S., Widmer, H., d'Heurle, F. M., and Nowick, A. S., *Appl. Phys. Lett.* **3,** 201 (1963).
68. Nowick, A. S., and Mader, S., *IBM J. Res. Develop.* **9,** 358 (1965).
69. Clarke, J. K. A., and Rafter, E. A., *Z. Phys. Chem. (Frankfurt am Main)* **67,** 169 (1969).
70. van der Plank, P., and Sachtler, W. M. H., *J. Catal.* **12,** 35 (1968).
71. Kemball, C., *Proc. Roy. Soc., Ser. A* **214,** 413 (1952).

72. Crawford, E., Roberts, M. W., and Kemball, C., *Trans. Faraday Soc.* **58,** 1761 (1962).
73. Moss, R. L., Gibbens, H. R., and Thomas, D. H., *J. Catal.* **16,** 181, (1970).
74. Moss, R. L., Duell, M. J., and Thomas, D. H., *Trans. Faraday Soc.* **59,** 216 (1963).
75. Suhrmann, R., Wedler, G., Wilke, H. G., and Reusmann, G., *Z. Phys. Chem. (Frankfurt am Main)* **26,** 85 (1960).
76. Pearson, W. B., "A Handbook of Lattice Spacings and Structures of Metals and Alloys," Vols. 1 and 2. Pergamon, London, 1964 and 1967.
77. Coles, B. R., *J. Inst. Metals* **84,** 346 (1955–1956).
78. Rachinger, W. A., *J. Sci. Instrum.* **25,** 254 (1948).
79. Byrne, J. J., and Clarke, J. K. A., *J. Catal.* **9,** 166 (1967).
80. Fukano, Y., *J. Phys. Soc. Jap.* **16,** 1195 (1961).
81. Richards, J. L., and McCann, W. H., *J. Vac. Sci. Technol.* **6,** 644 (1969).
82. Light, T. B., and Wagner, C. N. J., *J. Appl. Crystallogr.* **1,** 199 (1968).
83. Dubejko, M., and Olszewski, S., *Phys. Status Solidi* **16,** 399 (1966).
83a. Bouwman, R., Thesis, University of Leiden, Leiden, The Netherlands, (1970).
84. Zhavoronkova, K. N., Boreskov, G. K., and Nekipelov, V. N., *Dokl. Phys. Chem.* **177,** 859 (1967).
85. Hardy, W. A., and Linnett, J. W., *Trans. Faraday Soc.* **66,** 447 (1970).
86. Hutchins, G. A., *in* "Developments in Applied Spectroscopy" (E. L. Grove and A. J. Perkins, eds.), Vol. 7A, p. 80. Plenum, New York, 1969.
87. Marshall, D. J., and Hall, T. A., *J. Phys. D: Appl. Phys. 1,* 1651 (1968).
88. Alessandrini, E. I., and Kuptsis, J. D., *J. Vac. Sci. Technol.* **6,** 647 (1969).
89. Sewell, P. B., and Cohen, M., *Appl. Phys. Lett.* **11,** 298 (1967).
90. Sewell, P. B., Mitchell, D. F., and Cohen, M., *in* "Developments in Applied Spectroscopy" (E. L. Grove and A. J. Perkins, eds.), Vol. 7A, p. 61. Plenum, New York, 1969.
91. Harris, L. A., *J. Appl. Phys.* **39,** 1419 (1968).
92. Weber, R. E., and Peria, W. T., *J. Appl. Phys.* **38,** 4355 (1967).
93. Haas, T. W., Grant, J. T., and Dooley, G. J., *J. Vac. Sci. Technol.* **7,** 43 (1970).
94. Palmberg, P. W., and Rhodin, T. N., *J. Appl. Phys.* **39,** 2425 (1968).
95. Weber, R. E., and Johnson, A. L., *J. Appl. Phys.* **40,** 314 (1969).
95a. Ertl, G., and Kuppers, J., *Surface Sci.* **24,** 104 (1971).
95b. Ono, M., Takasu, Y., Nakayama, K., and Yamashina, T., *Surface Sci.* **26,** 313 (1971).
95c. Dobson, P. J., and Gibson, M. J., Personal communication (1971).
96. Montgomery, H., Pells, G. P., and Wray, E. M., *Proc. Roy. Soc., Ser. A* **301,** 261 (1967).
97. Hoare, F. E., *in* "Electronic Structure and Alloy Chemistry of the Transition Elements" (P. A. Beck, ed.), p. 29 Wiley (Interscience), New York, 1963.
98. Vuillemin, J. J., and Priestley, M. G., *Phys. Rev. Lett.* **14,** 307 (1965).
99. Coles, B. R., and Taylor, J. C., *Proc. Roy. Soc. Ser. A* **267,** 139 (1962).
100. Dugdale, J. S., and Guénault, A. M., *Phil. Mag.* **13,** 503 (1966).
101. Eggs, J., and Ulmer, K., *Z. Phys.* **213,** 293 (1968).
102. Dowden, D. A., *J. Res. Inst. Catal., Hokkaido Univ.* **14,** 1 (1966).
103. Bond, G. C., *Discuss. Faraday Soc.* **41,** 200 (1966).
104. Tamm, P. W., and Schmidt, L. D., *J. Chem. Phys.* **51,** 5352 (1969).
105. Whalley, L., Davis, B. J., and Moss, R. L., *Trans. Faraday Soc.* **66,** 3143 (1970).

106. Grimley, T. B., *in* "Molecular Processes on Solid Surfaces" (T. Drauglis, R. D. Gretz, and R. I. Jaffee, eds.), p. 299. McGraw-Hill, New York, 1969.
107. Dowden, D. A., *J. Chem. Soc., London* p. 242 (1950).
108. Dowden, D. A., and Reynolds, P. W., *Discuss. Faraday Soc.* **8**, 184 (1950).
109. Best, R. J., and Russell, W. W., *J. Amer. Chem. Soc.* **76**, 838 (1954).
110. Hall, W. K., and Emmett, P. H., *J. Phys. Chem.* **63**, 1102 (1959).
111. Hall, W. K., Cheselske, F. J., and Lutinski, F. E., *Actes 2nd Congr. Int. Catal., Paris, 1960* **2**, 2199 (1961).
112. Emmett, P. H., "Catalysis, Then and Now," Part I. Franklin, Englewood, New Jersey, 1965.
113. Gharpurey, M. K., and Emmett, P. H., *J. Phys. Chem.* **65**, 1182 (1961).
114. Sachtler, W. M. H., Dorgelo, G. J. H., and Jongepier, R., *in* "Basic Problems in Thin Film Physics" (R. Niedermayer and H. Mayer, eds.), p. 218. Vandenhoeck & Ruprecht, Göttingen, 1966.
115. van der Plank, P., and Sachtler, W. M. H., *J. Catal.* **7**, 300 (1967).
116. Hall, W. K., *J. Catal.* **6**, 314 (1966).
117. Clarke, J. K. A., and Byrne, J. J., *Nature (London)* **214**, 1109 (1967).
118. Byrne, J. J., Carr, P. F., and Clarke, J. K. A., *J. Catal.* **20**, 412 (1971).
119. Carr, P. F., and Clarke, J. K. A., *J. Chem. Soc. A* p. 985 (1971).
120. McMahon, E., Carr, P. F., and Clarke, J. K. A., *J. Chem. Soc. A* p. 2012 (1971).
121. Jongepier, R., and Sachtler, W. M. H., *J. Res. Inst. Catal., Hokkaido Univ.* **16**, 69 (1968).
121a. Plunkett, T. J., and Clarke, J. K. A., *J. Chem. Soc., Faraday Trans.* (in press).
122. Brenner, S. S., and McKinney, J. T., *Surface Sci.* **20**, 411 (1970).
123. Takeuchi, T., Tezuka, Y., and Takayasu, O., *J. Catal.* **14**, 126 (1969).
124. Clarke, J. K. A., and Spooner, T. A., *J. Phys. D: Appl. Phys.*, **4**, 1196 (1971).
125. Bond, G. C., "Catalysis by Metals," pp. 320, 248, 424. Academic Press, New York, 1962.
126. Whalley, L., Thomas, D. H., and Moss, R. L., *J. Catal.* **22**, 302 (1971).
127. Couper, A., and Eley, D. D., *Discuss. Faraday Soc.* **8**, 172 (1950).
128. Eley, D. D., and Luetic, P., *Trans. Faraday Soc.* **53**, 1483 (1957).
129. Daglish, A. G., and Eley, D. D., *Actes 2nd Congr. Int. Catal., Paris, 1960* **2**, 1615 (1961).
130. Eley, D. D., and Knights, C. F., *Proc. Roy. Soc., Ser. A* **294**, 1 (1966).
131. Bortner, M. H., and Parravano, G., *Advan. Catal. Relat. Subj.* **9**, 424 (1957).
132. Schwab, G.-M., and Gossner, K., *Z. Phys. Chem. (Frankfurt am Main)* **16**, 39 (1958).
133. Kowaka, M., *Nippon Kinzoku Gakkaishi* **23**, 655, 659 (1959).
134. Rienäcker, G., and Müller, H., *Z. Anorg. Allg. Chem.* **357**, 255 (1968).
135. Moss, R. L., and Thomas, D. H., *J. Catal.* **8**, 162 (1967).
136. Voge, H. H., and Adams, C. R., *Advan. Catal. Relat. Subj.* **17**, 151 (1967).
137. Twigg, G. H., *Trans. Faraday Soc.* **42**, 284 (1946).
138. Twigg, G. H., *Proc. Roy. Soc., Ser. A* **188**, 92, 105, 123 (1946).
139. Kenson, R. E., and Lapkin, M., *J. Phys. Chem.* **74**, 1493 (1970).
140. Kummer, J. T., *J. Phys. Chem.* **60**, 666 (1956).
141. Wilson, J. N., Voge, H. H., Stevenson, D. P., Smith, A. E., and Atkins, L. T., *J. Phys. Chem.* **63**, 463 (1959).
142. Hoare, F. E., Matthews, J. C., and Walling, J. C., *Proc. Roy. Soc., Ser. A* **216**, 502 (1953).
143. Kemball, C., and Patterson, W. R., *Proc. Roy. Soc., Ser. A* **270**, 219 (1962).

144. Patterson, W. R., and Kemball, C., *J. Catal.* **2,** 465 (1963).
145. Brennan, D., Hayward, D. O., and Trapnell, B. M. W., *Proc. Roy. Soc., Ser. A* **256,** 81 (1960).
146. Ostrovskii, V. E., and Kul'kova, N. V., *Dokl. Phys. Chem.* **161,** 354 (1965).
147. Gonzalez, O. D., and Parravano, G., *J. Amer. Chem. Soc.* **78,** 4533 (1956).
148. Couper, A., and Metcalfe, A., *J. Phys. Chem.* **70,** 1850 (1966).
149. Rienäcker, G., and Engels, S., *Z. Anorg. Allg. Chem.* **336,** 259 (1965).
150. Grimley, T. B., *Advan. Catal. Relat. Subj.* **12,** 1 (1960).
151. Raub, E., Beeskow, H., and Menzel, D., *Z. Metallk.* **50,** 428 (1959).
152. Barton, J. C., Green, J. A. S., and Lewis, F. A., *Trans. Faraday Soc.* **62,** 960 (1966).
153. Tverdovskii, I. P., and Stetsenko, A. I., *Dokl. Akad. Nauk SSSR,* **84,** 997 (1952).
154. Hoare, J. P., *J. Electrochem. Soc.* **107,** 820 (1960).
155. Bond, G. C., "Catalysis by Metals," Academic Press, New York, 1962.
156. Budworth, D. W., Hoare, F. E., and Preston, J., *Proc. Roy. Soc., Ser. A* **257,** 250 (1960).
157. Manuel, A. J., and St. Quinton, J. M. P., *Proc. Roy. Soc., Ser. A* **273,** 412 (1963).
158. McKee, D. W., and Norton, F. J., *J. Phys. Chem.* **68,** 481 (1964).
159. McKee, D. W., and Norton, F. J., *J. Catal.* **3,** 252 (1964).
160. McKee, D. W., and Norton, F. J., *J. Catal.* **4,** 510 (1965).
161. Kemball, C., *Advan. Catal. Relat. Subj.* **11,** 223 (1959).
162. Darling, A. S., Mintern, R. A., and Chaston, J. C., *J. Inst. Metals* **81,** 125 (1952–1953).
163. Münster, A., and Sagel, K., *Z. Phys. Chem. (Frankfurt am Main)* **14,** 296 (1958).
164. Reynolds, J. E., Averbach, B. L., and Cohen, M., *Acta Met.* **5,** 29 (1957).
165. Starostina, T. S., Khasin, A. V., Boreskov, G. K., and Plyasova, L. M., *Dokl. Phus. Chem.* **190,** 49 (1970).

Heat-Flow Microcalorimetry and Its Application to Heterogeneous Catalysis

P. C. GRAVELLE

Département de Chimie-Physique
Institut de Recherches sur la Catalyse
CNRS, Villeurbanne, France

I. Introduction	191
II. Principle of Heat-Flow Calorimetry	194
III. Some Heat-Flow Microcalorimeters That Can Be Used in Heterogeneous Catalysis Research	196
A. The Calvet Microcalorimeter	197
B. The Petit Microcalorimeter	201
C. The Petit–Eyraud Microcalorimeter	203
IV. Theory of Heat-Flow Calorimetry	206
A. A Simplified Theory—The Tian Equation	206
B. General Equations for the Heat Transfer in a Heat-Flow Calorimeter	211
V. Analysis of the Calorimetric Data	214
A. Methods of Recording Calorimetric Data	215
B. Data Correction and Determination of Thermokinetics	218
VI. Measurement of the Differential Heats of Gas–Solid Interactions	226
A. Description of the Associate Equipment. The Volumetric Line. The Calorimetric Cells	227
B. Calibration of the Calorimeter and Preliminary Experiments	232
VII. Applications of Heat-Flow Microcalorimetry in Heterogeneous Catalysis	237
A. Differential Heats of Adsorption. The "Energy Spectrum" of the Catalyst Surface	238
B. Differential Heats of Interaction between a Reactant and Preadsorbed Species. Reaction Mechanisms	246
C. Differential Heats of the Catalytic Reaction. Modifications of the Catalyst Surface	254
VIII. Conclusions	259
References	261

I. Introduction

The heat of adsorption generally evolved when a gas contacts the surface of a solid is a very important characteristic of the adsorbate–

adsorbent interaction, as a cursory glance at the first pages of most (if not all) textbooks on adsorption or heterogeneous catalysis indeed shows. Its fundamental importance and its early presentation in comprehensive treatises are explained by the fact that the heat of adsorption is obviously related to the energy of the bonds between the adsorbed particle and the adsorbent, although the relation is far from simple (*1, 2*), and thus to the *nature* of the bonds and to the *chemical reactivity* of the adsorbed species.

The magnitude of the heat of adsorption may be, for instance, considered as a simple, if not perfect, criterion to distinguish between physical adsorption and chemisorption [see inter alia (*3–5*)]. In some favorable cases, the satisfactory agreement between experimental heats of adsorption and heats calculated from theoretical models for the adsorption process has been considered as an argument for supporting the proposed models (*6*). The strength of the oxygen bond at the surface of the oxides of the fourth period, determined directly from calorimetric data or evaluated from the reaction rate of oxygen homomolecular isotopic exchange, has been correlated to the catalytic activity of these oxides in many catalytic reactions (*7*). The magnitude of the differential heat of adsorption of oxygen on nickel oxide catalysts has been related to the reactivity of the adsorbed oxygen species with respect to carbon monoxide (*8*).

The preceding list of examples, which is by no means exhaustive, confirms that the determination of heats of adsorption is of both fundamental and practical importance. However, in contrast with this basic importance which "cannot be overemphasised" (*9*), data on heats of adsorption, and particularly on calorimetric heats of irreversible adsorption processes, are relatively incomplete, as the careful perusal of any textbook on adsorption or heterogeneous catalysis will show.

The cause of the surprising scarcity of calorimetric data is certainly not that adsorption calorimetry is a new technique: heats of adsorption of hydrogen, oxygen, nitrogen, etc., on carbon had been already measured at low temperatures by Dewar in 1904 (*10*). But, although adsorption calorimetry has been known and used for many years, and although numerous investigators have not spared their skill, science, and ingenuity, it must be acknowledged that this technique has not yet reached the final point in its development. Construction, calibration, and operation of adsorption calorimeters are still debated among specialists (*11*), and it is clear that even in the case of the determination of heats of adsorption at the surface of metal films, which since the work of Beeck (*12*) has been particularly studied, no general agreement has been reached regarding the details of the experimental techniques (*11, 13*). Difficulties and uncertainties of the calorimetric techniques are certainly one of the sources of the numerous discrepancies that appear in the published results (*13*). Moreover, they explain why adsorption calorimetry is not as widely used as other, more recent, techniques.

The lack of calorimetric data is particularly evident in the case of the adsorption of gases on oxides or on oxide-supported metals, i.e., on solids similar to most industrial catalysts. Moreover, adsorption calorimeters are generally used at temperatures that are much lower than those usually found in industry, and it would be difficult indeed to adapt most usual adsorption calorimeters for the measurement of heats of adsorption of gases on industrial catalysts at elevated temperatures. The present success of gas chromatographic techniques for determining heats of reversible adsorption may be explained by the gap between the possibilities of the usual adsorption calorimeters and the requirements of industrial catalysis research.

A survey of the literature shows that although very different calorimeters or microcalorimeters have been used for measuring heats of adsorption, most of them were of the adiabatic type, only a few were isothermal, and until recently (*14, 15*), none were typical *heat-flow calorimeters*. This results probably from the fact that heat-flow calorimetry was developed more recently than isothermal or adiabatic calorimetry (*16, 17*). We believe, however, from our experience, that heat-flow calorimeters present, for the measurement of heats of adsorption, qualities and advantages which are not met by other calorimeters. Without entering, at this point, upon a discussion of the respective merits of different adsorption calorimeters, let us indicate briefly that heat-flow calorimeters are particularly adapted to the investigation (1) of slow adsorption or reaction processes, (2) at moderate or high temperatures, and (3) on solids which present a poor thermal diffusivity. Heat-flow calorimetry appears thus to allow the study of adsorption or reaction processes which cannot be studied conveniently with the usual adiabatic or pseudoadiabatic, adsorption calorimeters. In this respect, heat-flow calorimetry should be considered, actually, as a new tool in adsorption and heterogeneous catalysis research.

Finally, most previous calorimetric studies in this field have been devoted to adsorption processes only, and very seldom were these studies extended to the investigation of complete catalytic reactions. The work of Garner and his collaborators in Bristol (*18*) is a notable exception. Heat-flow calorimetry is particularly convenient for such studies (*19*).

The aims of this article are therefore:

(1) to describe the general principles of heat-flow calorimetry;

(2) to point out the specific advantages of heat-flow microcalorimeters for the determination of heats of adsorption; and

(3) to illustrate the use of this calorimetric technique in the study of heterogeneous catalysis reactions.

II. Principle of Heat-Flow Calorimetry

All calorimeters are composed of an inner vessel (the "calorimeter" vessel, A in Fig. 1), in which the thermal phenomenon under study is produced, and of a surrounding medium (shields, thermostat, etc., B in Fig. 1). Depending upon the intensity of the heat exchange between the inner vessel and its surroundings, three main types of calorimeters may be distinguished theoretically as indicated in Fig. 1.

When there is no heat exchange between the inner vessel and its surroundings (*adiabatic calorimeter*, 1 in Fig. 1), the temperature of the calorimeter vessel varies when heat is liberated or absorbed. The quantity of heat produced or absorbed may be calculated from this temperature change, if the heat capacity of the inner vessel and of its contents is known.

When the heat exchange between the inner vessel and its surroundings, maintained at a constant temperature T_0, occurs at an infinitely large rate (*isothermal calorimeter*, 2 in Fig. 1), the temperature of the inner vessel also remains constant. The heat produced or absorbed is generally evaluated from the intensity of a physical modification occuring at a constant temperature in the surrounding medium (phase transformation).

Finally, in the case of *heat-flow calorimeters* (3 in Fig. 1), the inner vessel and its surroundings are connected by a heat conductor (C in Fig. 1).

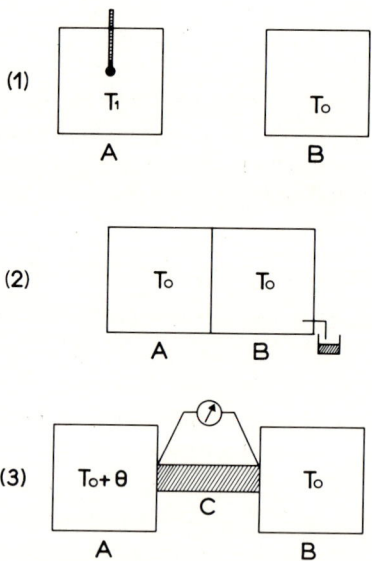

FIG. 1. Theoretical models of adiabatic (1), isothermal (2), and heat-flow (3) calorimeters.

When heat is liberated or absorbed in the calorimeter vessel, a thermal flux is established in the heat conductor and heat flows until the thermal equilibrium of the calorimetric system is restored. The heat capacity of the surrounding medium (heat sink) is supposed to be infinitely large and its temperature is not modified by the amount of heat flowing in or out. The quantity of heat flowing along the heat conductor is evaluated, as a function of time, from the intensity of a physical modification produced in the conductor by the heat flux. Usually, the temperature difference θ between the ends of the conductor is measured. Since heat is transferred by conduction along the heat conductor, calorimeters of this type are often also named "conduction calorimeters" ([20a]).

The ideal adiabatic or isothermal calorimeters that have been described have, of course, no real existence. The ideal adiabatic calorimeter should be, for instance, perfectly thermally insulated from its surroundings so that heat developed inside the calorimeter is totally retained therein. In practice, it is possible to restrain the thermal leaks, by the use of evacuated jackets for instance, as in Beeck-type adsorption calorimeters ([13]), but the thermal head between the inner vessel and its surroundings makes thermal leaks, through radiation or conduction, inevitable. It must be noted, however, that the usual isothermally jacketed and imperfectly insulated calorimeter (the "isoperibol" calorimeter after the denomination of O. Kubaschewski and R. Hultgren) ([20b]) may be considered as perfectly adiabatic at the initial time of the experiment. When the rate of the phenomenon under investigation is large enough, the maximum temperature change of the inner vessel, and thus the amount of heat, may be calculated from the extrapolation at time zero, on a semilogarithmic plot, of the observed temperature changes as a function of time. This simple application of Newton's cooling law has been used to determine heats of adsorption of gases at the surface of metal films ([21]).

Usual isothermal calorimeters also differ from the theoretical model that has been described: a temperature gradient, for instance, however small it may be, must exist within the calorimeter for heat to be transferred from the inner vessel to the heat sink.

It appears therefore that during the operation of all usual calorimeters, temperature gradients are developed between the inner vessel and its surroundings. The resulting thermal head must be associated, in all cases, to heat flows. In isoperibol calorimeters, heat flows (called thermal leaks in this case) are minimized. Conversely, they must be facilitated in isothermal calorimeters. All heat-measuring devices could therefore be named "heat-flow calorimeters." However, it must be noted that in isoperibol or isothermal calorimeters, the consequences of the heat flow are more easily determined than the heat flow itself. The temperature decrease which is recorded in a pseudoadiabatic calorimeter results from the heat exchange between the inner cell and its surroundings. Likewise, the phase transformation which occurs in the heat sink of an isothermal calorimeter is also the consequence of the heat flow between the inner vessel and its surroundings. We believe therefore, in agreement with the definition given above, that the term "heat-flow calorimeter" should be used exclusively in connection with calorimeters in which:

(1) a path is clearly defined mechanically for the flux of heat (or a constant fraction of it) between the inner vessel and the heat sink (a heat conductor in the case of conduction calorimeters), and

(2) the intensity of the heat flux may be measured, along this path, as a function of time (22).

Since heat exchange between the calorimeter vessel and the heat sink is not hindered in a heat-flow calorimeter, the temperature changes produced by the thermal phenomenon under investigation are usually very small (less than 10^{-4} degree in a Calvet microcalorimeter, for instance) (23). For most practical purposes, measurements in a heat-flow calorimeter may be considered as performed under isothermal conditions.

III. Some Heat-Flow Microcalorimeters That Can Be Used in Heterogeneous Catalysis Research

Many calorimetric measurements find applications in heterogeneous catalysis research. Heats of immersion, wetting, and solution, for instance, have been used in this field (24). This article, however, is mainly concerned with the calorimetric determination of heats produced by the adsorption or the reaction of gases at the catalyst surface. For this reason, no attempt is made to describe, in this section, all heat-flow calorimeters. The Benzinger microcalorimeter (17), designed primarily to measure heats of reaction, dilution, or mixing of liquid reagents, and in particular, heats of reaction in aqueous solution, is not described here, though it is a typical heat-flow calorimeter. On the other hand, all heat-flow microcalorimeters described in this section have been applied to the study of gas–solid interactions. They present a common characteristic required for this kind of investigation, namely, that the calorimeter vessel, in all cases, may be connected easily to an external vacuum line.

Heats of chemisorption may vary from a few kilocalories per mole of adsorbed gas to 100 kcal mole^{-1} (or more). It is certainly not necessary to employ a very sensitive calorimeter for their determination and, actually, many chemisorption processes have been studied in calorimeters with a low sensitivity. However, it is necessary to introduce a large quantity of gas in such calorimeters in order that the adsorbate–adsorbent interaction evolves a large amount of heat which may be measured accurately. The large amount of gas covers, of course, a large fraction of the adsorbent surface. Now, it is known that the surface of most adsorbents or catalysts is not energetically homogeneous. The determination of the surface heterogeneity, which is of both practical and theoretical importance, may be achieved by the repeated introduction into the calorimeter of very small doses or increments of gas which cover very progressively the active surface sites. Heats produced by the adsorption of very small doses of gas are usually called *differential heats of adsorption*. Adsorption of each dose of gas produces 0.1–1 cal, and the chemisorption process may last from a few minutes to several hours. The determination of differential heats of adsorption necessitates therefore the use of a very sensitive calorimeter. Such instruments, usually capable of measuring microcalories (10^{-6} cal), are named *microcalorimeters*.

A. The Calvet Microcalorimeter

The Calvet microcalorimeter *(16)* is an improved version of the first heat-flow calorimeter described by Tian in 1924 *(25)*. In this micro-

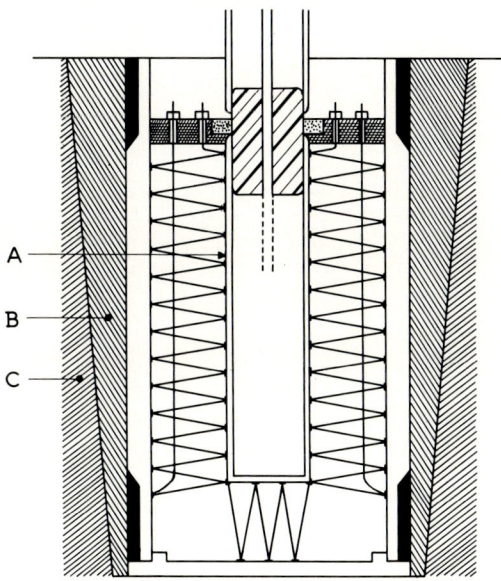

Fig. 2. Vertical section of a Tian–Calvet microcalorimetric element *(16)*: inner vessel (A) and hollow truncated cone (B) wedged in the heat sink (C). Reprinted from Calvet and Prat *(23)* with permission of Dunod.

calorimeter, as in most conduction calorimeters, a thermoelectric pile serves to transfer heat from the calorimeter vessel to the surrounding heat sink and to measure the resulting heat flux. A schematic representation of a typical Tian–Calvet calorimetric element is given in Fig. 2. The outside wall of the inner vessel (A) (a thin cylindrical silver cavity covered by a very thin insulating layer of mica) is completely surrounded with identical thermoelectric junctions, each separated from one another by equal intervals. The couples are affixed normal to the calorimeter wall. The reference junctions of the couples are attached to the inside cylindrical wall of a hollow truncated cone (B) which is wedged into a cavity of the same shape in the surrounding thermostated block (heat sink, C).

In heat-flow calorimeters, it is particularly important, as already indicated in Section II, that the heat sink remain, throughout the experiment, at a constant temperature. The construction of the heat sink and thermostat in the Calvet apparatus is shown in Fig. 3. The calorimetric element fits into a conical socket (A), cut in a cylindrical block of aluminium (B). The block is positioned between the bases of two truncated cones (C and C′), placed within a thick metal cylinder (D). The metal cylinder is, in

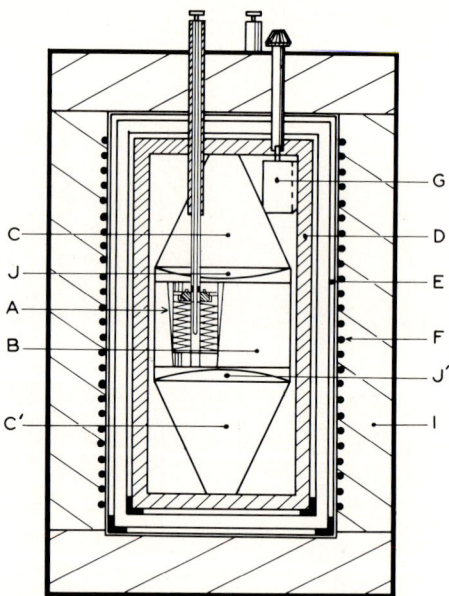

Fig. 3. Vertical section of the Calvet microcalorimeter (16): microcalorimetric element (A); the metal block (B); metallic cones (C and C′); thick metal cylinder (D); thermostat consisting of several metal canisters (E); electrical heater (F); switch (G); thermal insulation (I); and thermal lenses (J and J′). Reprinted from Calvet and Prat (23) with permission of Dunod.

turn, enclosed by a series of metal canisters (E), and the whole assembly is lodged within an insulated thermostat jacket (F and I).

All thermal perturbations, which may be issued from the thermostat jacket, are in the first place considerably damped by the multiple canisters, as demonstrated by Tian in 1923 (*26*). They are then diffused in the thick metal cylinder, thence to be conducted toward the metal block by the metal cones. The cones act as collimators, transforming the heat flux from the cone ends into a parallel vertical flux through the metal block. Thermal lenses of poor conducting steel located between the cones and the metal block (J and J' in Fig. 3) help to achieve the perfect radial symmetry of the thermal perturbations in the metal block. These perturbations, though very small, would nevertheless produce the erratic change of the emf delivered by the thermoelectric pile, were a single calorimetric element located in the metal block. In order to increase the stability of the recorded emf, two identical calorimetric elements are therefore symmetrically disposed in the metal block and the respective thermoelectric piles are connected in opposition. The process under investigation is carried out in one of the calorimeter vessels (the "active" element), the other serving as the tare or "reference" element. The active and reference thermoelectric piles being connected in opposition, the resultant emf is attributed solely to the process under study in the active cell. The twin calorimetric elements must be matched with particular care, because, for instance, a small difference in the thermal lag within these elements considerably reduces the sensitivity of the calorimeter by decreasing the emf stability (*27*). When these requirements are met, the stability of the zero reading, in the absence of production of heat in the active cell, is extremely good. Deflections of the record base line of less than 1 mm/day (corresponding to an apparent heat flux of 10^{-5} μW/sec) are the general rule rather than the exception (*28*). In numerous instances, Calvet microcalorimeters have been used successfully for experiments lasting a month or more. It appears, therefore, that the Calvet microcalorimeter is particularly suitable for the study of slow or even very slow thermal phenomena.

This instrument may be also used to investigate rapid heat evolutions. However, in this case, the thermal flow along the thermocouple wires may cause the transient heating (or cooling) of the thermoelectric junctions in contact with the heat sink and, thereby, induce erroneous emf changes. This cause of systematic errors may be eliminated by "compensating" the thermal effect generated in the active cell by Joule heating or Peltier cooling in the active element, according to the process, endothermic or exothermic, taking place. In constructing the microcalorimetric element, the thermoelectric couples are wired in such a way that by a switching arrangement, a part of the couples may be used for Peltier compensation by passing through them a current of known intensity for a measured time, while the second part of the couples is used for detecting the noncompensated part of the heat effect. Total compensation is not sought, but Peltier cooling is intended to balance the major part of the heat produced. Compensation of the heat effect is especially necessary in case of rapid and large heat evolutions in the active cell. For slow and small heat evolutions, compensation is less necessary and is not usually attempted.

Calvet and Persoz (29) have discussed at length the question of the sensitivity of the Calvet calorimeter in terms of the number of thermocouples used, the cross section and the length of the wires, and the thermoelectric power of the couples. On the basis of this analysis, the microcalorimetric elements are designed to operate near maximum sensitivity. The present-day version of a Tian–Calvet microcalorimetric element, which has been presented in Fig. 2, contains approximately 500 chromel-to-constantan thermocouples. The microcalorimeter, now commercially available, in which two of these elements are placed (Fig. 3) may be used from room temperature up to 200°C.

A high-temperature model of the Calvet microcalorimeter has also been designed. Its range of applicability extends to temperatures in excess of 600°C, i.e., it exceeds the temperature range of most industrial catalytic processes. Platinum and platinum–rhodium wires are then used in the construction of the thermoelectric piles (273 couples in a microcalorimetric element). A schematic representation of the high-temperature Calvet microcalorimeter is given in Fig. 4. The design of the thermostat and of the heat sink (a metal block located between the bases of two metal cones) is very similar to that of the ordinary Calvet microcalorimeter (Fig. 3). Dimensions of all parts except the twin microcalorimetric elements are, however, smaller, so that the whole calorimetric assembly may be housed in a muffle furnace which completely surrounds it. The switch which allows variation of the number of thermocouples in the detecting circuit and use, when necessary, of Peltier compensation, is located out of the main body in a separate thermostated box. Finally, it must be noted that very recently, the Calvet calorimeter has been adapted also to the measurement of heats of physical adsorption at low (liquid nitrogen) temperatures (30).

Because of the salient features of its design (multijacket thermostat, large heat sink), the Calvet microcalorimeter is a large and heavy instrument. Moreover, the attainment of a steady state (stable zero reading) at the beginning of an experiment is time-consuming, particularly at high temperatures (28). Although the objective of the Calvet design is sound, less heavy systems based on electrically controlled shields are now equally effective and more rapid in response. Such shields are employed in the

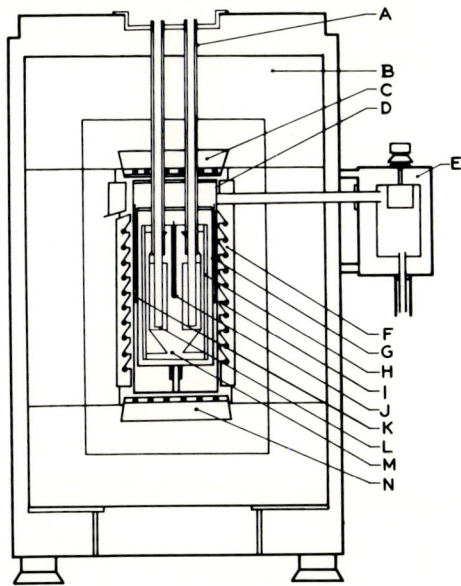

Fig. 4. Vertical cross section of a high-temperature Calvet calorimeter (16): cell guides (A); thermal insulation (B); top (C) and bottom (N) electrical heaters; thermostat consisting of several metal canisters (D, G, and H); switch (E); electrical heater (F); thermometers (I, J, and K); microcalorimetric element (L); and heat sink (M).

construction of the heat-flow microcalorimeters, which are described in the next sections. The central part, only, of these calorimeters will be therefore described.

B. The Petit Microcalorimeter (31)

Calvet and Guillaud (32) noted in 1965 that in order to increase the sensitivity of a heat-flow microcalorimeter, thermoelectric elements with a high factor of merit must be used. (The factor of merit f is defined by the relation: $f = \epsilon^2/\rho c$, where ϵ is the thermoelectric power of the element, ρ its electrical resistivity, and c its thermal conductivity.) They remarked that the factor of merit of thermoelements constructed with semiconductors (doped bismuth tellurides usually) is approximately 19 times greater than the factor of merit of chromel-to-constantan thermocouples. They described a Calvet-type microcalorimeter in which 195 semiconducting thermoelements were used instead of the usual thermoelectric pile.

In recent years, other heat-flow microcalorimeters equipped with commercially available semiconducting thermoelements have been described

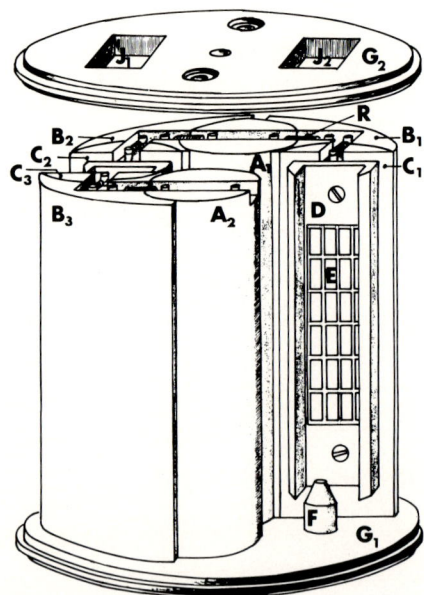

Fig. 5. The Petit microcalorimeter (*31*): vertical axles (A_1 and A_2); mobile arms (B_1–B_3); flux-meter holders (C_1–C_3); cell guide (D); thermoelectric element (E); cell-positioning block (F); top and bottom flanges (G_1 and G_2); portholes (J_1 and J_2); and springs (R).

(*31, 33*). The Petit microcalorimeter which is presented in Fig. 5 appears to be particularly suitable for chemisorption studies.

In this microcalorimeter, the heat sink is not a massive metal block but is divided into several parts which are mobile with respect to each other. Each thermoelectric element (E) and a cell guide (D) are affixed to a fluxmeter holder (C). The holder (C) is mobile with respect to a massive arm (B) which, in turn, rotates around a vertical axle (A). All parts of the heat sink are made of brass. Surfaces in contact are lubricated by silicone grease. Four thermoelectric elements (E) are mounted in this fashion. They enclose two parallelepipedic calorimetric cells, which can be made of glass (cells for the spectrography of liquids are particularly convenient) or of metal (in this case, the electrical insulation is provided by a very thin sheet of mica). The thermoelectric elements surrounding both cells are connected differentially, the Petit microcalorimeter being thus a twin differential calorimeter.

The purpose of the particular arrangement of the heat sink in the Petit microcalorimeter is to ensure an excellent and reproducible contact, at any temperature, between the surface of the thermoelectric elements and the outside walls of the calorimetric cells (*31*) and, moreover, to avoid scratching the soft surface of the thermoelements through repeated use of the calorimeter. When calorimetric cells are to be introduced into the microcalorimeter, all arms (B) are rotated in such a way that the gap between the opposite fluxmeter holders (C) is widened. After the cells have been introduced, the arms are allowed to rotate back to their former position and springs (R) hold the thermoelectric elements tight against the outside walls of the cells. The reverse procedure is used to remove the cells from the calorimeter.

It is clear that in this microcalorimeter, only a fraction of the outside wall of the inner vessel is covered by thermoelectric elements. Consequently, only a part of the total heat flux emitted by the cell is detected. This may be the cause of a systematic error which, however, can be avoided if the heat transfer via the thermoelectric elements constitutes a constant fraction of the total, irrespective of the process taking place in the calorimeter cell. The present version of the Petit microcalorimeter can be used only at moderate temperatures ($<100°C$), mainly because some components of the thermoelectric elements would be damaged at higher temperatures.

C. The Petit–Eyraud Microcalorimeter

This instrument was originally proposed for quantitative differential thermal analysis (DTA) (*34*) and it has proved indeed to be very suitable

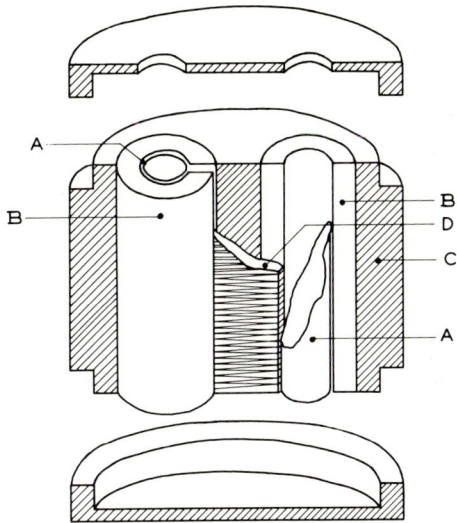

Fig. 6. The Petit–Eyraud calorimeter: calorimeter cells (A); cylinders made of insulating material (B); metal block (C); and plate of alumina supporting the thermoelectric pile (D). Reprinted from (*34*) with permission of Gauthier-Villars.

for this kind of investigation (*35*). The Petit–Eyraud calorimeter may also be used, however, at a constant temperature, particularly for studying gas–solid interactions (*36*). A schematic representation of an early model of this microcalorimeter is given in Fig. 6. Two thin-walled metal cylinders (A) are placed coaxially in two massive cylinders of a poor heat-conducting material (B). The cylinders (B) are wedged in two cylindrical sockets machined in a brass block (C). A parallelipipedic plate of alumina (D) supports a series of about 50 thermoelectric couples which, in equal number, are in thermal contact, through an electrical insulator, with each calorimetric cell (A). The process under investigation is carried out in one of the calorimeter cells, the other serving as a tare.

Several features of the early model (Fig. 6) have been modified in the present-day, high-temperature version of this calorimeter (Fig. 7) (*37*). Depending upon the temperature range envisaged, the block is made of refractory steel, alumina, or beryllium oxide and is machined to house the calorimeter itself. The thermoelectric pile (about 50 platinum to platinum–rhodium thermocouples) is affixed in the grooves of an alumina plate (A), which is permanently cemented to two cylindrical tubes of alumina (B). Cylindrical containers of platinum (C) ensure the uniformity of the temperature distribution within the calorimeter cells.

The Petit-Eyraud apparatus is a differential calorimeter but it is not a twin calorimeter. The reference cell serves also as a heat sink of limited heat capacity, since it collects, at least transiently, the heat flowing along

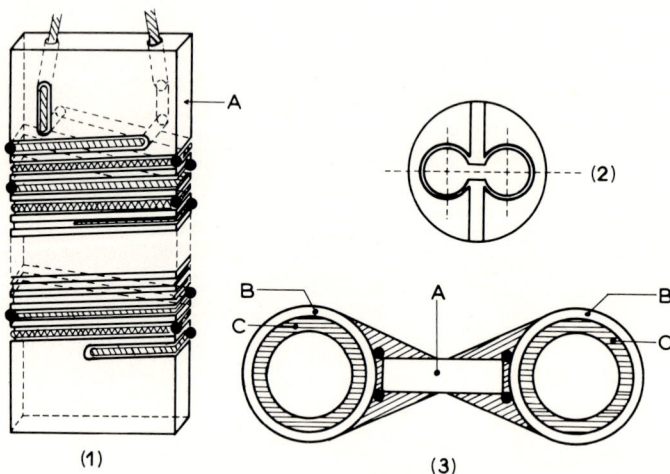

Fig. 7. The Eyraud calorimeter: the thermoelectric pile (1); the heat sink (2); horizontal section of the calorimeter (3); plate of alumina (A); calorimeter cells (B); and platinum cylinders (C). Reprinted from (*37*) with permission.

TABLE I

Intrinsic Sensitivities of Some Heat-Flow Microcalorimeters

Calorimeter	Ref.	Thermoelectric element	Temp. range (°C)	Intrinsic sensitivity[a] (μV/mW)
Calvet, standard model	(16)	Chromel to constantan couples	30–200	~60 (30°C)
Calvet, high-temperature version	(16)	Platinum to platinum–rhodium couples	30–800	~3 (500°C)
Petit	(31)	Semiconductors	30–100	~100 (30°C)
Eyraud	(37)	Platinum to platinum–rhodium couples	30–800	~0.04 (500°C)

[a] Sensitivities were determined at the temperatures in parentheses.

the thermocouple wires. In order not to modify appreciably the temperature of the reference cell and, thereby the temperature of the thermopile reference junctions, it is important to minimize the heat flux along the thermocouple wires. Since no Peltier cooling is possible in the usual versions of this calorimeter, heat evolutions in the active cell must be of a small intensity. This requirement limits ultimately the precision of the measurements. Moreover, the fraction of the total heat flow which is transferred via the thermopile is particularly small in this calorimeter. This feature may be the source of errors, as already pointed out in the preceding section. The Petit–Eyraud microcalorimeter is nevertheless a useful instrument which, because of its relatively simple and rugged design, may be constructed in many research laboratories.

The *intrinsic sensitivity* of a heat-flow calorimeter is defined as the value of the steady emf that is produced by the thermoelectric elements when a unit of thermal power is dissipated continuously in the active cell of the calorimeter (*38*). In the case of microcalorimeters, it is conveniently expressed in microvolts per milliwatt (μV/mW). This ratio, which is characteristic of the calorimeter itself, is particularly useful for comparison purposes. Typical values for the intrinsic sensitivity of the microcalorimeters that have been described in this section are collected in Table I, together with the temperature ranges in which these instruments may be utilized. The intrinsic sensitivity has, however, very little practical importance, since it yields no indication of the maximum amplification that may be applied to the emf generated by the thermoelements without developing excessive noise in the indicating device.

IV. Theory of Heat-Flow Calorimetry

No theory can possibly take into account the arrangement of a real heat-flow calorimeter in all its details. Theoretical models of heat-flow calorimeters, which are necessarily simplified versions of the actual instruments, will therefore be used in the following calculations. It must be remarked that because of the limitations of the theory, no absolute measurements can be made with a heat-flow calorimeter, nor with any calorimeter. It is possible, however, to compare successive measurements with precision. A calorimetric study necessarily involves the calibration of the calorimeter and, upon this operation, depends the accuracy of the whole series of measurements.

A. A Simplified Theory—The Tian Equation (16)

The relation between the emf of the thermoelectric pile and the heat flux from the calorimeter cell will be first established. Let us suppose (Fig. 8) that the process under investigation takes place in a calorimeter vessel (A), which is completely surrounded by n identical thermoelectric junctions, each separated from one another by equal intervals. The thermocouples are attached to the external surface of the calorimeter cell (A), which constitutes the internal boundary (E_{int}) of the pile and to the inside wall of the heat sink (B), constituting the external boundary (E_{ext}) of the thermoelectric pile. The heat sink (B) is maintained at a constant temperature (θ_e).

Moreover, let each junction occupy, at the internal boundary, an area S and let the average uncovered area surrounding each junction be S'. Let θ_j be the temperature of a given junction j at the internal boundary.

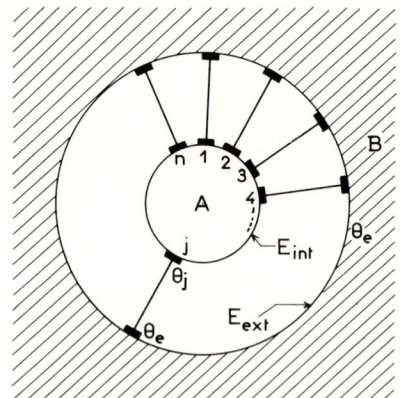

Fig. 8. Theoretical model of a heat-flow calorimeter.

Since the total area $S + S'$ is small, it is permissible to accept that θ_j measures the temperature of the junction and of its adjacent uncovered area. The temperature of any other thermoelectric junction and of its adjacent uncovered area may, however, be different from θ_j.

The fraction of the heat flux from an area $(S + S')_j$, conducted by the thermocouple wire, is:

$$\phi_j = \lambda \Phi_j = c(\theta_j - \theta_e), \tag{1}$$

where c is the thermal conductivity of the wire. The other fraction $(1 - \lambda)\Phi_j$ of the heat flux is transferred from the surface by other means (conduction, radiation, convection). It may be assumed that heat flows encountered in microcalorimetric work vary within a relatively small range and that λ is consequently constant and independant of Φ.

The total heat flux emitted by the area $(S + S')_j$ is therefore:

$$\Phi_j = (c/\lambda)(\theta_j - \theta_e). \tag{2}$$

For n identical thermoelectric junctions, the total heat flux is

$$\Phi = (c/\lambda) \sum_{j=1}^{n} (\theta_j - \theta_e). \tag{3}$$

However, the emf generated in a thermocouple is given by

$$e_j = \epsilon_0 (\theta_j - \theta_e), \tag{4}$$

where ϵ_0 is the thermoelectric power of the thermocouple. Since all the couples are connected in series, the total emf produced by n thermocouples is

$$E = \epsilon_0 \sum_{j=1}^{n} (\theta_j - \theta_e). \tag{5}$$

The combination of Eqs. (3) and (5) gives

$$E = (\epsilon_0 \lambda / c) \Phi, \tag{6}$$

so that the emf generated by the thermoelectric pile is proportional to the total heat flux from the internal boundary, irrespective of temperature gradients at the internal boundary or within the calorimeter cell.

It is possible to define such a temperature (θ_i) that if all the thermoelectric junctions at the internal boundary were maintained at this temperature, the total emf would be identical to the emf actually generated by the thermoelectric pile:

$$E = \epsilon_0 \sum_{j=1}^{n} (\theta_j - \theta_e) = n\epsilon_0 (\theta_i - \theta_e). \tag{7}$$

Similarly, Eq. (3) may be written

$$\Phi = p(\theta_i - \theta_e), \tag{8}$$

where p is the total heat power lost by the calorimetric vessel when the temperature difference between the internal and external boundaries is unity. Comparison of Eqs. (6)–(8) shows that

$$p = nc/\lambda. \tag{9}$$

The heat transfer coefficient p is an important characteristic of a heat flow calorimeter.

Now the thermal balance of the inner cell of a heat-flow calorimeter may be established. Let the thermal power developed in the cell, at time t, be called W.

Part of the thermal power is transferred to the heat sink, the corresponding flux being

$$\Phi = p(\theta_i - \theta_e) = p\theta, \tag{8}$$

if $\theta = (\theta_i - \theta_e)$.

Part of the thermal power remains, however, in the cell and increases its temperature. As a first approximation, let us assume that the temperature increase, during time dt, is uniform $(d\theta)$ throughout the cell which has a uniform heat capacity μ. The amount of thermal power which remains in the cell is therefore equal to $\mu \, (d\theta/dt)$, and finally,

$$W = p\theta + \mu \frac{d\theta}{dt}. \tag{10}$$

This fundamental equation of heat-flow calorimeters is called the *Tian equation*.

From Eq. (7), it is clear that the recorder indication Δ, a galvanometer deflection for instance, is proportional to θ,

$$\Delta = g\theta. \tag{11}$$

The proportionality constant g includes such parameters as the number of thermoelectric couples in the pile, their thermoelectric power, and the gain of the amplification device. It is supposed, moreover, that the response of the recording line is considerably faster than the thermal lag in the calorimeter. The Tian equation may also be written therefore:

$$W = \frac{p\Delta}{g} + \frac{\mu}{g} \frac{d\Delta}{dt}. \tag{12}$$

The record trace $[\Delta = f(t)$, usually called *thermogram*] obtained when the evolution of heat stops in the calorimeter cell ($W = 0$) is readily deduced from Eq. (12) by integrating

$$\frac{p}{g}\Delta + \frac{\mu}{g}\frac{d\Delta}{dt} = 0. \tag{13}$$

The exponential decrease of Δ follows the simple Newton's cooling law

$$\Delta = \Delta_0 \exp[-(p/\mu)t]. \tag{14}$$

The reciprocal value of the constant p/μ has a dimension of $[t]$ and is called the calorimeter *time constant* τ. This is another important characteristic of a heat-flow calorimeter:

$$\tau = \mu/p. \tag{15}$$

The value of the time constant depends upon the calorimeter itself p and upon the heat capacity of the calorimeter cell and of its contents μ. Typical, but necessarily approximate, values of the time constant for some heat-flow microcalorimeters are given in Table II.

By combining Eqs. (12) and (15), the Tian equation may be written:

$$W = \frac{p}{g}\left(\Delta + \tau \frac{d\Delta}{dt}\right). \tag{16}$$

The quantity of heat which is developed by the process under investigation is determined by integrating the Tian equation [Eq. (12), for instance],

$$Q_t^{t'} = \frac{p}{g}\int_t^{t'} \Delta\, dt + \frac{\mu}{g}\int_t^{t'} d\Delta. \tag{17}$$

When a Joule heating or a Peltier cooling are used to compensate part of the heat absorbed or liberated in the calorimeter cell, Eq. (17) gives

TABLE II

The Time Constant of Some Heat-Flow Microcalorimeters

Calorimeter	Time constant τ (sec)
Calvet (16)	~400
Petit (31)	~140
Eyraud (37)	~130

exclusively the noncompensated part. The total heat is given then by

$$Q_t^{t'} = \frac{p}{g}\int_t^{t'} \Delta\, dt + \frac{\mu}{g}\int_t^{t'} d\Delta + \int_t^{t'} W_{\text{Joule or Peltier}}\, dt. \qquad (18)$$

The integration limits t and t' may be so selected that the complete thermal effect under study is included within these limits. Then, the second integral in Eq. (17) cancels out because a stable zero reading is recorded before the initiation and after the completion of the experiment ($d\Delta = 0$). The total heat produced during the experiment is therefore given by

$$Q_{\text{total}} = \frac{p}{g}\int_{\text{experiment}} \Delta\, dt, \qquad (19)$$

i.e., by the *area under the thermogram*. The proportionality coefficient p/g is determined by calibration experiments (see Section VI.B).

It must be noted that the heat capacity of the calorimeter cell and of its contents μ, which appears in the second term of Tian's equation [Eq. (12)], disappears from the final expression giving the total heat [Eq. (19)]. This simply means that all the heat produced in the calorimeter cell must eventually be evacuated to the heat sink, whatever the heat capacity of the inner cell may be. Changes of the heat capacity of the inner cell or of its contents influence the *shape* of the thermogram but not the *area* limited by the thermogram. It is for this reason that heat-flow microcalorimeters, with a high sensitivity, are particularly convenient for investigating adsorption processes at the surface of poor heat-conducting solids similar in this respect to most industrial catalysts.

As already indicated, Tian's equation supposes (1) that the temperature of the external boundary of the thermoelectric element θ_e, and consequently of the heat sink, remains constant; and (2) that the temperature θ_i of the inner cell is uniform at all times. The first condition is reasonably well satisfied when the heat capacity of the heat sink is large and when the rate of the heat flux is small enough to avoid the accumulation of heat at the external boundary. The second condition, however, is physically impossible to satisfy since any heat evolution necessarily produces heat flows and temperature gradients. It is only in the case of slow thermal phenomena that the second condition underlying Tian's equation is approximately valid, i.e., that temperature gradients within the inner cell are low enough to be neglected. The evolution of many thermal phenomena is indeed slow with respect to the time constant of heat-flow calorimeters (Table II) and, in numerous cases, it has been shown that the Tian equation is valid (*16*).

This equation, however, cannot be used to analyze the thermograms produced by fast heat evolutions. The equations, which describe in the general case the thermal lag in a heat-flow calorimeter are presented in the next section.

B. General Equations for the Heat Transfer in a Heat Flow Calorimeter

The basic principle of heat-flow calorimetry is certainly to be found in the linear equations of Onsager which relate the temperature or potential gradients across the thermoelements to the resulting flux of heat or electricity (*16*). Experimental verifications have been made (*39–41*) and they have shown that the Calvet microcalorimeter, for instance, behaves, within 0.2%, as a linear system at 25°C (*41*). A heat-flow calorimeter may be therefore considered as a *transducer* which produces the linear transformation of any function of time [$f(t)$, the input, i.e., the thermal phenomenon under investigation] into another function of time [$g(t)$, the response, i.e., the thermogram]. The problem is evidently to define the corresponding linear operator.

Now, if the response of a linear system to a unit impulse $h(t)$ is known, the system is fully characterized, i.e., the response $g(t)$ to an arbitrary input $f(t)$ can be found in terms of $h(t)$ [see, e.g. (*42*)]. This can be expressed either by the convolution

$$g(t) = h(t) * f(t), \qquad (20)$$

or, if the unit impulse is compatible with Laplace transforms, by the product

$$G(p) = H(p) \cdot F(p), \qquad (21)$$

where $G(p)$ and $F(p)$ are respectively the Laplace transforms of $g(t)$ and $f(t)$ and where

$$H(p) = \int_0^\infty e^{-pt} h(t)\, dt. \qquad (22)$$

The calorimeter response to a unit impulse must therefore be determined. This may be achieved by solving the Fourier equation [Eq. (23)] for a theoretical model of a heat-flow calorimeter and for this particular heat evolution.

$$\Delta\theta = \frac{1}{h} \frac{\partial\theta}{\partial t}, \qquad (23)$$

where θ is the temperature and h the thermal diffusivity.

The boundary conditions for the differential equation [Eq. (23)] depend upon the theoretical model of a heat-flow calorimeter which is adopted in the calculations.

Laville (*43*) has supposed that the calorimeter is composed of a heat-conducting body (the internal boundary in Fig. 8) which receives, on a fraction (S_1) of its surface at temperature θ_1, a heat flux $\phi(t)$ generated within the calorimeter cell. Another fraction of its surface S_2, at temperature θ_2, emits a heat flux which diffuses towards the heat sink at temperature θ_3.

The corresponding boundary conditions are

$$K \frac{d\theta_1}{dn_1} = \phi(t), \qquad K \frac{d\theta_2}{dn_2} + p(\theta_2 - \theta_3) = 0, \qquad (24)$$

where K is the thermal conductivity of the heat-conducting body, n_1 and n_2 are any space variables, p is the heat transfer coefficient, and $\phi(t)$ is, in the calculations, a unit flux.

In the calculations proposed by Camia (*44*), a heat pulse is produced within the calorimeter cell, which is initially in thermal equilibrium. The heat pulse diffuses through the heat-conducting body toward the heat sink which is maintained at a constant temperature θ_3.

In this case, the boundary conditions are

$$d\theta_1/dn_1 = 0, \qquad \theta_3 = \text{constant}. \qquad (25)$$

In both calculations, the boundary conditions are linear with respect to θ and its first-order derivatives. The solution of the Fourier equation, with respect to the space variables, may be developed in a series of orthogonal functions, which are exponential with respect to the time variable [for the solution of similar problems, see (*45*)]. The time-dependance of the temperature distribution along a single space variable r, resulting from a unit pulse, is therefore given by

$$\theta(r, t) = \sum_{i=1}^{\infty} \theta_i(r) \exp(-t/\tau_i). \qquad (26)$$

The dimension of the coefficients τ_i is $[t]$. They are characteristic of the calorimeter.

The calorimeter response (the emf–time curve or the thermogram) is, of course, proportional at any time to the temperature difference which exists between two definite values of the space variable r_1 and r_2 where the active and reference junctions of the thermoelement are located:

$$h(t) = \theta(r, t)_{r_1} - \theta(r, t)_{r_2}. \qquad (27)$$

The impulse response of the calorimeter is therefore given by

$$h(t) = \sum_{i=1}^{\infty} a_i \exp(-t/\tau_i), \tag{28}$$

where the coefficients a_i are also characteristic of the calorimeter itself.

The combination of Eqs. (28) and (22) gives the Laplace transform of the impulse response $H(p)$ which allows us to solve Eq. (21). By the inverse transformation, the relation which gives the output of the linear system $g(t)$ (the thermogram) to any input $f(t)$ (the thermal phenomenon under investigation) is obtained. This general equation for the heat transfer in a heat-flow calorimeter may be written (40, 46):

$$f(t) = \frac{1}{\sum_i a_i \tau_i} \left[g(t) + \sum_i \tau_i \frac{dg(t)}{dt} + \sum_{i \neq j} \tau_i \tau_j \frac{d^2 g(t)}{dt^2} + \cdots \right.$$

$$\left. + \sum_{i \neq j \neq k} \tau_i, \tau_j, \ldots, \tau_k \frac{d^k g(t)}{dt^k} + \tau_1, \tau_2, \ldots, \tau_k, \ldots, \tau_n \frac{d^n g(t)}{dt^n} \right], \tag{29}$$

where $i = 1, 2, \ldots, n$ and $i < j < \cdots k \cdots < n$.

In the particular case where $i = 1$, the general equation (29) becomes

$$f(t) = \frac{1}{a_1 \tau_1} \left[g(t) + \tau_1 \frac{dg(t)}{dt} \right]. \tag{30}$$

This simplified equation is equivalent to Tian's equation [Eq. (16)], and it appears that τ_1 is indeed the time constant τ of the calorimeter. Thence, the successive coefficients τ_i in Eq. (29) may be called the calorimeter time constants of 1st, 2nd, ..., ith order. When the Tian equation applies correctly, all time constants τ_i except the first τ may be neglected. Since the value of the coefficients τ_i of successive order decreases sharply [the following values, for instance, have been reported (40): $\tau_1 = 144$ sec, $\tau_2 = 38.5$ sec, $\tau_3 = 8.6$ sec, $\tau_4 \simeq 1$ sec], this approximation is often valid, and the linear transformation of many thermal phenomena produced by the thermal lag in the calorimeter may actually be represented correctly by Eqs. (16) or (30). It has already been shown (Section IV.A) that the total heat produced in the calorimeter cell is then proportional to the area limited by the thermogram.

When the evolution of the thermal phenomenon is fast, a rather large number of coefficients τ_i must be considered in Eq. (29) in order to define correctly the linear operator. The total amount of heat produced is still easily determined since it can be shown that proportionality of the areas under the input and output functions is a general property of linear systems, *whatever the input function may be (47)*. However, the calorimetric system may deviate from linearity, and thus the simple proportionality between the area under the thermogram and the total amount of heat may no longer be valid, in the case of very intense thermal effects or of pulselike phenomena, the frequency of which exceeds the cutoff frequency characteristic of the calorimeter. It has been shown experimentally, for instance, that the quantity of heat evolved in a very short time is proportional to the maximum ordinate of the sharp peak which is then recorded and not to its area *(16)*. This result shows that in the case of very fast exothermic phenomena, the maximum temperature increase of the calorimeter vessel is a measure of the quantity of heat evolved. Consequently, measurements in a heat-flow calorimeter may be made not only under quasi-isothermal conditions (as they are very often) but also under quasi-adiabatic conditions. Before assuming the proportionality between the area under the thermogram and the total heat evolved, it is therefore important to carefully determine the range of linearity of the calorimeter response. This may be done by application of the simple laws which govern all linear systems. If the thermal inputs $f_1(t)$ and $f_2(t)$ produce, respectively, the thermograms $g_1(t)$ and $g_2(t)$, a thermal input equal to $f_1(t) + f_2(t)$ must produce a thermogram which is the sum of $g_1(t)$ and $g_2(t)$; likewise, $nf_1(t)$ must yield $ng_1(t)$. The thermal inputs $f(t)$ are conveniently generated by means of Joule heatings *(39)*.

In the range of linearity, Eq. (29) correctly represents the heat transfer within the calorimeter. It should be possible, then, by means of this equation to achieve the deconvolution of the thermogram, i.e., knowing $g(t)$ (the thermogram) and the parameters in Eq. (29), to define $f(t)$ (the input). This is evidently the final objective of the analysis of the calorimeter data, since the determination of the input $f(t)$ not only yields the total amount of heat produced, but also defines completely the kinetics of the thermal phenomenon under investigation.

V. Analysis of the Calorimetric Data

The differential equations [Eqs. (10) and (29)], which represent the heat transfer in a heat-flow calorimeter, indicate explicitly that the data obtained with calorimeters of this type are related to the kinetics of the thermal phenomenon under investigation. A thermogram is the representation, as a function of time, of the heat evolution in the calorimeter cell, but this representation is distorted by the thermal inertia of the calorimeter *(48)*. It could be concluded from this observation that in order to improve heat-flow calorimeters, one should construct instruments, with a small thermal lag and, consequently, short time constants. Equation (15) shows, however, that in order to decrease the time constant, the heat-transfer coefficient p should be increased. This would evidently be detrimental to the sensitivity of the instrument. In Tables I and II, comparison of the data for Calvet and Petit calorimeters shows, for instance, that the use of semiconducting thermoelements with a high factor of merit, in the Petit calorimeter, does not increase the intrinsic sensitivity to a large extent (Table I), mainly because the corresponding heat-transfer coefficient is larger than the transfer coefficient in the Calvet calorimeter, as indeed the respective time constants show (Table II). Two kinds of heat-flow calorimeters could therefore be defined (16): (1) "integrator" instruments with a high sensitivity but a rather large thermal inertia, and, conversely, (2) "ballistic" instruments with a rather poor sensitivity but a small thermal inertia. In microcalorimetric work where a high sensitivity is often of prime importance, calorimeters with a rather large thermal lag must be used, though the continuing progress in the construction of amplification devices progressively makes this requirement less rigid. Equation (15) shows, moreover, that the time constant of a given calorimeter may be varied by changing the heat capacity of the calorimeter cell contents: kinetics of the heat evolution are less distorted if a smaller amount of material is placed in the inner cell.

In any case, thermal inertia may be decreased but never completely removed (48). Data from heat-flow calorimeters must therefore be recorded in such a way that (1) the total amount of heat produced by the phenomenon under investigation is measured with precision, and (2) the correction of the thermograms is easily performed.

A. Methods of Recording Calorimetric Data

The equations describing the heat transfer in a heat-flow calorimeter [Eqs. (10) and (29)] have been established with the help of theoretical models. In a real calorimeter, the response of the instrument (the thermogram) corresponds to a fraction only of the total heat produced in the calorimeter cell, because thermal leaks, through conduction, convection, or radiation, are unavoidable. It is therefore essential for meaningful results that heat transfer via the thermoelements constitutes a *constant fraction* of the total heat, irrespective of the process taking place in the calorimeter cell. In particular, this fraction must remain constant during the whole course of the experiment: it should not vary with the modification of the thermal power evolved in the cell nor with the progress of the thermal phenomenon in the material placed in the cell. During microcalorimetric work, the first condition is nearly always satisfied, as the linear plot of the intrinsic sensitivity of the calorimeter versus the thermal power dissipated in the cell indeed shows (*37*, *41*, *49*). The second condition is more difficult to meet, especially in calorimeters of Petit–Eyraud type, i.e., when the thermoelements cover but a small fraction of the inner cell surface (Fig. 6). The heat flux emitted by the calorimeter cell maintains, however, throughout the experiment, a radial symmetry when the reactants located near the axis of the cell are surrounded by a thick sheath made of a material with a good thermal conductivity, such as a metal, which ensures, at all times, the uniformity of the radial temperature distribution within the calorimeter cell. The fraction of the total heat which is transferred via the thermoelements must also remain the same during calibration and actual experiments. The arrangement of the calorimeter cell must therefore be very similar in both cases.

Although most heat-flow calorimeters are multipurpose instruments, it is clear that for each particular type of experiment, the inner calorimeter cell must be especially designed and carefully tested. The reliability of the calorimetric data and, thence, the precision of the results depend, to a large extent, upon the arrangement of the inner cell. Typical arrangements for adsorption studies are described in the next section (Section VI.A).

Now, the precision of the calorimetric results depends also upon the nature of the amplifier and indicating devices on the recording line [for a general review, see (*50*, *51*)]. Analog recorders are often used, since they allow, at any time, an easy check of the calorimeter operation. A cheap, but efficient, recording line is composed, for instance, of a galvanometer and of a spot-following recorder. The galvanometer has indeed much to recommend it in terms of sensitivity and noise-to-signal ratio. Moreover, one of the advantages of using an electro-optical system is certainly that it cuts off any noise of electronic origin, issued from the recording units, from being eventually fed back to the calorimeter thermoelements. Special care must be given, however, to the galvanometer installation, to protect it from vibrations, and to the electrical and thermal insulation of the leads connecting the calorimeter and the galvanometer (*41*).

The main drawback of the galvanometer–spot follower system is that the sensitivity level of the line is defined ab initio and cannot be modified except by increasing or decreasing the number of thermoelements in the circuit. A more versatile, but more expensive, system is provided by the addition of a dc amplifier to a recording voltmeter.

As already indicated (Section IV.A), the quantity of heat evolved in the calorimeter cell is measured, in the case of usual heat evolutions, by the area limited by the thermogram. The integration of the calorimetric curves is, therefore, often needed. This may be achieved by means of integrating devices which may be added to the recorder. From our experience, however, many instruments designed for the integration of gas chromatographic peaks do not present the high level of stability which is needed in microcalorimetric work, especially in the case of prolonged experiments (several weeks). Manual integration by means of high precision planimeters is then a tedious, but necessary, operation.

A very efficient but more expensive method is to integrate the thermograms by means of a computer. This operation, which involves of course the digitization of the data, is to be recommended especially when not only the total heat but also the time-dependence of the heat evolution are to be deduced from the calorimetric data. It must be noted that digital methods are inherently more accurate than analog methods, as indeed a comparison of both methods shows (40). In many cases, however, analysis of the data is made, off-line, by means of a time-shared computer service. Therefore, the digital information must be stored, on punched paper tape, for instance. In these cases, we believe that a simultaneous analog record of the experiment is still needed, in order to control, at any time, the progress of the experiments which are often very long.

Figure 9 gives a diagrammatic representation of a system which may be used to record, simultaneously, analog and digital data from two calorimeters (52). The calorimeter signals are first amplified and recorded, in an analog form, either by a dc amplifier and an analog voltmeter or by a galvanometer followed by a spot-recorder. Digitization is then achieved

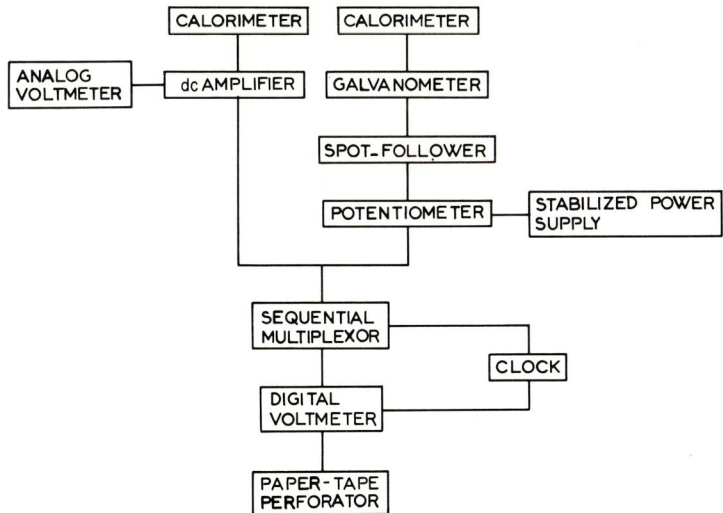

Fig. 9. Diagram of system designed to record and store simultaneously the data from two calorimeters (52).

by means of a digital voltmeter (or any analog-to-digital converter). The linear potentiometer (in Fig. 9) delivers, at any time, an emf which is proportional to the ordinate of the analog records. The sequential multiplexor and the digital voltmeter are used to sample the information from the different calorimeters and to feed it, in a predetermined sequence, to the paper-tape perforator. Samples may be stored at a rate of 6/sec and, at least, four separate recording lines (or calorimeters) may be connected to a single sequential multiplexor–digital voltmeter system before the maximum capacity of the tape punch is reached.

Although the handling of data by the computer is rapidly gaining in popularity, it should not be understood that recording lines such as the system described on Fig. 9 are necessary for all experiments with a heat-flow calorimeter. When the evolution of heat in the calorimeter cell is slow and when, consequently, Tian's equation [Eq. (12)] applies correctly, the analysis of the data certainly does not necessitate the use of a computer. Most of our previous studies (19) have indeed been conducted without the help of these very convenient, but expensive, machines. Another advantage of the system presented on Fig. 9 should, therefore, be pointed out: the digital part of any recording line in the system may evidently be disconnected when digital data storage is not necessary. But, it should be noted that this operation does not perturb the analog part of the recording line and, therefore, the calorimetric experiments may proceed without the interruption caused, for instance, by recalibration tests.

B. Data Correction and Determination of Thermokinetics

The development of the theory of heat-flow calorimetry (Section VI) has demonstrated that the response of a calorimeter of this type is, because of the thermal inertia of the instrument, a distorted representation of the time-dependence of the evolution of heat produced, in the calorimeter cell, by the phenomenon under investigation. This is evidently the basic feature of heat-flow calorimetry. It is therefore particularly important to profit from this characteristic and to correct the calorimetric data in order to gain information on the *thermokinetics* of the process taking place in a heat-flow calorimeter.

Now, the simple Tian equation

$$f(t) = \frac{1}{a_1 \tau_1} \left[g(t) + \tau_1 \frac{dg(t)}{dt} \right] \tag{30}$$

shows that when the heat evolution is very slow, i.e., when $\tau_1 [dg(t)/dt]$ is very small compared to $g(t)$, the second term in Eq. (30) may be neglected: the thermogram $g(t)$ is then directly proportional to the thermal input $f(t)$, the *"thermogenesis,"* after Calvet (*16*). This approximation is indeed acceptable in many experiments and is very useful to compare, at least qualitatively, the rates of different thermal phenomena (*53*, *54*), or to detect, from the profile of the thermogram, the occurrence of distinct, but quasi-simultaneous, heat evolutions (*55*).

However, for a quantitative kinetic analysis of the calorimetric data, corrections must be applied, because the thermal inertia of the calorimeter, which cannot be completely removed, causes in many cases the "blurring" of the thermogenesis. Several methods have been proposed, with different levels of sophistication.

1. *Manual Correction (56)*

From Tian's equation [Eq. (30)], it appears that in order to transform the calorimeter response $g(t)$ into a curve proportional to the thermal input $f(t)$, it is sufficient to add, algebraically, to the ordinate of each point on the thermogram $g(t)$, a correction term which is the product of the calorimeter time constant τ_1, by the slope of the tangent to the thermogram at this particular point. This may be achieved manually by the geometrical construction presented on Fig. 10.

Let us consider a point M on a thermogram C, and the tangent MT to the calorimetric curve C, at this particular point. If the segment MH, parallel to the time axis, is equal to the first time constant τ_1 of the calo-

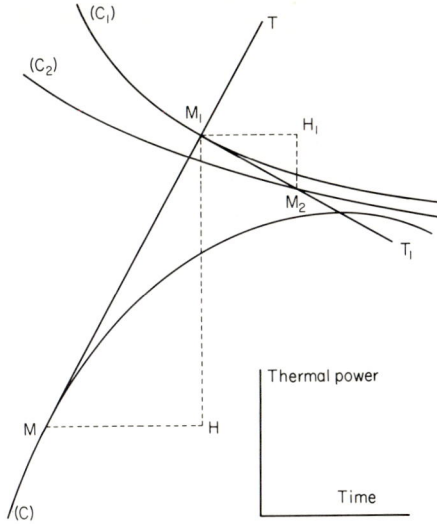

FIG. 10. Manual correction of a thermogram (see text). Reprinted from (*57*) with permission.

rimeter, the correction term $\tau_1[dg(t)/dt]$, is clearly given by HM_1. This operation may be repeated for any point on the thermogram C, and it yields the corrected curve C_1, which is a representation of the thermal input $f(t)$, when the Tian equation applies correctly.

The value of the calorimeter time constant τ ($=\tau_1$), may be determined from the cooling curve which is recorded, for instance, when a Joule heating, which produced a constant deviation Δ_0 of the base line (Fig. 11), is suddenly stopped (16). The comparison of Eqs. (14) and (15) shows that the cooling curve is represented by

$$\Delta = \Delta_0 e^{-t/\tau} \quad \text{or} \quad \ln \Delta/\Delta_0 = -t/\tau. \tag{31}$$

The time $t_{1/2}$, which corresponds to the half-maximum deviation $\Delta = \Delta_0/2$, is given by

$$t_{1/2}/\tau = \ln 2 = 0.69 \quad \text{or} \quad \tau = t_{1/2}/0.69. \tag{32}$$

The time constant τ is therefore very simply related to $t_{1/2}$. Moreover, it can be shown also (13) that the time constant represents the time necessary for attaining thermal equilibrium—and a stable base line—if the calorimeter cooling was linear and followed the tangent in the beginning of the cooling (Fig. 11).

It should be recalled, at this point, that the value of the time constant is related to the heat capacity of the calorimeter cell and of its contents [Eq. (15)]. For meaningful results, it is therefore essential that the arrangement of the inner cell remain identical for both the Joule heating and the thermal phenomenon under investigation. Strictly speaking, the time constant would be unchanged if it were possible to keep the thermal paths completely identical in both cases. This condition is, of course, very difficult to meet.

Finally, it has been assumed that the heat transfer in the calorimeter

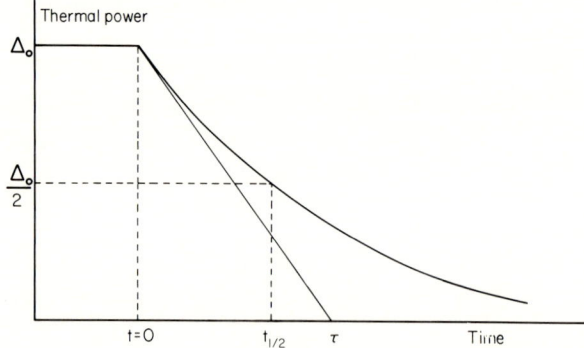

FIG. 11. Thermogram recorded after switching off a constant-power Joule heating.

is correctly represented by Tian's equation [Eq. (30)] and that, consequently, the cooling curve (Fig. 11) decreases exponentially, according to Newton's law [Eq. (31)]. As already discussed (Section IV), this approximation is valid when the evolution of heat is slow with respect to the time constant. In the case of rapid heat evolutions, a more complete equation for the heat transfer must be utilized [Eq. (29)]. This equation may also be conveniently represented by the following system of linear equations (40, 46):

$$g_1(t) = g(t) + \tau_1 \frac{dg(t)}{dt},$$

$$g_2(t) = g_1(t) + \tau_2 \frac{dg_1(t)}{dt}, \qquad (33)$$

$$\vdots$$

$$g_n(t) = g_{n-1}(t) + \tau_n \frac{dg_{n-1}(t)}{dt},$$

where $f(t) = \lim_{n \to \infty}[g_n(t)]$. From this set of equations, it appears that the corrected curve C_1 in Fig. 10, which is proportional to the thermal input $f(t)$, in the case of a slow phenomenon [Eq. (30)], is related only to $g_1(t)$ in the case of a rapid heat evolution. A closer approximation of $f(t)$ is obtained therefore if the manual correction is repeated with the use of the time constants of successive orders ($\tau_1, \tau_2, \ldots$). Two such corrections are presented in Fig. 10 (curves C_1 and C_2). The time constants of increasing order may be determined successively, according to Eq. (26), by the analysis, on a semilogarithmic plot, of the response of the calorimeter to a unit impulse (40). Thermal paths, within the calorimeter cell, should be identical, of course, for the unit impulse and for the thermal input under study. The detailed analysis of the impulse response can yield the values of three or four time constants of increasing order (τ_1–τ_3 or τ_4) (40, 57). However, owing to practical difficulties, only two successive manual corrections can be made. In many cases, this small number of corrections is sufficient and a correct representation of the thermokinetics of the phenomenon is obtained (57).

The manual method of data correction, which has been described, though efficient and theoretically sound, is however very tedious and requires skilled experimenters. For this reason, various other methods have been proposed to correct automatically the calorimetric data. In some methods ("on-line" correction), raw data may be processed immediately, i.e., during the course of the experiment itself. In other methods ("off-line" correction), it is more convenient to store raw data, in a digital form, in order to correct it by means of a computer.

2. "On-Line" Correction

The successive corrections necessary to transform the thermogram $g(t)$ into $f(t)$, according to Eqs. (33), may be done by adding to the amplification and recording line a suitable analog RC network (40, 46, 56). The electrical circuit, presented on Fig. 12, transforms any voltage function $g_{k-1}(t)$ (the input) into a new function of time $g_k(t)$ (the output) which is related to $g_{k-1}(t)$ by

$$g_k(t) = \frac{r_k}{R_k}\left[g_{k-1}(t) + R_k C_k \frac{dg_{k-1}(t)}{dt}\right], \qquad (34)$$

provided that r_k is small compared to R_k and large compared to the impedance of the input circuit (58). If the respective values of R_k and C_k, the resistance and capacitance of the parallel RC circuit, are so chosen that their product is equal to τ_k, the time constant of the kth order of the calorimeter, this circuit may achieve the kth linear transformation of the calorimeter response. The complete correction of the thermogram is automatically obtained if n similar circuits, in which the respective $R \cdot C$ products are equal to the time constants of successive order, are connected in series. Such a network which allows four successive corrections of the thermograms is presented in Fig. 13 (40). The main drawback of this system is that the intensity of the output decreases as the number of successive corrections increases, particularly when the time constants are large, and that, consequently, the noise of electronic origin becomes progressively more important. High quality amplifiers are therefore needed. It must be noted also that the values of all the components in the successive parallel RC circuits are determined for a particular arrangement of the calorimeter cell. Any modification of the experimental conditions implies therefore a new determination of the time constants and the readjustment of all RC circuits. However, this method of data correction has found some practical applications (57, 59).

It should not, of course, be necessary to correct the thermograms if the

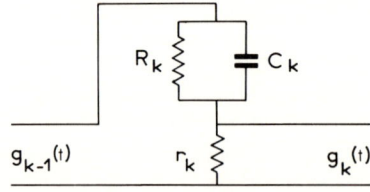

Fig. 12. Parallel RC circuit for the automatic correction of thermograms. Reprinted from (40) with permission.

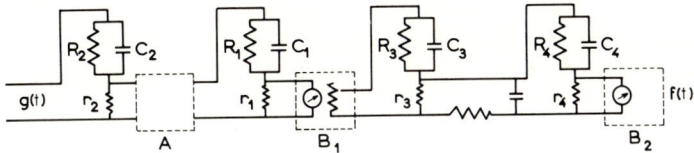

FIG. 13. RC network for four successive corrections of the thermograms: resistances (R and r); capacitances (C); amplifier (A); and recorders (B). Reprinted from (*40*) with permission.

true kinetics of the phenomenon under study were not distorted by the thermal inertia of the calorimeter. A very promising method, which indeed derives from the original Tian's method (*25*), has been proposed recently (*60, 61*) whereby the thermal inertia of a heat-flow calorimeter may be decreased considerably without decreasing simultaneously its sensitivity. This method is based on the *exact* compensation by a Peltier cooling of the heat evolved in the calorimeter cell. It is, therefore, completely different from the method of *approximate* Peltier compensation, presented by Calvet et al. (*16*), to avoid excessive transient heating of the reference junctions of the thermoelements. The emf generated by part of the thermoelectric elements, surrounding the calorimeter cell, is amplified and used to regulate the current flowing in the Peltier junctions (the second part of the thermoelectric elements) in contact with the calorimeter cell. The electrical circuits are adjusted in such a way that, at any time, the heat flow via the thermoelements is negligible. The Peltier current which is then recorded is proportional to all thermal perturbations at the internal boundary of the calorimeter, and, therefore, its time-dependence is a correct representation of the kinetics of the heat evolution within the calorimeter cell. Moreover, it should be remarked that, because of this arrangement, the calorimeter becomes, simultaneously, adiabatic (no heat transfer between the internal and external boundaries) and isothermal (all heat evolved is compensated). Electrical energy is, however, transferred via the thermoelements. Since the main sources of thermal inertia are located within the external and internal boundaries, the exact Peltier compensation considerably decreases the time constants of the calorimeter response. Actually, all inertia would be removed if it were possible to compensate, by a Peltier cooling, all heat at the exact location where it is evolved. This is, of course, impossible, and the residual thermal lag finds its origin in heat transfers within the calorimeter cell itself.

The efficiency of this method has been demonstrated for several types of heat-flow calorimeters. The rather long time constant of a Calvet-type calorimeter (200 sec), for instance. is decreased to $\simeq 10$ sec, when exact Peltier cooling is used (*61*). Similarly, the time constant of calorimeters equipped with semiconducting thermoelements (230, 75, and 12 sec) are decreased to 58, 10, and 0.7 sec, respectively (*60*). The heat waves generated by a cyclic phenomenon (frequency, 1 Hz) are, for instance, correctly separated in a calorimeter with a reduced time constant (0.7 sec), whereas they are "blurred" when no Peltier compensation is used (*60*). This method has already found some practical applications in differential thermal analysis (*33*). However, it is not possible to compensate, by a Peltier cooling, the thermal leaks which necessarily occur. It is therefore particularly important for a correct application of this method that the fraction of the total heat which is transferred via the thermoelements, or compensated when Peltier cooling is produced, remain constant throughout the calibration and actual experiments. Finally, it should be remarked that exact Peltier compensation produces the decrease of thermal inertia, but it is not actually a method for data correction. For this reason, the analysis of the calorimeter response to a unit impulse according to Eq. (26) is not then needed, and one of the difficulties of most data-correcting methods is therefore avoided.

3. *"Off-Line" Correction*

The large amount of calorimetric data, which can be conveniently stored in a digital form, may, of course, be used in a computer to solve the general equation for the heat transfer in a heat-flow calorimeter (Section IV.B):

$$g(t) = f(t) * h(t). \tag{20}$$

Equation (20) may be written explicitly (*42*) as

$$g(t) = \int_0^t f(\tau) \cdot h(t - \tau) \, d\tau. \tag{35}$$

The problem is apparently simple and may be expressed in the following way: knowing $g(t)$, the thermogram, and $h(t)$, the calorimeter response to a unit impulse, solve Eqs. (20) or (35) and determine $f(t)$, the thermokinetics of the phenomenon taking place in the calorimeter. However, the digital information which is used in the computer does not allow the continuous integration of Eq. (35). Both functions $g(t)$ and $h(t)$ are indeed stored and manipulated as series of discrete steps (samples). For a computer's convenience, Eq. (35) must therefore be written

$$g(k) = \sum_{i=0}^{k} f(i) \cdot h(k - i), \tag{36}$$

where $f(i) = (f(t))_{t=iT}$, T being the sampling period. This linear system

[Eq. (36)] may also be presented in the equivalent matrix form:

$$\begin{bmatrix} g(0) \\ g(1) \\ \vdots \\ g(k) \\ \vdots \end{bmatrix} = \begin{bmatrix} h(0) & 0 & \cdots & 0 & \cdots \\ h(1) & h(0) & & \vdots & \\ \vdots & \vdots & & \vdots & \\ h(k) & h(k-1) & \cdots & h(0) & \cdots \\ \vdots & \vdots & & \vdots & \end{bmatrix} \begin{bmatrix} f(0) \\ f(1) \\ \vdots \\ f(k) \\ \vdots \end{bmatrix}. \quad (37)$$

It seems, therefore, that the impulse response $h(t)$—or $h(0)$, $h(1)$, ..., $h(k)$, ...,—being known, the determination of $g(0)$, $g(1)$, ..., $g(k)$, ..., is sufficient to calculate $f(0), f(1), ..., f(k), ...,$ by means of the preceding system of linear equations [Eqs. (36) or (37)]. This is indeed correct mathematically. But, the value of both functions, $h(t)$ and $g(t)$, is very small for the first samples [$f(0)$, ..., and $g(0)$, ...,] and the relative error of their determination is large (62). Since all successive samples from 0 to k appear in the linear equation of kth order, as indeed Eq. (37) shows, errors cumulate and they produce the oscillation, with a progressively larger amplitude, of the function $f(t)$. The system represented by Eqs. (36) or (37) is unstable (63).

It has been shown recently, however, that these equations may be solved (62), by means of the state functions theory (64) and/or the time-domain matrix methods (63). Figure 14 shows, for instance, that the computer calculations allow us to determine, with a good approximation, the time-dependence of thermal phenomena taking place in the calorimeter, although all significant details of their kinetics are completely "blurred" on the thermogram (62). This method has been recently used to correct

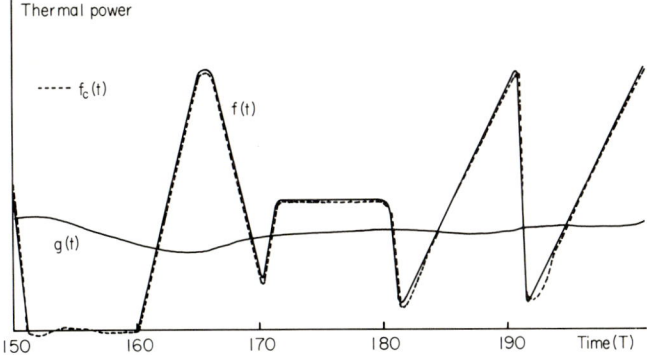

FIG. 14. Automatic correction of the thermograms by the state-function theory and the time-domain matrix methods: the input $f(t)$; the thermogram $g(t)$; the corrected thermogram $f_c(t)$; T is the sampling period (20 sec). Reprinted from (62) with permission.

the thermograms recorded during the calorimetric study of a catalyzed reaction (the decomposition of nitrous oxide on nickel oxide) (*52*).

Finally, the difficulty of most data-processing methods should not be overlooked. In all methods, but the exact Peltier compensation method, the deconvolution of the thermogram necessitates the determination of the calorimeter impulse response. This may be phrased differently by saying that the removal of the distortion caused, on the thermogram, by thermal inertia requires the measurement of the calorimeter time constants. This is, indeed, in many cases a formidable task (*65*), because during the calibration experiments, a defined thermal unit phenomenon (Dirac pulse or Heaviside step) must be produced by using the exact arrangement of the inner cell that is needed for the actual experiments. In both cases, thermal paths should be identical. Very often, such calibration experiments are very difficult to make with the required precision, and, therefore, detailed data processing should not be attempted.

VI. Measurement of the Differential Heats of Gas–Solid Interactions

The adsorption of dn_s moles of gas, at a constant temperature, is accompanied usually by the transfer of an amount of heat dQ to the surrounding constant temperature shield. The magnitude of dQ depends, of course, upon the conditions prevailing during the adsorption process, but, in all cases, differential heats of adsorption are defined by

$$q = dQ/dn_s. \tag{38}$$

The usual, "differential heat of adsorption," q_d is determined for an isothermal process during which the volumes V_G and V_S of, respectively, the gas and adsorbent phases and the area, A, of the adsorbent remain constant, the adsorbent and chamber being inert (*66*):

$$q_d = (dQ/dn_s)_{V_G, V_S, A}. \tag{39}$$

True differential heats of adsorption may be determined from equilibrium data when adsorption is thermodynamically reversible. However, when this process is not reversible, a calorimeter must be employed, and the so-called "differential heats," which are then measured, refer actually to the average heats evolved during the adsorption of small doses of gas:

$$(q_d)_{cal} = (\Delta Q/\Delta n_s)_{V_G, V_S, A}. \tag{40}$$

It is for this reason that stepped curves are convenient to represent accurately the evolution of the calorimetric heats of adsorption $(q_d)_{cal}$ expressed in kilocalories per mole, versus the quantity of adsorbed gas, expressed, for instance, by unit weight or by unit surface area of the adsorbent (see, for instance, Figs. 19 and 22).

The determination of these curves requires not only the measurement of small amounts of heat in a microcalorimeter, but also the simultaneous determination of the corresponding quantity of adsorbed gas. Volumetric measurements are to be preferred to gravimetric measurements for these determinations because it would be very difficult indeed to ensure a good, and reproducible, thermal contact between a sample of adsorbent, hanging from a balance beam, and the inner cell of a heat-flow calorimeter.

A. Description of the Associate Equipment. The Volumetric Line. The Calorimetric Cells

Figure 15 gives a diagrammatic representation of a volumetric line which is used in connection with a high-temperature Calvet microcalorimeter (67). Other volumetric lines which have been described present the same general features (15, 68). In the case of corrosive gases or vapors, metallic systems may be used (69). In all cases, a sampling system (A in Fig. 15) permits the introduction of a small quantity of gas (or vapor) in a calibrated part of the volumetric line (between stopcocks R_1 and R_0 in Fig. 15) where its pressure Pi is measured (by means of the McLeod gage B in Fig. 15). The gas is then allowed to contact the adsorbent placed in the calorimeter cell C (by opening stopcock R_0 in Fig. 15). The heat evolution is recorded and when it has come to completion, the final equi-

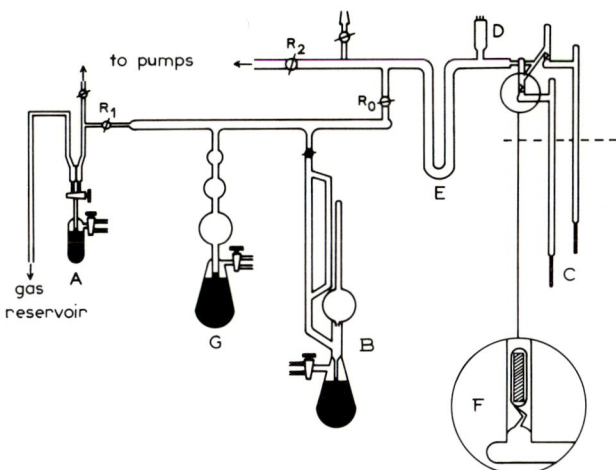

Fig. 15. Volumetric line used in connection with a Calvet microcalorimeter (67).

librium pressure of the gas, Pf, is measured. From the two pressure measurements Pi and Pf, the quantity of gas which has reacted with the adsorbent must be determined and it is compared to the amount of heat which has been simultaneously evolved. Stopcock R_0 is then closed again, another dose of gas is introduced, and the whole procedure described above is repeated. The same operations are successively carried out until saturation of the adsorbent is reached or until a predetermined final pressure is attained.

The volumetric line, presented on Fig. 15, is connected to a classical high vacuum system. Pressures in the lowest range which may be attained in the line ($\sim 10^{-6}$ Torr) are measured with the ionization gage D. In order to protect the sample from mercury or grease vapors, a cold trap E is placed between the adsorption cell C and the last stopcock or mercury reservoir. A break seal, F in Fig. 15, allows the connection of the sample, maintained under vacuum after its preparation or pretreatment in a separate apparatus, to the volumetric line without the use of greased joints. Such an arrangement is particularly necessary when the adsorbent, and thence the calorimeter, are maintained at a high temperature ($\sim 500°C$). Of course, nothing prevents the use of ultrahigh-vacuum systems in connection with heat-flow calorimeters, and we believe, indeed, that it would be desirable to repeat, with calorimeters of this type, some experiments concerning the determination of heats of adsorption on clean solid surfaces which have been made previously under pseudoadiabatic conditions.

The usual precautions concerning the volumetric determination of adsorbed amounts must evidently be taken [for a recent review on volumetric measurements, see (70)]. The measurement of the initial volume (NTP) of the dose, from Pi, is straightforward since the volume of the line, between R_0 and R_1, maintained at a constant temperature, may be calibrated by means of the gas buret, G in Fig. 15. The quantity of gas remaining in the volumetric line at the end of the adsorption of a given dose is less easily calculated, because different parts of the line (the trap, the adsorption cells) are maintained at different temperatures. However, it is possible to determine experimentally, during calibration experiments, an expansion coefficient k which is defined as the ratio of the pressures Pi and Pf measured before and after the opening of stopcock R_0 (Fig. 15), in the absence of adsorbent in the adsorption cell. This coefficient allows the calculation of the quantity of gas remaining in the whole volumetric line after the adsorption of a dose, provided that the temperature distribution along the line is the same during calibration and adsorption experiments and that the final pressures Pf are similar in both cases. The value of the expansion coefficient, however, does not change significantly when the

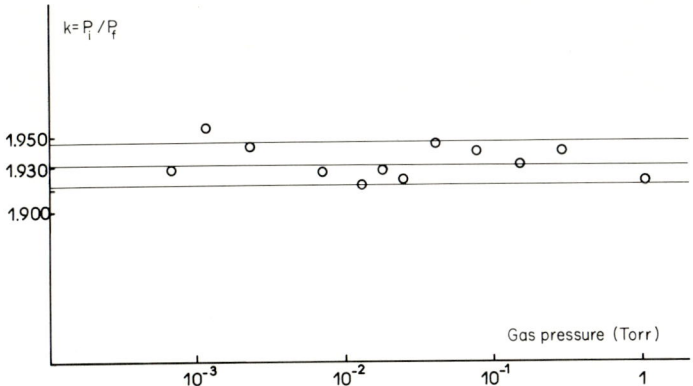

FIG. 16. Variation of the expansion coefficient k as a function of the gas pressure in the volumetric line (67).

pressure Pf in the line varies within the usual limits ($\sim 10^{-4}$–~ 2 Torr). Figure 16 shows, for instance, the pressure-dependence of the expansion coefficient k in the case of a volumetric line used in connection with a Calvet calorimeter heated at 450°C and a trap cooled at liquid nitrogen temperature, the gas being oxygen or carbon monoxide (67).

The importance of a careful construction and calibration of the volumetric line must be emphasized, because the relative error on the volumetric measurements is often larger than the relative error on the corresponding microcalorimetric measurements. Moreover, it is not exclusively for this reason that the temperature of the volumetric line must be kept constant (within 1 or 2°): small modifications of the temperature of the line and, thence, of the partial pressure of a gas in the line may indeed produce adsorption–desorption phenomena which are detected by a sensitive heat-flow microcalorimeter. An example of such phenomena is presented in Fig. 17. Thermograms A and B were recorded during the slow interaction, at 200°C, of doses of oxygen (~ 0.02 cm^3 NTP) with the surface of samples of nickel oxide, prepared by the dehydration, *in vacuo*, of nickel hydroxide at 200°C and containing, in both cases, similar amounts of preadsorbed oxygen (~ 4.5 cm^3 O$_2$/gm) (71). In the case of thermogram A, the trap (E in Fig. 15) in the volumetric line was cooled at liquid nitrogen temperature, whereas in the case of thermogram B, it was not cooled. Although the general profile of both thermograms is similar, fluctuations do appear in thermogram B which are not apparent in thermogram A. It was observed, moreover, that these fluctuations were synchronized with the cooling–heating cycles of the air-conditioning unit regulating the temperature in the laboratory (within 2°). Cooling the trap at liquid

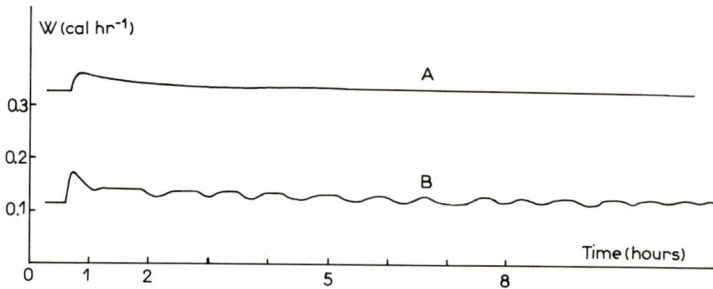

FIG. 17. Thermograms recorded during the adsorption of doses of oxygen at the surface of nickel-oxide samples containing preadsorbed oxygen, the cold trap being cooled (A) or not cooled (B) (*71*).

nitrogen—or dry ice—temperature immediately stopped, however, the fluctuations on the thermogram. A mass-spectroscopic analysis of the condensate in the cold trap has demonstrated that it was composed essentially of water. The fluctuations in thermogram B reveal, therefore, that water is desorbed continuously at 200°C from the incompletely dehydrated sample of nickel oxide. In the absence of a cold trap, a partial pressure of water vapor slowly builds up during the oxygen adsorption experiment, and the cyclic modifications of the temperature of the volumetric line result in periodic desorptions or adsorptions of water molecules. The heat produced by these successive phenomena is, of course, recorded and it appears as a periodic fluctuation superimposed on the thermogram produced by the main heat evolution, i.e., the oxygen adsorption.

The adsorption cell (C in Fig. 15) which contains the adsorbent must be placed in the inner cell of the calorimeter and a good thermal contact must be established between the sample and the sensing elements of the calorimeter. The mechanical contact between the volumetric line and the calorimeter occurs, therefore, in the calorimeter cell itself. Thence, any relative movement or vibrations between these parts of the apparatus must be strictly avoided. This necessitates the very careful installation of the whole apparatus, especially if experiments of long duration are to be made.

The adsorbent—a powder generally, but it could be a metal or oxide film—is placed in a glass tube (the adsorption cell C in Fig. 15) which is connected to the volumetric and vacuum lines. The bottom part of the tube, which contains the adsorbent and is located in the calorimeter cell, is made of thin-walled (0.2–0.3 mm) blown tubing (A in Fig. 18). In order to avoid the slow diffusion of gases through a thick layer of adsorbent (see Section VII.A), the sample is often placed in the annular space between the inner wall of the adsorption cell and the outer wall of a cylinder made of glass,

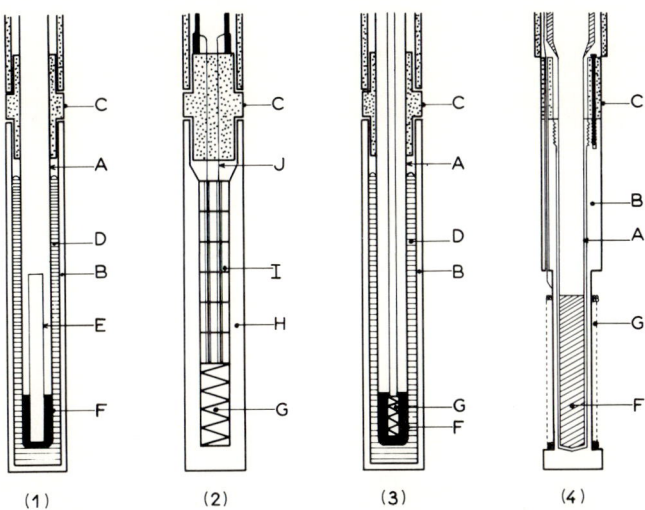

FIG. 18. Calorimeter cells equipped for adsorption studies at moderate (1) or high (4) temperatures: calibration cell (2); adsorption cell with an electrical heater (3); thin-walled glass tube (A); metallic container (B); stopple made of insulating material (C); mercury or gallium (D); cylinder made of glass or platinum gauze (E); sample of adsorbent (F); electrical calibration heater (G); stainless-steel cylinder (H); insulating beads (I); thin gold wires (J) (55, 67). Reprinted from (55) with permission.

or platinum, gauze (15, 55). All materials used for the construction of the adsorption cell must be selected not only for their convenience but also for their inactivity with respect to the reagents used in the experiments. The adsorption cell is immersed in a metallic (stainless-steel) container (B in Fig. 18), through a stopple made of a poor heat-conducting material (Teflon, asbestos, or alumina, C in Fig. 18). The thermal contact between the adsorption cell and the metal container is ensured by mercury, at room temperature, or gallium (or silicone oil), in the intermediate temperature range (up to ~200°C) (15, 55). At still higher temperatures (up to 500°C), no liquid thermal joint can be used, and the adsorption cell must be blown directly into the thick-walled metal container in order to push-fit in it [cell (4), Fig. 18] (67). The metallic container B, in its turn, must push-fit into the internal cavity of the calorimeter cell, the silver cylinder in contact, through a thin layer of mica, with the active thermoelectric junctions in the case of the Calvet microcalorimeter, for instance (see Section III.A). Several containers must therefore be accurately machined when the calorimeter is to be used in a wide temperature range. Because of the different thermal dilatation of the inner calorimeter cell and of the metallic container, the latter, though carefully adjusted at a particular temperature in the range, would, indeed, either move too freely or damage the calorimeter cell at other temperatures.

B. Calibration of the Calorimeter and Preliminary Experiments

In order to minimize the fluctuations of the emf of the thermoelements which arise from the imperfect temperature control in the thermostat, twin or differential systems are used in most heat-flow microcalorimeters (Section III). These systems have a maximum efficiency when the thermal responses, i.e., the time-constants, of both calorimetric elements are identical (*16*). The arrangements of both calorimetric cells must therefore be very similar. For this reason, two adsorption cells are symmetrically connected to the volumetric line, through identical break seals (Fig. 15). They are immersed in carefully matched metallic containers, filled, if necessary, with the same quantity of mercury or gallium, which fit similarly in the cavities of the twin elements. The adsorbent is placed, of course, in one adsorption cell, the second one being empty. The dissymmetry produced by the sample is, however, in most cases, unimportant because the weight (50–300 mg) and, thence, the heat capacity of the sample are small compared to the weight and the heat capacity of the whole calorimeter cell. The symmetry of the twin or differential system and the efficiency of the arrangement of the cells are easily controlled by recording the emf produced in the absence of heat evolution in the calorimeter cells. If the symmetry of the twin calorimetric elements were perfect, the emf delivered by the thermoelectric piles in opposition would be unmeasurably small. In most cases with Calvet microcalorimeters, the emf, though small, is however measurable, but it remains very stable during long periods of time (several weeks) when the arrangement in both calorimetric cells is identical (*28*).

All heat evolutions which occur simultaneously, in a similar manner, in both twin calorimetric elements connected differentially, are evidently not recorded. This particularity of twin or differential systems is particularly useful to eliminate, at least partially, from the thermograms, secondary thermal phenomena which would otherwise complicate the analysis of the calorimetric data. The introduction of a dose of gas into a single adsorption cell, containing no adsorbent, appears, for instance, on the calorimetric record as a sharp peak because it is not possible to preheat the gas at the exact temperature of the calorimeter. However, when the dose of gas is introduced simultaneously in both adsorption cells, containing no adsorbent, the corresponding calorimetric curve is considerably reduced. Its area (0.5–3 mm^2, at 200°C) is then much smaller than the area of most thermograms of adsorption ($\sim$300 mm^2), and no correction for the gas-temperature effect is usually needed (*55*).

The adsorption cell, connected to the volumetric line out of the calorimeter, constitutes a path for thermal leakage and, for instance, heat is transferred continuously from the calorimetric cell to the outside via the gas phase in the adsorption cell. This phenomenon is especially pronounced when the calorimeter is heated at a high temperature. The thermal balance which is attained in the absence of heat evolutions in the calorimetric elements and, thence, the position of the base line on the record depend upon the quantity of heat which is continuously removed from the cell. Changes of the gas pressure, which occur in the course of adsorption experiments, would produce, therefore, the modification of the thermal equilibrium in the calorimetric element and the deflection of the base line if these pressure changes occurred in a single adsorption cell. Such permanent deflections of the base line have, indeed, been observed, when the gas pressure is modified in a single cell containing no adsorbent. However, no permanent deflection of the base line is detected, even when the temperature in the calorimeter is as high as 500°C, when the pressure is modified simultaneously in both adsorption cells containing no adsorbent, provided that the pressure is maintained below 2 Torr (*55*, *67*).

Now, it is necessary to calibrate the calorimeter in order to analyze quantitatively the recorded thermograms and determine the amount of heat evolved by the interaction of a dose of gas with the adsorbent surface. The use of a standard substance or of a standard reaction is certainly the most simple and reliable method, though indirect, for calibrating a calorimeter, since it does not require any modification of the inner cell arrangement. [For a recent review on calibration procedures, see (*72*).] No standard adsorbent–adsorbate system has been defined, however, and the direct electrical calibration must therefore be used. It should be remarked, moreover, that the comparison of the experimental heat of a catalytic reaction with the known change of enthalpy associated with the reaction at the same temperature provides, in some favorable cases, a direct control of the electrical calibration (see Section VII.C).

The electrical calibration should require the accurate measurement of the quantity of electrical energy needed to reproduce the thermogram characteristic of the reaction or adsorption under study. Precise matching of the reaction and electrical calibration recordings is not necessary, however, when the calorimeter behaves as a linear system: the total heat produced is then, in all cases, proportional to the area under the thermogram (*47*). It is therefore particularly important to carefully determine the range of linearity of any given heat-flow calorimeter (*39*). When the calorimeter can be identified with a linear system, and especially when Tian's equation is applicable, as indeed it is in many adsorption experiments, the aim of the electrical calibration is then, simply, to determine the value of the calibration ratio which measures the amount of electrical energy that is needed to produce a steady unit deflection of the record baseline. This ratio is usually expressed in microwatts per millimeter.

The calibration cell used at moderate temperatures (30–200°C) is presented in Fig. 18, cell (2) (*55*). The heater (∼600 ohms), which is made of manganin wire wound on a small rod of alumina, is pushed into a cylindrical cavity bored into a cylindrical bar of stainless steel, in such a way that the heater in the calibration cell and the sample of adsorbent in the adsorption cell occupy, in both cases, the same location in the calorimeter cell. The steel cylinder push-fits into the calorimeter cell. The heater ends are brazed to thin gold wires (0.07 mm), electrically insulated by alumina beads, which pass through a thick layer of poor heat-conducting material (Teflon) on top of the steel cylinder, and are then affixed to large-section leads connected to the outside circuit. The purpose of this arrangement is to reduce thermal leakage along the large section leads.

It must be noted that although the calibration cell is very different from the adsorption cell [Fig. 18, cells (1) and (2)], the heat capacity of both cells is not very different, as the similar values of the time constant of the calorimeter containing one cell or the other indeed show (350 sec in the case of the calibration cell and 400 sec in the case of the adsorption cell) (*55*). This is explained by the fact that in both cases, the calorimeter cell is almost completely filled with a metal. However, the glass tube which is immersed in the calorimeter cell and the pressure changes which occur in the course of the adsorption experiments may be the sources of variable thermal leaks. The importance of these leaks was appreciated by means of the following control experiments.

A heater was introduced in the adsorption cell, containing a powder inert with respect to the gas phase [Fig. 18, cell (3)]. Calibration tests were carried out by means of this heater, different pressures having been successively maintained in the adsorption cell. The results of the calibration tests are presented in Table III, for the case of a Calvet microcalorimeter heated at 200°C equipped with a galvanometer and a spot-follower (*55*). It appears from these results that the calibration ratio does not vary significantly when the pressure of the gas in the adsorption cell changes from ∼10^{-5} to 2 Torr. Moreover, the heater in the calibration cell and the heater in the adsorption cell yield similar values of the calibration ratio. Thermal leaks along the glass tube immersed in the calorimeter cell may therefore be neglected, and the value of the calibration ratio determined by these tests may be used to analyze the thermograms recorded in the course of adsorption experiments, during which the gas pressure may vary from 10^{-6} to ∼2 Torr. Finally, the precision (∼1%) of the calorimetric results, for experiments of this type, is limited by the deviations of the ratio W/Δ between thermal power and the corresponding deflection of the base line (Table III).

When the calorimeter is heated at temperatures exceeding ∼200°C, it

TABLE III

Electrical Calibration of a Calvet Microcalorimeter Heated at 200°C[a]

Type of cell	Gas pressure (Torr)	Calibration ratio (μW/mm)	Mean deviation of results (μW/mm)	
			Constant thermal power	Variable thermal power
Calibration[b]	—	2.356	0.002	0.016
Adsorption[c]	8×10^{-6}	2.356	0.007	—
	2×10^{-3}	2.357	0.007	—
	6×10^{-1}	2.358	0.003	—
	2.5	2.354	0.004	—

[a] Gravelle et al. (55).
[b] Shown as (2) in Fig 18.
[c] Shown as (3) in Fig. 18.

becomes even more important than at lower temperatures to maintain identical arrangements in the calorimeter cell for both calibration tests and adsorption experiments. For this reason, a high-temperature cell was constructed in which the adsorption cell and the calibration heater are permanently located [Fig. 18, cell (4)] (67). As already indicated, the thin-walled glass tube which contains the adsorbent is then directly blown into the metallic container B. The glass tube which connects the adsorption cell into the metallic container to the volumetric line has a large section, and its outer diameter is but slightly smaller than the inner diameter of the cylindrical hole bored through the metal block for introducing the cells. The purpose of this arrangement is to reduce air convection around the tube and thus to minimize possible fluctuations of the emf produced by this secondary heat transfer. It is difficult to construct a calibration heater which can be used permanently at high temperatures ($\sim$500°C) and which is, at the same time, electrically insulated and in good thermal contact with the metallic container in the calorimeter cell. A commercially available thermocouple wire, electrically insulated by magnesia and encased in a steel sheathing, has been found suitable for this purpose. Several successive layers of this wire are wound in an annular space cut into the metallic container in such a way that the heater height is not very different from that of the sample in the adsorption cell. The heater is permanently connected to the external circuit. Calibration tests have been carried out with this heater. However, the heat evolved in the adsorption cell must flow through the adsorbent, the walls of the glass tube, and the metallic container before it is transferred via the thermoelements to the heat sink, whereas the heat path is much shorter in the case of calibration tests. The validity of these tests has therefore been controlled by means of another heater introduced in the adsorption cell containing the adsorbent (internal heater). Electrical calibration tests have been carried out, using successively or simultaneously, the calibration heater (then called "external heater") and the internal heater. The results (67) are presented on Table IV for the case of a high-temperature Calvet calorimeter, heated at 384°C, equipped with a dc amplifier and a galvanometric recorder (a base line deflection of 1 mm is equivalent to 0.1 μV). The values of the calibration ratio are, in all cases, very similar. These results demonstrate, in particular, that the leads to the internal heater do not produce measurable thermal leaks. As in the case of similar tests at 200°C (Table III), the precision of the calorimetric results ($\sim 2\%$) is limited by the variations of the ratio W/Δ between thermal power and the corresponding deflection of the base line.

It is clear that the calibration of a calorimeter and the preliminary experiments which have been described are operations of paramount importance. In the case of the apparatus that we use, they have shown that corrections are often not necessary, and that the area of the thermogram is in many cases directly proportional to the amount of heat evolved during a adsorption phenomenon, provided that the gas pressure is maintained below 2 Torr. It may not be so with all adsorption calorimeters, especially if a large sensitivity is needed or if the symmetry of the twin system is not perfect. However, the calibration tests and preliminary experiments which have been described can be used to determine eventually the necessary corrections. Moreover, it should not be forgotten that the

TABLE IV

Electrical Calibration of a Calvet Microcalorimeter Heated at 384°C[a]

Heater	Calibration ratio (μW/mm)	Mean deviation of results for variable thermal power (μW/mm)	Time constant (sec)
Calibration	29.9	0.4	$\sim$220
External	29.9	0.4	$\sim$220
Internal	29.7	0.4	$\sim$480

[a] Reymond (67).

characteristics of a heat-flow calorimeter, or even of a volumetric line, may change with time. Calibration tests should be repeated frequently and, of course, every time that the apparatus is, even slightly, modified. Newly constructed calibration cells must always be compared to older ones, in the same calorimeter, before they are used for actual calibration tests. Finally, the measurement of the differential heats of adsorption of oxygen on nickel oxide, NiO (200) (*19*) has been used regularly as a convenient and reproducible test for comparing the performances of all heat-flow microcalorimeters in our laboratory (see Section VII.A).

The main time constant τ of the calorimeter may be calculated by the methods indicated previously (Section V.B) from the thermograms recorded during electrical calibration. Moreover, Joule heating is certainly the most convenient method for producing, in an adsorption calorimeter, the unit heat impulses that are needed for the study of the calorimeter response and, then, for the processing of the calorimetric data. The very different values of the calorimeter time constant (Table IV), calculated from Joule heatings in the external and internal heaters, both of them being simultaneously in the high-temperature calorimeter cell, confirm however that the determination of the calorimeter response is a very difficult problem indeed. In all cases where only qualitative information on the kinetics are required, the time constant obtained by means of a Joule heating in the *internal* heater, located within the adsorption cell in the adsorbent itself, may be used for an approximate manual correction of the thermograms. On the other hand, when a more complete data correction is needed, it is essential that heat should be transferred via the same paths during both calibration by unit impulses and experiment. This condition cannot be completely met when calibration tests and experiments are different physical phenomena: Joule heatings and adsorption processes, for instance. In order to minimize the differences between heat paths, the heat sources (heater and adsorbent) should be both located near the center of the calorimeter cell and they should occupy a relatively small volume (*52*).

VII. Applications of Heat-Flow Microcalorimetry in Heterogeneous Catalysis

The mechanism of any catalytic reaction may be divided in a number of steps: adsorption of at least one reactant, surface interactions between reactants, desorption of the products, etc. Each of these steps is associated with a modification of the enthalpy of the system. But the total energy released, or absorbed, during all successive steps of the actual reaction mechanism must be identical to the change of enthalpy associated with the noncatalyzed reaction at the same temperature, provided that the catalyst is not modified in the course of its action. It appears, therefore, that at least in some favorable cases, it is possible to study separately, in a calorimeter, the different steps of the catalytic reaction, and thus, to deduce probable reaction intermediates or reaction mechanisms from the calorimetric data. Although the efficiency of this method, based on very simple thermodynamic arguments, has been clearly demonstrated by Garner and his collaborators [for a review, see (*18*)], its importance does not seem to have found yet a general recognition, and in most cases, the use of calorimeters in this field has been limited to the investigation of adsorption processes.

It is true, however, that many catalytic reactions cannot be studied conveniently, under given conditions, with usual adsorption calorimeters of the isoperibol type, either because the catalyst is a poor heat-conducting material or because the reaction rate is too low. The use of heat-flow calorimeters, as has been shown in the previous sections of this article, does not present such limitations, and for this reason, these calorimeters are particularly suitable not only for the study of adsorption processes but also for more complete investigations of reaction mechanisms at the surface of oxides or oxide-supported metals. The aim of this section is therefore to present a comprehensive picture of the possibilities and limitations of heat-flow calorimetry in heterogeneous catalysis. The use of Calvet microcalorimeters in the study of a particular system (the oxidation of carbon monoxide at the surface of divided nickel oxides) has moreover been reviewed in a recent article of this series (*19*).

A. Differential Heats of Adsorption. The "Energy Spectrum" of the Catalyst Surface

Differential heats of adsorption yield, in many instances, very relevant information concerning not only the type, or types, of bond between the adsorbed particles and the adsorbent but also the heterogeneity of the catalyst surface. Comparison of Eqs. (39) and (40) shows, however, that a closer approximation to true differential heats is obtained when smaller doses of gas are admitted in the adsorption cell. Clearly, a very sensitive calorimeter is required, particularly when adsorption occurs slowly.

Calvet microcalorimeters are particularly convenient for such studies. Figure 19 shows, for instance, the evolution of the differential heat of adsorption of oxygen, measured at 30°C with a Calvet calorimeter, as a function of the total amount of oxygen adsorbed on the surface of a sample (100 mg) of nickel oxide, NiO(200) (*19, 73*). The volume of the first

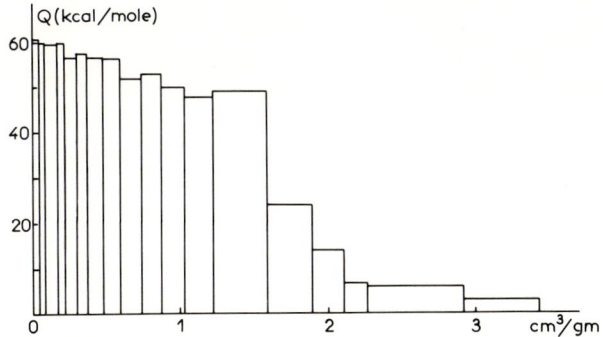

FIG. 19. Differential heats of adsorption of oxygen on nickel oxide, NiO (200), at 30°C. Reprinted from (*19*) with permission.

increments of adsorbed gas ($\sim$0.1 cm³ gm⁻¹) corresponds indeed to a very small increase of surface coverage [$\Delta\theta = \sim 2 \times 10^{-3}$, if as usual (*74*), it is assumed that one oxygen atom can be adsorbed on each surface nickel ion]. Similar doses could be used during most adsorption experiments. However, in many cases, larger doses are admitted in the adsorption cell in order to reduce the duration of the experiments.

The thermograms recorded during the adsorption of the first doses of oxygen on NiO(200) are similar to curve A in Fig. 20. They indicate that the oxide surface presents a high reactivity toward the gas. However, after the maximum ordinate, the heat flow decreases more progressively with time than would be expected if cooling of the calorimeter cell was occurring exclusively. Therefore, although an important fraction of the dose reacts rapidly with the sample, a part of the dose, which cannot be assessed

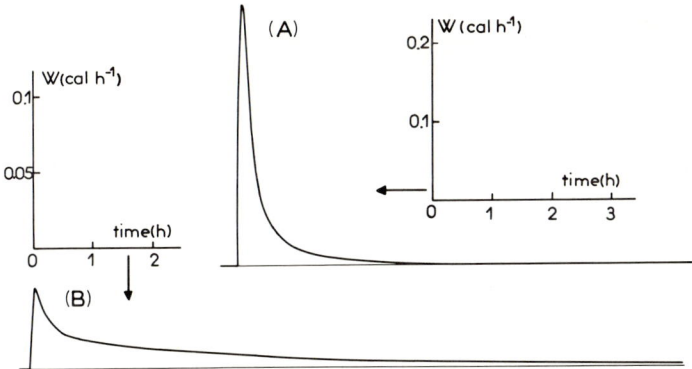

FIG. 20. Thermograms recorded during the adsorption of doses of oxygen on nickel oxide at 30°C, the equilibrium pressure of the gas being smaller (A) or larger (B) than 10⁻³ Torr. Reprinted from (*19*) with permission.

quantitatively from these results, is stabilized slowly on the surface. This observation has indeed been repeated very often in the case of chemisorptions on oxides, and it may be wondered if this is not a general feature of many "fast" chemisorption processes.

It is, of course, not necessary to use a heat-flow microcalorimeter in order to determine the heat released by rapid adsorption phenomena. Dell and Stone (74), for instance, using an isoperibol calorimeter of the Garner–Veal type, found an initial heat of 54 ± 4 kcal mole^{-1} for the adsorption of oxygen on nickel oxide at 20°C. The agreement with the value (60 ± 2 kcal mole^{-1}) in Fig. 19 is remarkably good, particularly if it is considered that very different methods were used for the preparation of the nickel-oxide samples (19, 74).

However, when the oxygen equilibrium pressure over the sample in the calorimeter cell increases to $\sim 10^{-3}$ Torr, the profile of the thermograms changes distinctly (Fig. 20, curve B). Although a fraction of the dose still reacts rapidly with the surface, the adsorption of the main part of the dose proceeds very slowly and a heat-flow microcalorimeter is now required to follow its progress. Integration of the thermograms shows that the slow adsorption process is associated with lower heats (5–20 kcal mole^{-1}) than the fast process (50–60 kcal mole^{-1}) (Fig. 20). These calorimetric results suggest, therefore, that two kinds of adsorbed species are formed when oxygen contacts nickel oxide at 30°C. The adsorption isotherm and electrical conductivity measurements have indeed confirmed the calorimetric results (19, 73).

The formation of different surface species during a single adsorption process may therefore be detected by means of heat-flow microcalorimetry. The results are particularly convincing when the various kinds of adsorbed species are characterized not only by different heats but also by different rates of formation. Moreover, when the adsorption isotherms corresponding to the various kinds of adsorbed species are quite distinct, the formation of a particular species may occur preferentially during part of the adsorption experiment, as in the preceding example (Fig. 20, curves A and B). The detection of the various kinds of adsorbed species is then particularly easy. This is not, however, a necessary condition, and processes occurring quasi-simultaneously in the calorimeter cell may still be detected and qualitatively separated when the corresponding heat-evolutions are quite different. It has been shown, for instance, that the thermogram presented in Fig. 21, curve 1, which was recorded during the reaction of a dose of carbon monoxide with a sample of nickel oxide at 200°C (55), is the sum of curve 2 (the desorption of generated carbon dioxide), curve 4 (the formation of metal germs), and curve 5 (the reaction at the metal-oxide interface). It must be noted, in passing, that heat-flow calorimetry has thus confirmed

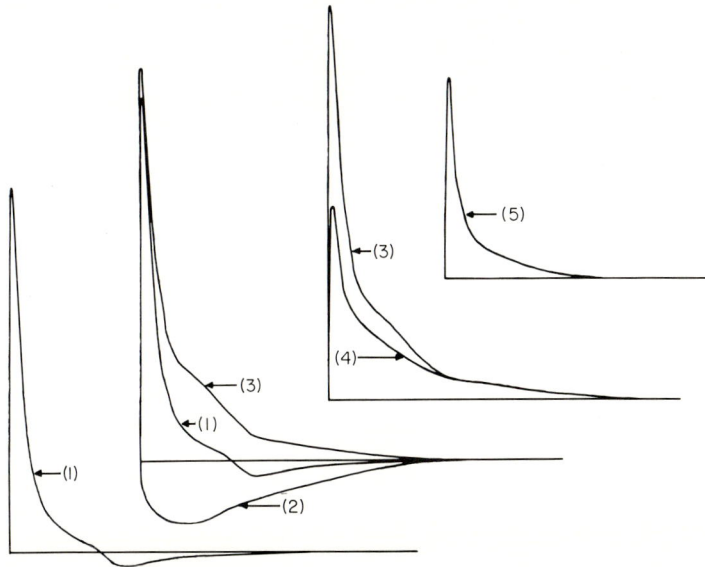

Fig. 21. Qualitative analysis of a thermogram recorded during the reduction of nickel oxide by a dose of carbon monoxide at 200°C. The thermogram (1) may be considered as the sum of curves (2), (4), and (5). Curve (3) is the difference between curves (1) and (2). Reprinted from (55) with permission.

directly the development of an interface reaction during a reduction process (55).

It must be acknowledged, however, that the determination of the number of the different surface species which are formed during an adsorption process is often more difficult by means of calorimetry than by spectroscopic techniques. This may be phrased differently by saying that the "resolution" of spectra is usually better than the resolution of thermograms. Progress in data correction and analysis should probably improve the calorimetric results in that respect. The complex interactions with surface cations, anions, and defects which occur when carbon monoxide contacts nickel oxide at room temperature are thus revealed by the modifications of the infrared spectrum of the sample (75) but not by the differential heats of the CO-adsorption (76). Any modification of the nickel-oxide surface which alters its defect structure produces, however, a change of its "*energy spectrum*" with respect to carbon monoxide that is more clearly shown by heat-flow calorimetry (77) than by IR spectroscopy.

It appears, therefore, that although it is often difficult to determine the nature of the adsorbed species, or even to distinguish between the different kinds of adsorbed species from the calorimetric data, the change of the

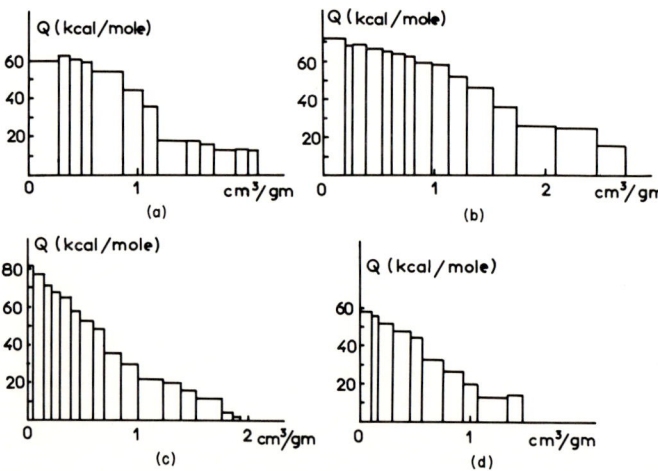

FIG. 22. Differential heats of adsorption of oxygen at 30°C on four different samples of pure and doped nickel oxide. (a) NiO (200), (b) NiO(Li) (250), (c) NiO (250), (d) NiO(Ga) (250). Reprinted from (8) with permission. Copyright © 1969 by Academic Press, Inc. New York.

differential heats of adsorption with coverage pictures quite clearly the distribution of surface sites with respect to a given adsorbate and their varying reactivity on given adsorbents. This point may be illustrated by Fig. 22 which shows the Q–θ curves for the oxygen adsorption at 30°C on a series of pure and doped nickel oxides (8). These curves demonstrate quite explicitly that the number, the reactivity, and the distribution of surface sites are significantly modified, with respect to oxygen, when the composition or the pretreatment of the samples are changed. This information is particularly relevant to catalysis studies, since, as it was indeed shown in this particular example (8), only a fraction of the "energy spectrum" of the surface sites may actually be useful during the catalytic reaction.

It ought to be verified, however, in all cases, that the experimental Q–θ curve truly represents the distribution of surface sites with respect to a given adsorbate under specified conditions. The definition of differential heats of adsorption [Eq. (39)] includes, in particular, the condition that the surface area of the adsorbent A remain unchanged during the experiment. The whole expanse of the catalyst surface must therefore be accessible to the gas molecules during the adsorption of all successive doses. The adsorption of the gas should not be limited by diffusion, either within the adsorbent layer (external diffusion) or in the pores (internal diffusion). Diffusion, in either case, restricts the accessibility to the adsorbent surface.

When it occurs, the adsorption on reactive sites, located in shielded areas, may therefore occur after less reactive sites, better exposed, have reacted. Diffusion may thus cause the "smoothing out" of significant details in the energy spectrum and the Q–θ curves, determined in the presence of diffusion phenomena, indicate less surface heterogeneity than actually exists on the adsorbent surface.

External diffusion is especially to be feared in the case of adsorption calorimeters which contain a large quantity of adsorbent. Stone and his colleagues, for instance have described a calorimeter in which special tubes for gas entry must be placed within the adsorbent layer in order to eliminate external diffusion (78). During most experiments with heat-flow microcalorimeters, the quantity of adsorbent is small (50–200 mg). Moreover, the thickness of the adsorbent layer in the annular space between the inside wall of the adsorption cell and the outside wall of the cylinder of glass gauze (Fig. 18) never exceeds $\sim$2 mm. Under these conditions, external diffusion is unlikely and its absence may indeed be verified by the similitude of the Q–θ curves for the adsorption of a given gas, at a given temperature, on samples from the same batch which are characterized by very different weights. The comparison of Figs. 22, curve 1, and 23, for instance, shows that the increase of the sample weight from 125 mg (Fig. 22, curve 1) to 550 mg (Fig. 23) does not modify significantly the Q–θ curve for the adsorption of oxygen at 30°C on nickel-oxide samples prepared from the same batch of hydroxide. The sample used in the similar experiment reported in Fig. 19 was prepared from a different batch of hydroxide. The comparison of Figs. 22, curve 1, or 23 and 19 therefore gives an indication of the reproducibility of the sample preparation.

The most reliable method for detecting the influence of internal diffusion upon the profile of Q–θ curves would be to determine calorimetrically and to compare the differential heats of adsorption of a given gas on the surface of similar samples with different porosities. But it would be very difficult

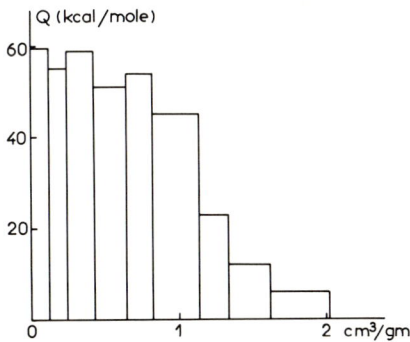

FIG. 23. Differential heats of adsorption of oxygen on nickel oxide at 30°C (sample weight, 550 mg).

indeed to ascertain whether the modification of the adsorbent texture does or does not alter simultaneously the reactivity, the number, or the distribution of its surface sites. Therefore, this method, though theoretically sound, is not recommended in the general case, and internal diffusion is certainly less easily controlled than external diffusion. The comparison of the heats evolved during the adsorption of large or small doses of gas on the same adsorbent may, however, provide some clues.

Let suppose that very reactive and less reactive sites are distributed at random on the surface of the porous adsorbent. On the one hand, when a large dose of gas contacts the previously evacuated sample, the number of gas molecules usually exceeds the number of very reactive sites readily accessible to the gas, and chemisorption occurs also on less reactive sites. The same situation exists during the adsorption of the next large doses and consequently, because of diffusional limitations, the experimental heats do not picture correctly the energy spectrum of the surface sites. On the other hand, when a small dose of gas contacts the previously evacuated sample, the number of molecules in the dose may be small compared to the number of readily accessible reactive sites. In such a case, adsorption should occur preferentially on reactive sites. When diffusion limits the accessibility of surface sites, the heat released by the adsorption of small doses should therefore be higher than the heat evolved during the adsorption of large doses. The similarity of the initial heats ($\theta = 0$) of adsorption of oxygen on nickel oxide at 30°C, which is observed when small (Fig. 19) or large (Fig. 22, curve 1) doses of gas are adsorbed on similar samples, may be considered as an indication that chemisorption, in this case, is not limited by diffusion phenomena. As the admission of small doses is repeated, adsorption proceeds either preferentially on the most active sites or simultaneously on all sets of surface sites depending upon the gas pressure, diffusional limitations, and the adsorption isotherms corresponding to each set of surface sites. Reactive sites react usually at a lower gas pressure than less reactive sites. This is indeed observed in the case of the adsorption of oxygen on nickel oxide at 30°C: surface sites characterized by high heats of adsorption (50–60 kcal/mole) (Fig. 19) react with oxygen when the equilibrium gas pressure is smaller than 10^{-3} Torr, whereas adsorption on less reactive sites (corresponding heats of adsorption: 5–20 kcal/mole) proceeds at much higher gas pressures (73). The admission of small doses which prevents the rapid increase of the equilibrium pressure, therefore favors the reaction of the gas with the most active surface sites, even when they are located in shielded areas. The use of small doses of gas is therefore beneficial in experiments of adsorption calorimetry, particularly when the surface sites are distributed at random on a porous adsorbent.

Finally, the presence or the absence of diffusional limitations may be verified, in some cases, by indirect methods. For instance, the Q–θ curve on Fig. 19 has been determined by the successive introductions of doses of molecular oxygen into the adsorption cell containing a sample of nickel oxide maintained at 30°C. Now, surface oxygen species may also be formed, on the surface of nickel oxide by the decomposition at 30°C of small quantities of nitrous oxide:

$$Ni^{2+} + N_2O(g) \rightarrow N_2(g) + O^-(ads) + Ni^{3+}. \qquad (41)$$

Reaction (41) has indeed been confirmed by pressure measurements and mass spectroscopic analyses (73, 79). The experimental heats for reaction (41) may be introduced in the following thermochemical cycle:

$$Ni^{2+} + N_2O(g) \rightarrow N_2(g) + O^- (ads) + Ni^{3+}, \qquad (41)$$

$$\tfrac{1}{2} O_2(g) + N_2(g) \rightarrow N_2O(g), \quad (-20 \text{ kcal}), \qquad (42)$$

$$\overline{Ni^{2+} + \tfrac{1}{2} O_2(g) \rightarrow O^- (ads) + Ni^{3+},} \qquad (43)$$

in order to calculate the energy of adsorption [Eq. (43)] of the oxygen species created by the decomposition of nitrous oxide. The results are presented on Fig. 24. The calculated heats (Fig. 24) are similar to the differential heats determined experimentally during the adsorption of molecular oxygen on a similar sample of nickel oxide (Fig. 19). The calculated heats, corresponding to the decomposition of nitrous oxide, however, decrease more progressively with an increased coverage by adsorbed oxygen than the experimental heats. An explanation of this phenomenon may be that pairs of sites are needed for the dissociation of molecular oxygen to occur [Eq. (43)], whereas isolated sites may be also active in the nitrous oxide decomposition. The similarity of the Q–θ curves (Figs. 19 and 24) suggest

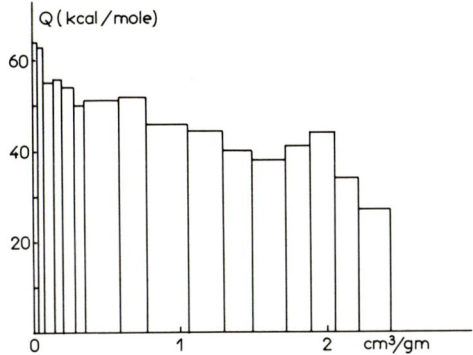

FIG. 24. Differential heats of adsorption of oxygen on nickel oxide [NiO(200)], as calculated from the experimental heats of decomposition of N_2O at 30°C. Reprinted from (19) with permission.

that surface oxygen species created in both cases present the same bond with the surface and, thereby, the same structure (*73*). But the similarity of the Q–θ curves indicates moreover that adsorption of oxygen on the most reactive surface sites of these samples of nickel oxide is not perturbed by diffusion because the changes of the gas pressure during the course of these experiments are very different. Whereas the oxygen pressure increases slowly but regularly during the adsorption of molecular oxygen, no increase of the gas pressure occurs during the decomposition of nitrous oxide if the sample is replaced under vacuum after the reaction of each dose of nitrous oxide (*73*), or the gas pressure increases very rapidly indeed if all generated nitrogen accumulates in the volumetric line (*79*).

B. Differential Heats of Interaction between a Reactant and Preadsorbed Species. Reaction Mechanisms

The adsorption of at least one reactant is the first step of the mechanism of any catalytic reaction. This step is followed by surface interactions between adsorbed species or between a gaseous reactant and adsorbed species. In many cases, these interactions may be detected by the successive adsorptions of the reactants in different sequences. Heat-flow microcalorimetry can be used with profit for such studies (*19*).

Preadsorption of oxygen at 30°C on the surface of a gallium-doped nickel oxide produces, for instance, a considerable increase of the differ-

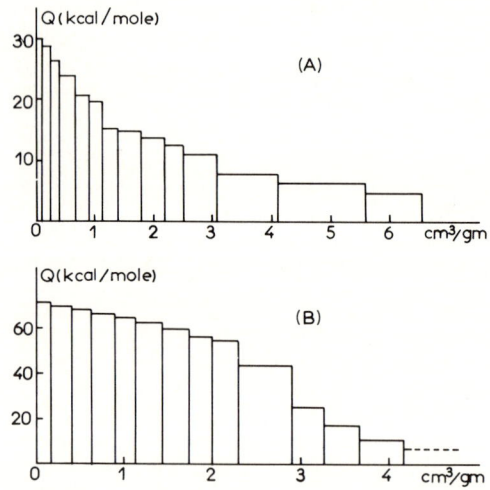

Fig. 25. Differential heats of adsorption of carbon monoxide at 30°C on fresh (A) or oxygenated (B) samples of a gallium-doped nickel oxide. Reprinted from (*53*) with permission *J. Chim. Phys.*

ential heats of adsorption of carbon monoxide at 30°C on this solid (53). The Q–θ curves for the adsorption of carbon monoxide on fresh or oxygenated samples are presented in Fig. 25, curves A and B. The very different aspect of these curves shows clearly that the calorimetric study of the chemisorption of carbon monoxide at 30°C constitutes a sensitive test for detecting oxygen species adsorbed on nickel oxide (80). The amount of carbon monoxide, adsorbed on the oxygen-precovered sample (Fig. 25, curve B) which produces heats exceeding the initial heat of adsorption of carbon monoxide on the fresh sample (30 kcal/mole, Fig. 25, curve A) is close to twice the amount of preadsorbed oxygen (Fig. 22, curve 4). It seems therefore that one molecule of carbon monoxide interacts with every preadsorbed oxygen ion:

$$CO(g) + O^- (ads) + Ni^{3+} \rightarrow CO_2(g) + Ni^{2+}. \qquad (44)$$

Equation (44) has indeed been confirmed by electrical conductivity measurements and by the detection of carbon dioxide condensed in the cold trap (53).

The stoichiometry of an interaction between gas molecules and preadsorbed species may thus be deduced from the modifications of the Q–θ curves for a given reactant which are produced by the presence of preadsorbed species on the solid. The results are, of course, particularly conclusive when the differential heats of adsorption of small doses of reactant are measured in a sensitive calorimeter. But, such a qualitative analysis of the calorimetric data, though very useful, does not allow definite conclusions. In the preceding example, for instance, a fraction of carbon dioxide may remain adsorbed on the solid:

$$CO (g) + O^- (ads) + Ni^{3+} \rightarrow CO_2 (ads) + Ni^{2+}. \qquad (45)$$

The calorimetric data may be used, however, in a more quantitative manner, to verify separately, by means of thermochemical cycles, the different hypotheses. In cycle 1 (Table V), the experimental heats recorded during (1) the adsorption of oxygen on the sample of gallium-doped nickel oxide (Fig. 22, curve 4) and (2) the subsequent adsorption of carbon monoxide (Fig. 25, curve B) are added. The sum should be equal to 68 kcal/mole, the heat of the homogeneous combustion of carbon monoxide (Eq. 46) if the hypothesis which is included in the cycle [the formation of *gaseous* carbon dioxide, Eq. (44)] is correct. Calculations may be achieved for any surface coverage, but in all cases, the stoichiometry of the interactions must be taken into account. In Table V, cycle 1, for instance, is tested for two surface coverages (column a, initial coverage, 0, for both oxygen and carbon monoxide; column b, 0.9 cm^3 O$_2$/gm and 1.8 cm^3 CO/gm). Cycle 2 (Table V), in which the second hypothesis is included

TABLE V

Thermochemical Cycles Testing the Formation of Gaseous (Cycle 1) or Adsorbed (Cycle 2) Carbon Dioxide by the Interaction of Carbon Monoxide with Oxygen Preadsorbed on Gallium-Doped Nickel Oxide[a]

Cycle 1		a	b
(43)	$\frac{1}{2}O_2(g) + Ni^{2+} \rightarrow O^-(ads) + Ni^{3+}$	+29 (0)[b]	+12 (0.9)
(44)	$CO(g) + O^-(ads) + Ni^{3+} \rightarrow CO_2(g) + Ni^{2+}$	+73 (0)	+60 (1.8)
(46) = (43) +(44)	$CO(g) + \frac{1}{2}O_2(g) \rightarrow CO_2(g)$	+102	+72

Cycle 2		a	b
(43)	$\frac{1}{2}O_2(g) + Ni^{2+} \rightarrow O^-(ads) + Ni^{3+}$	+29 (0)	+12 (0.9)
(45)	$CO(g) + O^-(ads) + Ni^{3+} \rightarrow CO_2(ads) + Ni^{2+}$	+73 (0)	+60 (1.8)
(47)	$CO_2(ads) \rightarrow CO_2(g)$	−28 (0)	−24 (1.8)
(46) = (43) + (45) + (47)	$CO(g) + \frac{1}{2}O_2(g) \rightarrow CO_2(g)$	+74	+48

[a] Gravelle et al. (53).
[b] Numbers not in parentheses refer to the heat of adsorption or interaction, in kilocalories per mole or kilocalories per atom depending on the stoichiometry of the proposed interaction; numbers in parentheses refer to the corresponding surface coverages in cubic centimeters per gram.

[the formation of *adsorbed* carbon dioxide, Eq. (45)], is tested for the same surface coverages (columns a and b). The heats of desorption of carbon dioxide which appear in cycle 2 are equal, in absolute value, to the experimental heats of adsorption of carbon dioxide for the same coverages (81).

Before analyzing the results of these, or similar, thermochemical cycles, the assumptions which have been made must be critically examined. Since the cycles are tested for different surface coverages, it is assumed first that the Q–θ curves represent correctly, in all cases, the distribution of reactive sites—the energy spectrum—on the surface of the adsorbent. This point has been discussed in the preceding section (Section VII.A). It is assumed moreover that, for instance, the first doses of carbon monoxide ($\theta = 0$) interact with oxygen species adsorbed on the most reactive surface sites ($\theta = 0$). This assumption, which is certainly not acceptable in all cases, ought to be verified directly. This may be achieved in separate experiments by adsorbing *limited* amounts of the different reactants in the same sequence on another sample of the same adsorbent. Figure 26 reports, for instance, the differential heats of interaction of carbon monoxide at 30°C with a sample of gallium-doped nickel oxide containing preadsorbed oxygen (0.4 cm^3 O$_2$/gm). The comparison of the Q-θ curves in Figs. 25, curve B, and 26 shows that preadsorption of a large ($\sim$1.4 cm^3 O$_2$/gm) of a small (0.4 cm^3 O$_2$/gm) quantity of oxygen does not influence significantly the initial value ($\theta = 0$) of the heat produced by the interaction of carbon monoxide with the sample. Moreover a sharp decrease of the differential heats is observed when the surface coverage by carbon monoxide ($\sim$0.8 cm^3 CO/gm) exceeds twice the quantity of preadsorbed oxygen. These results which confirm the stoichiometry of the interaction [Eqs. (44) or (45)] demonstrate moreover that oxygen and carbon monoxide in the first doses of these gases are adsorbed or react on the same surface sites. These results support, therefore, the calculations in cycles 1 and 2 (Table V). Similar experiments ought to be made in all cases where thermochemical cycles are tested for different surface coverages.

Now, the heat of the homogeneous combustion of carbon monoxide being 68 kcal/mole, cycle 1 (Table V) is balanced only for high surface coverages (column b), whereas cycle 2 is balanced only for a low coverage (column a). The calorimetric data indicate therefore that the interaction of carbon monoxide with oxygen preadsorbed on gallium-doped nickel oxide produces *adsorbed* carbon dioxide when it occurs on the most reactive surface sites. On the less reactive sites, the interaction yields *gaseous* carbon dioxide. If it is supposed that separate adsorption experiments picture correctly the adsorptions and surface interactions which occur during the catalytic reaction, these results indicate that during the catalytic combustion of carbon monoxide at 30°C, the most reactive sites on a gallium-doped nickel oxide will eventually be inhibited by carbon dioxide,

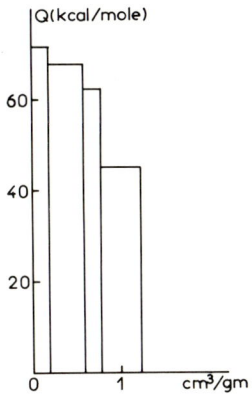

Fig. 26. Differential heats of interaction of carbon monoxide at 30°C with a sample of gallium-doped nickel oxide, containing a limited amount (0.4 cm^3 O$_2$ gm^{-1}) of preadsorbed oxygen.

the reaction product, and that on less reactive sites, interactions (43) and (44) may be the steps of a probable reaction mechanism.

The analysis of the calorimetric data by means of thermochemical cycles yields, therefore, very significant information concerning probable reaction mechanisms and the reactivity of surface sites. In some cases, it has been possible to determine, by this method, which part of the energy spectrum of the surface sites is actually active during the catalytic process (8). This method requires, however, that the Q–θ curves are determined accurately, and, thence, that a sensitive calorimeter is used. It should be noted, for instance, that neither cycle 1 nor cycle 2 in Table V are balanced if the *integral* heats of the successive adsorptions of oxygen (30 kcal/mole) and carbon monoxide (47 kcal/mole) and the integral heat of the adsorption of carbon dioxide (18 kcal/mole) are used in the calculations. Moreover, the agreement of a cycle with thermodynamic data constitutes only a

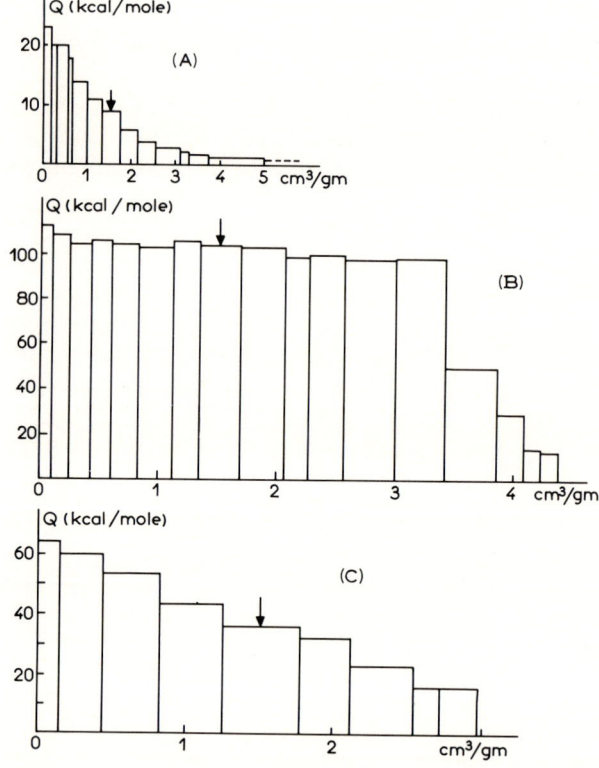

FIG. 27. Differential heats versus coverage for the successive adsorptions, at 30°C, of carbon monoxide (A), oxygen (B), and, again, carbon monoxide (C) on the surface of lithium-doped nickel oxide. Reprinted from (54) with permission J. Chim. Phys.

TABLE VI

Thermochemical Cycle Testing the Formation of Gaseous Carbon Dioxide at the End of the Adsorption Sequence $(CO-O_2-CO)^a$

(48)	$CO(g) \rightarrow CO(ads)$	$+7^b$
(49)	$Ni^{2+} + CO(ads) + O_2(g) \rightarrow CO_3^-(ads) + Ni^{3+}$	$+103$
(50a)	$Ni^{3+} + CO_3^-(ads) + CO(g) \rightarrow 2 CO_2(g) + Ni^{2+}$	$+35$
(46) = ½[(48) + (49) + (50a)]	$CO(g) + ½O_2(g) \rightarrow CO_2(g)$	$+72.5$

^{a} Gravelle *et al.* (*54*).
^{b} All heats are expressed in kilocalories per mole.

presumption that the hypothesis included in the cycle is correct: the balance of a single thermochemical cycle may be considered as fortuitous, and, indeed, in one particular case, the balance of a cycle was shown to be accidental (*54*). All possible hypotheses must be successively tested by means of thermochemical cycles, and, indeed, in most cases, all but one are rejected. Finally, the calorimetric results must be confronted with the results of other experimental techniques.

The analysis of the thermograms recorded during the interaction of the successive doses of the different reactants in the sequence may also yield very relevant informations. Through the use of different techniques, it has been shown, for instance, that the different steps of the mechanism of the CO oxidation, at room temperature, at the surface of pure [NiO (200)] (*19, 82*) or lithium-doped (*54*) nickel oxide, may be written:

$$CO(g) \rightarrow CO(ads), \quad (48)$$

$$CO(ads) + O_2(g) + Ni^{2+} \rightarrow CO_3^-(ads) + Ni^{3+}, \quad (49)$$

$$CO_3^-(ads) + CO(ads) + Ni^{3+} \rightarrow 2 CO_2(g) + Ni^{2+}, \quad (50)$$

The rate-limiting step of this mechanism is the interaction, in the adsorbed phase, between the carbonate species and adsorbed carbon monoxide (Eq. 50) (*82*). The successive steps of this reaction mechanism can be studied separately in a calorimeter, for instance, by the successive adsorptions of carbon monoxide, oxygen and, again, carbon monoxide. The differential heats corresponding to the successive adsorptions of the sequence $(CO-O_2-CO)$ at the surface of a lithium-doped nickel oxide are presented on Fig. 27 (*54*). The thermochemical cycle, presented on Table VI, shows that the differential heats recorded during all steps of the sequence for a surface coverage close to 1.5 cm³/gm (the corresponding doses are indicated by arrows on Fig. 27) are in agreement with the proposed mechanism

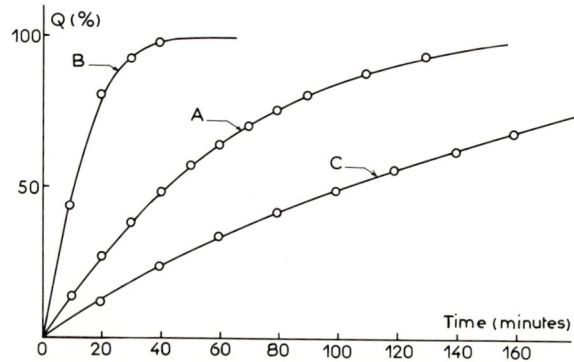

FIG. 28. Percentage of heat evolved as a function of time during the interaction of the doses indicated by arrows on curves A, B, and C in Fig. 27 (curve A in Fig. 28 corresponds to the dose on curve A in Fig. 27, etc.). Reprinted from (54) with permission J. Chim. Phys.

[Eqs. (48)–(50a)]. The thermograms recorded during the interaction of the doses indicated by arrows on Fig. 27 have been analyzed and corrected (their integration gave the differential heats on Fig. 27). The percentage of the heat evolved during the reaction of each of these doses is plotted on Fig. 28 as a function of time. Heat is evolved rapidly during the formation of the carbonate complex [Eq. (49), Fig. 28, curve B]. The decomposition of the complex species by carbon monoxide is slow [Eq. (50a), Fig. 28, curve C], and it should be observed that this interaction is even slower than the adsorption of carbon monoxide *for the same surface coverage* [Eq. (48), Fig. 28, curve A]. The calorimetric data therefore indicate that the slowest interaction is the decomposition of the carbonate complex by *adsorbed* carbon monoxide. This interaction yields gaseous carbon dioxide (Table VI). Identical conclusions have been deduced from the kinetic study of the reaction (82).

Heat-flow microcalorimetry may be used, therefore, not only to detect, by means of adsorption sequences, the different surface interactions between reactants which constitute, in favorable cases, the steps of probable reaction mechanisms, but also to determine the rates of these surface processes. The comparison of the adsorption or interaction rates, deduced from the thermograms recorded during an adsorption sequence, is particularly reliable, because the arrangement of the calorimetric cells remains unchanged during all the steps of the sequence. Moreover, it should be remembered that the curves on Fig. 28 represent the adsorption or interaction rates on a very small fraction of the catalyst surface which is, very probably, active during the catalytic reaction (Table VI). It is for these reasons that the calorimetric results may be compared directly to the results of the kinetic study of the reaction. The similar analysis of the thermograms recorded during adsorptions or interactions of reactants on other fractions of the catalyst surface could also yield information, for instance, on the rate of formation of reaction inhibitors. Finally the calorimetric study of different adsorption sequences has shown that in some cases (even in the case of so-called "simple" catalytic reactions), two reaction paths may be equally probable on the same catalyst (8, 83). The analysis of the thermograms is then particularly useful, since it may indicate which reaction mechanism actually governs the catalytic reaction.

The calorimetric method which has been outlined in this section is not applicable to the study of surface interactions or of reaction mechanisms which occur between reversibly adsorbed species. But, even in these unfavorable cases, heat-flow microcalorimetry may still yield useful information concerning either the nature of the adsorbed species, the distribution of sites, or the irreversible modifications which occur frequently on the catalyst surface during the course of the reaction.

It is evident, however, from the preceding examples that the detailed analysis of surface interactions by means of adsorption calorimetry is

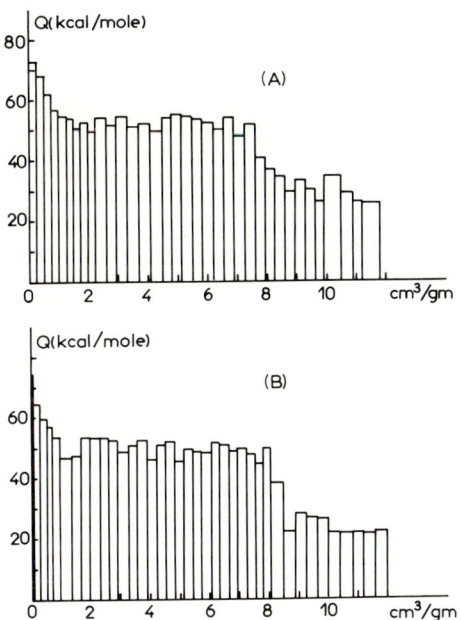

FIG. 29. Differential heats of interaction of carbon monoxide, at 200°C, with samples of nickel oxide containing excess oxygen, preadsorbed rapidly (A) or slowly (B).

time-consuming. The use of heat-flow microcalorimetry should follow rather than precede the use of more rapidly rewarding techniques. Finally, it may be wondered if in some cases the Q–θ curve, which is determined by the repeated introduction of small doses of gas during long periods of time, truly represents the distribution of sites as it exists when a large amount of gas contacts the adsorbent. The study of surface interactions may often give an answer to this question. Figure 29 represents, for instance, the evolution of the differential heats of the interaction of carbon monoxide, at 200°C, with two samples (A and B) of nickel oxide containing preadsorbed oxygen (71). Small doses of oxygen were adsorbed successively, during approximately two weeks, on sample B placed in a Calvet microcalorimeter maintained at 200°C. On the other hand, sample A, heated at 200°C, was left for only 24 hr in the presence of air (50 Torr). Both samples present the same reactivity toward carbon monoxide. These calorimetric results therefore demonstrate that the distribution of adsorbed oxygen is identical in both cases, and that it does not depend upon the particulars of the adsorption experiment.

C. Differential Heats of the Catalytic Reaction. Modifications of the Catalyst Surface

Under steady state conditions, the heat evolved by a catalyzed reaction is evidently equal, in absolute value, to the change of enthalpy associated at the same temperature with the same reaction in the homogeneous phase. Its experimental determination therefore presents very little interest unless a direct confirmation of the calorimeter calibration is sought (Section VI.B). However, the activities of fresh or aged catalysts are usually different. These modifications of catalytic activity, which present a great practical importance and which are called, depending upon the nature of the modification, stabilization, aging, poisoning, inhibition, or activation of the catalyst, are caused by surface interactions occurring during the catalytic reaction itself. Calorimetry may be used to study these surface interactions which, indeed, are also detected in many cases by the calorimetric study of adsorption sequences.

One of the conclusions deduced from the thermochemical cycle 2 in Table V, for instance, is that in the course of the catalytic combustion of carbon monoxide at 30°C, the most reactive surface sites of gallium-doped nickel oxide are inhibited by the reaction product, carbon dioxide. This conclusion ought to be verified directly by the calorimetric study of the reaction. Small doses of the stoichiometric reaction mixture ($CO + \frac{1}{2}O_2$) were therefore introduced successively in the calorimetric cell of a Calvet microcalorimeter containing a freshly prepared sample of gallium-doped

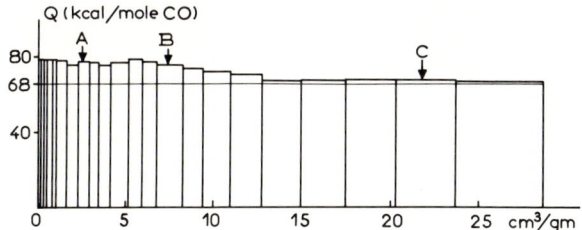

FIG. 30. Differential heats of reaction at 30°C of a stoichiometric mixture (CO + $\frac{1}{2}O_2$) at the surface of a freshly prepared sample of gallium-doped nickel oxide. Reprinted from (53) with permission J. Chim. Phys.

nickel oxide (53). At 30°C, the reaction of each dose is complete in a reasonable time and the final pressure is, in all cases, very small ($<10^{-4}$ Torr), all carbon dioxide being condensed in the cold trap. The heats released by the reaction of these doses, expressed per mole of consumed carbon monoxide, are presented as a function of the total volume of stoichiometric mixture on Fig. 30. The heats measured during the reaction of the first doses (78 kcal mole^{-1}) cannot be explained by the adsorption of the reactants (Fig. 22, curve 4 and Fig. 25, curve A). The catalytic reaction occurs, therefore, although the surface coverage is very small during the reaction of the first doses. Carbon dioxide is, indeed, condensed in the cold trap during the reaction of all doses. The heats of reaction of the first doses are higher than the heat of combustion of carbon monoxide (68 kcal/mole). However, the heats of reaction decrease progressively, and as the total volume of consumed reaction mixture becomes close to 15 cm^3/gm, they do not differ significantly (69 kcal/mole) from the normal heat of reaction.

The large heats, recorded at the beginning of the experiment, confirm that *part* of the generated carbon dioxide remains adsorbed on the catalyst surface. If all carbon dioxide had remained adsorbed on the surface during the reaction of the first doses, the registered heat of reaction should have been 96 kcal/mole:

$$CO(g) + \tfrac{1}{2}O_2(g) \rightarrow CO_2(g) + 68 \text{ kcal} \tag{46}$$

$$CO_2(g) \rightarrow CO_2(ads) + 28 \text{ kcal}$$
$$\text{(exptl. value)} \tag{47a}$$

$$CO(g) + \tfrac{1}{2}O_2(g) \rightarrow CO_2(ads) + 96 \text{ kcal} \tag{51}$$

The initial heats of reaction are lower (78 kcal/mole). This confirms that gaseous carbon dioxide is, in part, evolved during the reaction of the first doses. Therefore, inhibition of the most reactive sites of the catalyst surface

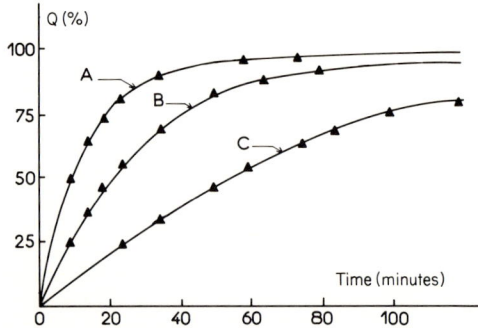

Fig. 31. Percentage of heat evolved, as a function of time, during the reaction of doses A, B, and C (Fig. 30) of stoichiometric mixture (CO + $\tfrac{1}{2}O_2$) at the surface of a sample of gallium-doped nickel oxide (curve A in Fig. 31 corresponds to dose A in Fig. 30, etc. Reprinted from (53) with permission J. Chim. Phys.

proceeds during the reaction, but progressively. In Fig. 31, the percentage of the heat released during the reaction of doses A, B, and C (Fig. 30), determined by the analysis of the corresponding thermograms, is plotted as a function of time. The heat of reaction is produced more slowly for dose C than for doses B or A. Therefore, the catalytic activity decreases as the number of doses of reaction mixture introduced in the volumetric line increases. The curve corresponding to dose C indicates a steady value of the activity, inhibition of the surface being then completed (heat of reaction, 69 kcal/mole, Fig. 30, compared with 68 kcal/mole for the enthalpy change of the homogeneous reaction). These calorimetric results

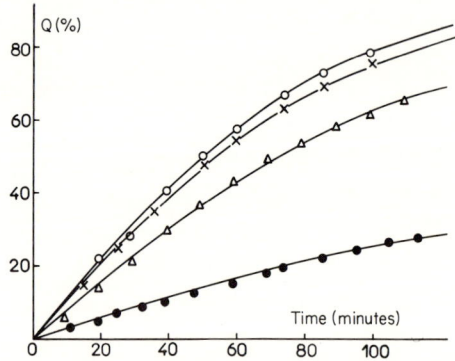

Fig. 32. Percentage of heat evolved as a function of time during the reaction of doses of stoichiometric mixture (CO + $\tfrac{1}{2}O_2$) at 30°C on four samples of pure and doped nickel oxide. (These curves characterize the steady activity of the different samples.) ● NiO (200), △ NiO(Li) (250), ○ NiO (250), × NiO(Ga) (250). Reprinted from (53) J. Chim. Phys. with permission.

demonstrate a progressive inhibition of the oxide surface by generated carbon dioxide remaining partially adsorbed. The same conclusion has already been deduced from the study of surface interactions. Heat-flow calorimetry may therefore be used to confirm directly some conclusions of the study of adsorption sequences.

It should be observed that curve C (Fig. 31) which characterizes the steady activity of the catalyst surface may be directly compared to activity curves obtained by other techniques. The evolutions of heat as a function of time produced by the reaction of doses of reaction mixture at the surface of four different nickel oxide catalysts are, for instance, plotted on Fig. 32. These curves are very similar indeed to the kinetic curves obtained with the same catalysts in a classical static reactor (Fig. 33) (8).

Heat-flow calorimetry may be used also to detect the surface modifications which occur very frequently when a freshly prepared catalyst contacts the reaction mixture. Reduction of titanium oxide at 450°C by carbon monoxide for 15 hr, for instance, enhances the catalytic activity of the solid for the oxidation of carbon monoxide at 450°C (84) and creates very active sites with respect to oxygen. The differential heats of adsorption of oxygen at 450°C on the surface of reduced titanium dioxide (anatase) have been measured with a high-temperature Calvet calorimeter (67). The results of two separate experiments on different samples are presented on Fig. 34 in order to show the reproducibility of the determination of differential heats and of the sample preparation.

When small doses of the reaction mixture ($CO + \frac{1}{2}O_2$) contact at 450°C a freshly reduced sample of anatase, heat is evolved rapidly. But, it has been observed that when the thermal balance of the calorimeter is apparently restored, part of the gas has not been consumed. Pressure measure-

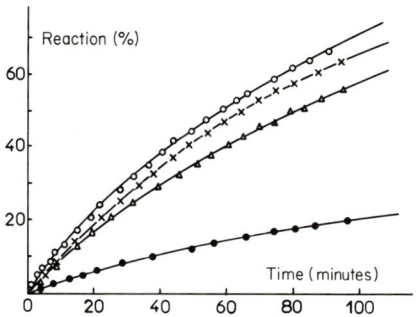

FIG. 33. Reaction yield as a function of time for carbon monoxide oxidation at room temperature on pure and doped nickel oxides. ● NiO (200), △ NiO(Li) (250), ○ NiO (250) × NiO(Ga) (250). Reprinted from (8) with permission. Copyright © 1969 by Academic Press, Inc., New York.

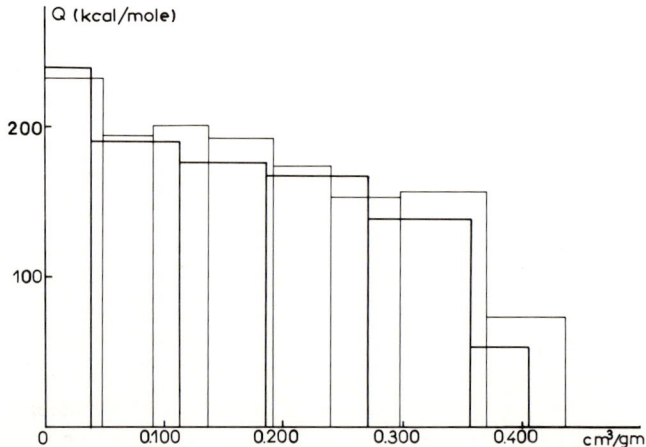

Fig. 34. Differential heats of adsorption of oxygen at 450°C on the surface of samples of titanium oxide (anatase) which have been previously reduced by carbon monoxide (p_{CO} = 70 Torr) at 450°C for 15 hr (67).

ments and mass spectrometric analyses have shown that the gas remaining in the volumetric line is pure carbon monoxide. Carbon monoxide however, reacts very slowly with the sample, but its reaction does not produce any measurable thermal effect. The differential heats recorded during the reaction of the first doses of reaction mixture (expressed per mole of consumed gas) are presented on Fig. 35. They are much higher than the normal heat of combustion of carbon monoxide (~68 kcal/mole CO), but

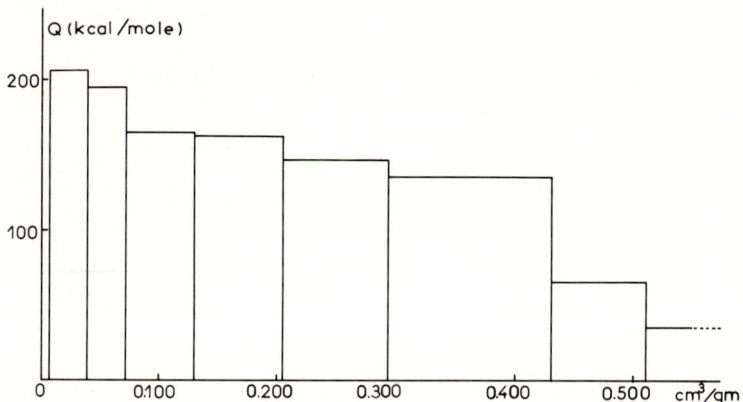

Fig. 35. Differential heats released by the interaction at 450°C of doses of the mixture (CO + ½O$_2$) with a sample of titanium oxide (anatase) which has been previously reduced by carbon monoxide (p_{CO} = 70 Torr) at 450°C for 15 hr. (67).

they are very similar indeed to the differential heats of adsorption of pure oxygen on the same titanium oxide at the same temperature (Fig. 34).

As the number of doses of reaction mixture admitted to the sample increases, the profile of the thermograms changes. Simultaneously, the heat evolved decreases to $\sim$70 kcal/mole CO. The catalytic reaction occurs. Its rate is much slower than the rate of adsorption of oxygen, but it is faster than the rate of the interaction of carbon monoxide with the catalyst surface. These results indicate, therefore, that the very reactive sites, toward oxygen, which are formed during the reducing pretreatment of the sample, are not active for the oxidation of carbon monoxide at 450°C. They react with all oxygen in the first small doses of the reaction mixture, and it is only when these sites are saturated by oxygen that the catalytic reaction proceeds on less reactive surface sites.

Finally, experimental procedures differing from that described in the preceding examples could also be employed for studying catalytic reactions by means of heat-flow calorimetry. In order to assess, at least qualitatively, but rapidly, the decay of the activity of a catalyst in the course of its action, the reaction mixture could be, for instance, either diluted in a carrier gas and fed continuously to the catalyst placed in the calorimeter, or injected as successive slugs in the stream of carrier gas. Calorimetric and kinetic data could therefore be recorded simultaneously, at least in favorable cases, by using flow or pulse reactors equipped with heat-flow calorimeters in place of the usual furnaces.

VIII. Conclusions

In the various sections of this article, it has been attempted to show that heat-flow calorimetry does not present some of the theoretical or practical limitations which restrain the use of other calorimetric techniques in adsorption or heterogeneous catalysis studies. Provided that some relatively simple calibration tests and preliminary experiments, which have been described, are carefully made, the heat evolved during fast or slow adsorptions or surface interactions may be measured with precision in heat-flow calorimeters which are, moreover, particularly suitable for investigating surface phenomena on solids with a poor heat conductivity, as most industrial catalysts indeed are. The excellent stability of the zero reading, the high sensitivity level, and the remarkable fidelity which characterize many heat-flow microcalorimeters, and especially the Calvet microcalorimeters, permit, in most cases, the correct determination of the Q–θ curve—the energy spectrum of the adsorbent surface with respect to a given adsorbate—by the repeated introduction of small doses of gas in the calorimeter cell containing the adsorbent, although some of these experiments last for many days. Heat-flow microcalorimeters have been used for chemisorption studies in a wide temperature range, extending from 30°C to $\sim$500°C, which includes the usual temperatures of most industrial catalytic processes.

Moreover, the use of heat-flow calorimetry in heterogeneous catalysis research is not limited to the measurement of differential heats of adsorption. Surface interactions between adsorbed species or between gases and adsorbed species, similar to the interactions which either constitute some of the steps of the reaction mechanisms or produce, during the catalytic reaction, the inhibition of the catalyst, may also be studied by this experimental technique. The calorimetric results, compared to thermodynamic data in thermochemical cycles, yield, in the favorable cases, useful information concerning the most probable reaction mechanisms or the fraction of the energy spectrum of surface sites which is really active during the catalytic reaction. Some of the conclusions of these investigations may be controlled directly by the calorimetric studies of the catalytic reaction itself.

Finally, the response of a heat-flow calorimeter, the thermogram, which is recorded during the experiment may be used to determine the time-dependence of the heat evolution, the thermokinetics of the phenomenon under investigation. This is very probably the most important application of heat-flow calorimetry. It has already found many practical uses in the field of adsorption and heterogeneous catalysis. But, because of the unceasing progress in data storage, correction, and analysis, this specific application of heat-flow calorimetry may become even more important in the near future. Late Professor Calvet remarked to us, twelve years ago that "it would be foolish to pretend explaining everything by means of heat-flow calorimeters, but, thermogenesis being the most common feature of all natural phenomena, heat-flow calorimetry will reveal, in many cases, new aspects of old experiments." Experiment after experiment, we have become convinced that this remark is totally applicable to the field of surface phenomena and related subjects, and we do believe that heat-flow calorimetry should be considered, at least for this reason, as an exciting new tool in heterogeneous catalysis research.

Acknowledgments

The author is grateful to Professor J. L. Petit and Dr. C. Brie, Institut National des Sciences Appliquées de Lyon, for helpful comments, and to Professor S. J. Teichner, Faculté des Sciences de l'Université de Lyon, for a critical reading of the manuscript.

References

1. De Boer, J. H., *Advan. Catal. Relat. Subj.* **8,** 17 (1956).
2. Ehrlich, G., *J. Chem. Phys.* **31,** 1111 (1959).
3. Young, D. M., and Crowell, A. D., "Physical Adsorption of Gases," p. 2. Butterworth, London, 1962.
4. Hayward, D. O., and Trapnell, B. M. W., "Chemisorption," p. 3. Butterworth, London, 1964.
5. Thomas, W. J., and Thomas, J. M., "Introduction to the Principles of Heterogeneous Catalysis," p. 15. Academic Press, New York, 1967.
6. Hayward, D. O., and Trapnell, B. M. W., "Chemisorption," p. 194. Butterworth, London, 1964.
7. Boreskov, G. K., *Proc. 3rd Int. Congr. Catal., Amsterdam, 1964* p. 163 (1965); Boreskov, G. K., Sazonov, V. A., and Popovskii, V. V., *Dokl. Akad. Nauk SSSR* **176,** 1331 (1967).
8. El Shobaky, G., Gravelle, P. C., and Teichner, S. J., *J. Catal.* **14,** 4 (1969).
9. Thomson, S. J., and Webb, G., "Heterogeneous Catalysis," p. 27. Oliver & Boyd, Edinburgh, 1968.
10. Dewar, J., *Proc. Roy. Soc., Ser. A* **74,** 122 (1904).
11. Wedler, G., King, D. A., Brennan, D., and Černý, S., *Discuss. Faraday Soc.* **41,** 104–112 (1966).
12. Beeck, O., *Rev. Mod. Phys.* **17,** 61 (1945).
13. Černý, S., and Ponec, V., *Catal. Rev.* **2,** 249 (1968).
14. Calvet, E., *C. R. Acad. Sci.* **232,** 964 (1951).
15. Gravelle, P. C., *J. Chim. Phys.* **61,** 455 (1964).
16. Calvet, E., and Prat, H., eds., "Microcalorimétrie, Applications Physicochimiques et Biologiques." Masson, Paris, 1956; Calvet, E., Prat, H., and Skinner, H. A., "Recent Progress in Microcalorimetry." Pergamon, Oxford, 1963.
17. Kitzinger, C., and Benzinger, T. H., *Z. Naturforsch. B* **10,** 375 (1955); Benzinger, T. H., and Kitzinger, C., *in* "Temperature, Its Measurement and Control in Science and Industry" (C. M. Herzfield, ed.), Part III, Sect. III, p. 43. Reinhold, New York, 1963.
18. Stone, F. S., *Advan. Catal.* **13,** 1 (1962).
19. Gravelle, P. C., and Teichner, S. J., *Advan. Catal.* **20,** 167 (1969).
20a. Skinner, H. A., *in* "Biochemical Microcalorimetry" (H. D. Brown, ed.), p. 1. Academic Press, New York, 1969; *20b.* Kubaschewski, O., and Hultgren, R., *in* "Experimental Thermochemistry" (H. A. Skinner, ed.), Vol. II, p. 51. Wiley (Interscience), New York, 1962.
21. Klemperer, D. F., and Stone, F. S., *Proc. Roy. Soc., Ser. A* **243,** 375 (1957).
22. Camia, F. M., Chabert, J., Gravelle, P. C., Laffitte, M., Macqueron, J. L., Petit, J. L., and Tachoire, P., *Rev. Gen. Therm.* **7,** 895 (1968).
23. Calvet, E., and Prat, H., "Récents Progrès en Microcalorimétrie," p. 8. Dunod, Paris, 1958.
24. Chessick, J. J., and Zettlemoyer, A. C., *Advan. Catal.* **11,** 263 (1959).
25. Tian, A., *C. R. Acad. Sci.* **178,** 705 (1924).
26. Tian, A., *J. Chim. Phys.* **20,** 132 (1923).
27. Zielenkiewicz, W., Zielenkiewicz, A., and Kurek, T., *Bull. Acad. Polon. Sci., Ser. Sci. Chim.* **16,** 95 (1968).
28. Gravelle, P. C., *Rev. Gen. Therm.* **8,** 873 (1969).
29. Calvet, E., and Persoz, M., *in* "Microcalorimétrie, Applications Physicochimiques et Biologiques" (E. Calvet and H. Prat, eds.), p. 100. Masson, Paris, 1956.

30. Rouquerol, J., *Colloq. Int. Cent. Nat. Rech. Sci.* **201** (in press).
31. Petit, J. L., and Brie, C., *C. R. Acad. Sci., Ser.* B **267,** 1380 (1968); Petit, J. L., personal communication.
32. Calvet, E., and Guillaud, C., *C. R. Acad. Sci.* **260,** 525 (1965).
33. Macqueron, J. L., Sinicki, G., and Bernard, R., *C. R. Acad. Sci., Ser.* B **266,** 1 (1968); Gery, A., Sinicki, G., Laurent, M., and Macqueron, J. L., *C. R. Acad. Sci., Ser.* B **266,** 113 (1968).
34. Petit, J. L., Sicard, L., and Eyraud, L., *C. R. Acad Sci.* **252,** 1740 (1961).
35. Roux, A., Richard, M., Eyraud, L., and Elston, J., *J. Phys. (Paris)* **25,** 51 (1964); Fetiveau, M., Fetiveau, Y., Richard, M., and Eyraud, L., *Bull. Soc. Chim. Fr.* p. 450 (1965); Berger, C., Richard, M., and Eyraud, L., *Bull. Soc. Chim. Fr.* p. 1491 (1965).
36. Eyraud, M., Ph.D. Thesis, No. 370, Univ. de Lyon, Lyon, 1965; Basset, J., Mathieu, M. V., and Prettre, M., *J. Chim. Phys.* **66,** 1264 (1969).
37. Eyraud, L., *in* "Thermodynamique Chimique" (J. Bousquet, ed.), p. 81. Masson, Paris, 1969.
38. Chabert, J., Petit, J. L., and Gravelle, P. C., *Rev. Gen. Therm.* **7,** 897 (1968).
39. Brie, C., Petit, J. L., and Gravelle, P. C., *C. R. Acad. Sci., Ser. B, 273,* 1 (1971).
40. Thouvenin, Y., Hinnen, C., and Rousseau, A., *Colloq. Int. Cent. Nat. Rech. Sci.* **156,** 65 (1967).
41. Chevillot, J. P., Goldwaser, D., Hinnen, C., Koehler, C., and Rousseau, A., *J. Chim. Phys.* **67,** 49 (1970).
42. Papoulis, A., "The Fourier Integral and its Applications," p. 81. McGraw-Hill, New York, 1962.
43. Laville, G., *C. R. Acad. Sci.* **240,** 1060 (1955).
44. Camia, F. M., "Traité de Cinétique Impulsionnelle," p. 18. Dunod, Paris, 1967.
45. Chapman, A. J., "Heat Transfer," 2nd Ed. Macmillan, New York, 1967.
46. Camia, F. M., *Journées Int. Transm. Chaleur, C. R., Paris, 1961* p. 703 (1962).
47. Brie, C., Petit, J. L., and Gravelle, P. C., *C. R. Acad. Sci., Ser. B, 273,* 305 (1971).
48. Camia, F. M., *Colloq. Int. Cent. Nat. Rech. Sci.* **156,** 83 (1967).
49. Bros, J. P., *Bull. Soc. Chim. Fr.* p. 2582 (1966).
50. Praglin, J., *in* "Biochemical Microcalorimetry" (H. D. Brown, ed.), p. 199. Academic Press, New York, 1969.
51. Pennington, S. N., and Brown, H. D., *in* "Biochemical Microcalorimetry" (H. D. Brown, ed.), p. 207. Academic Press, New York, 1969.
52. Brie, C., Ph.D. Thesis, No. 44, Univ. Claude Bernard, Lyon, France, 1971.
53. Gravelle, P. C., El Shobaky, G., and Teichner, S. J., *J. Chim. Phys.* **66,** 1760 (1969).
54. Gravelle, P. C., El Shobaky, G., and Teichner, S. J., *J. Chim. Phys.* **66,** 1953 (1969).
55. Gravelle, P. C., Marty, G., and Teichner, S. J., *Bull. Soc. Chim. Fr.* p. 1525 (1969).
56. Calvet, E., and Camia, F. M., *J. Chim. Phys.* **55,** 818 (1958).
57. Garrigues, J. C., Roux, R., and Valette, A., *Colloq. Int. Cent. Nat. Rech. Sci.* **156,** 357 (1967).
58. Olsen, G. H., "Electronics, a General Introduction for the Non-Specialist," p. 468. Butterworth, London, 1968.
59. Rinfret, M., and Lam, V. T., *J. Chim. Phys.* **65,** 1457 (1968).
60. Macqueron, J. L., Gery, A., Laurent, M., and Sinicki, G., *C. R. Acad. Sci., Ser.* B **266,** 1297 (1968).
61. Coten, M., Laffitte, M., and Camia, F. M., *Proc, 1st Int. Conf. Calorimetry Thermodyn., Warsaw, 1969* 67 (1971).

62. Brie, C., Guivarch, M., and Petit, J. L., *Proc. 1st Int. Conf. Calorimetry Thermodyn., Warsaw, 1969* 73 (1971).
63. Dorf, R. C., "Time-Domain Analysis and Design of Control Systems," Addison-Wesley, Reading, Massachusetts, 1964.
64. Schultz, D. G., and Melsa, J. L., "State Functions and Linear Control Systems," McGraw-Hill, New York, 1967.
65. Chevillot, J. P., Goldwaser, D., Hinnen, C., Koehler, C., and Rousseau, A., *J. Chim. Phys.* **67**, 56 (1970).
66. Young, D. M., and Crowell, A. D., "Physical Adsorption of Gases," p. 71. Butterworth, London, 1962.
67. Reymond, J. P., Ph.D. Thesis, No. 20, Univ. Claude Bernard, Lyon, France, 1971.
68. Della Gatta, G., and Fubini, B., *Cahi. Therm. Ser. B*, **1,** 72 (1971); Bastick, M., and Dupupet, G., *Cahi. Therm. Ser. B*, **1,** 44 (1971).
69. Abassin, J. J., Barberi, P., Guillouet, Y., Hartmanshenn, O., Lambard, J., Machefer, J., and Michel, J., Commissariat à l'Energie Atomique (France). *Report* No. 2932 (1966).
70. Knor, Z., *Catal. Rev*, **1,** 257 (1967).
71. Marty, G., Ph.D. Thesis, No. 715, Univ. de Lyon, Lyon, 1970.
72. Berger, R. L., *in* "Biochemical Microcalorimetry" (H. D. Brown, ed.), p. 221. Academic Press, New York, 1969.
73. Gravelle, P. C., Marty, G., and Teichner, S. J., *Proc. 1st Int. Conf. Calorimetry Thermodyn., Warsaw, 1969* 365 (1971).
74. Dell, R. M., and Stone, F. S., *Trans. Faraday Soc.* **50,** 501 (1954).
75. Courtois, M., and Teichner, S. J., *J. Catal.* **1,** 121 (1962).
76. Gravelle, P. C., and Teichner, S. J., *J. Chim. Phys.* **61,** 527 (1964).
77. Gravelle, P. C., Marty, G., El Shobaky, G., and Teichner, S. J., *Bull. Soc. Chim. Fr.* p. 1517 (1969).
78. Gale, R. L., Haber, J., and Stone, F. S., *J. Catal.* **1,** 32 (1962).
79. Gravelle, P. C., unpublished observations.
80. El Shobaky, G., Gravelle, P. C., and Teichner, S. J., *Colloq. Int. Cent. Nat. Rech. Sci.* **156,** 175 (1967).
81. Gravelle, P. C., *in* "La Catalyse au Laboratoire et dans l'Industrie" (B. Claudel, ed.), p. 88. Masson, Paris, 1967.
82. Coué, J., Gravelle, P. C., Ranc, R. E., Rué, P., and Teichner, S. J., *Proc. 3rd Int. Congr. Catal., Amsterdam, 1964* p. 748. (1965).
83. El Shobaky, G., Gravelle, P. C., and Teichner, S. J., *Advan. Chem. Ser.* **76,** 292 (1968).
84. Long, J., and Teichner, S. J., *Bull. Soc. Chim. Fr.* p. 2625 (1965); Vainchtock, M. T., Vergnon, P., Juillet, F., and Teichner, S. J., *Bull. Soc. Chim. Fr.* p. 2806, 2812 (1970).

Electron Spin Resonance in Catalysis

JACK H. LUNSFORD

Department of Chemistry
Texas A&M University
College Station, Texas

I. Introduction	265
II. Theoretical Basis	267
A. The Spin Hamiltonian	267
B. From g and Hyperfine Tensors to Molecular Structure	271
C. Relaxation Phenomena	279
III. Experimental Considerations	282
A. The Spectrometer	282
B. Quantitative Determination of Spin Concentration	286
C. Analysis of Polycrystalline Spectra	287
IV. Applications	295
A. Adsorbed Atoms and Molecules	295
B. Surface Defects	315
C. Transition Metal Ions	320
Appendix A. Quantum Mechanics for Spin Systems	326
Appendix B. Determination of Energy Levels from the Spin Hamiltonian	328
Appendix C. The g Tensor	332
Appendix D. The Hyperfine Tensor	336
References	340

I. Introduction

The electron spin resonance (ESR) technique has been extensively used to study paramagnetic species that exist on various solid surfaces. These species may be supported metal ions, surface defects, or adsorbed molecules, ions, etc. Of course, each surface entity must have one or more unpaired electrons. In addition, other factors such as spin–spin interactions, the crystal field interaction, and the relaxation time will have a significant effect upon the spectrum. The extent of information obtainable from ESR data varies from a simple confirmation that an unknown paramagnetic species is present to a detailed description of the bonding and orientation of the surface complex. Of particular importance to the catalytic chemist is the high sensitivity, which may offer an exclusive technique for studying low concentrations of active sites. In certain cases unexpected oxidation states have been detected, while in other work commonly proposed intermediates have been identified. For example, the presence of the oxygen anions O_2^- and O^- have often been inferred through mechanistic studies and conductivity measurements. Both of these species have now been identified by means of ESR spectroscopy.

A stimulating introduction to magnetic resonance spectroscopy applied to catalysis was written by O'Reilly (1) at a time when little experimental work had actually been carried out. Kokes (2) later described the technique and reviewed the literature up to about 1965. Adrian (3) has also discussed the application of ESR to surface studies. Although several excellent texts have been written on the general subject of electron spin resonance, none of these specifically treats problems related to heterogeneous catalysis. For a more qualitative introduction, the reader is referred to a book by Bersohn and Baird (4); whereas, for a more comprehensive treatment at the introductory level, books by Carrington and McLachlan (5) and by Bolton and Wertz (6) are recommended. The ESR spectra of inorganic radicals have been reviewed in a book by Atkins and Symons (7) and in an article by Symons (8). Other general review articles have been written by Symons (9) and by Morton (10).

The present article was designed both to be instructional and to review the rapidly expanding literature. The scope of this review has been limited to paramagnetic species that are reasonably well defined by means of their spectra. Several important areas of study have thus been omitted because the origin of the paramagnetic species can only be deduced in a very general sense. Coke formation, for example, has been studied by means of the ESR technique (11); yet, the spectrum of coke is not very useful in the identification of the free radicals at the surface. Likewise, the spectra of various palladium and other transition metal compounds are void of structure and the spectra are of questionable value in the identification of the surface complexes (12–15). A limited amount of work has been carried out on supported ferromagnetic materials, but this also has been neglected in the present review (16).

A working knowledge of the ESR technique requires an elementary background in certain aspects of quantum mechanics. Such information is well documented in texts on the subject; however, it is often difficult to find a concise review of the essential operations for spin systems. The limited material presented here, particularly in the appendices, is given solely to introduce (or refresh) the reader in the procedures for such operations. The material is presented strictly from a functional viewpoint, and no attempt has been made to justify the procedures from elementary principles. The choice of subject matter was judged according to its utility, based upon the author's own experiences.

II. Theoretical Basis

A. The Spin Hamiltonian

In ESR spectroscopy one is observing transitions from one energy state to another. For the experimental results to have meaning, it must be possible to relate these transitions to fundamental properties of the paramagnetic species. The various interactions which give rise to energy states of the spin system are written in an operator form which is called the spin Hamiltonian. This spin Hamiltonian includes all possible interactions between spins or between spins and the external magnetic field. It does not include coulombic and other interactions directly; however, these enter in through the g tensor, the hyperfine tensor, etc. In one sense the spin Hamiltonian may be considered to be an empirical equation which is used to describe the experimentally observed energy levels of the system; yet, each parameter in the equation can be, in principle, related to fundamental interactions in the paramagnetic species.

Perhaps it would be well to start with a simple example of the application of the spin Hamiltonian using a case in which the interactions are easily defined. In practice the situation is often reversed; that is, the experimental results indicate a particular spin Hamiltonian, and then one searches for species which are consistent with this Hamiltonian. In the present example it is convenient to begin with one of the simplest free radicals, the hydrogen atom. For this atom there are three spin interactions that must be considered: the interaction between the magnetic moment of the unpaired electron and the external magnetic field (electronic Zeeman term), the interaction between the magnetic moment of the proton and the external magnetic field (nuclear Zeeman term), and the interaction between the magnetic moment of the proton and the electron (hyperfine term). The resulting spin Hamiltonian which includes these three terms is

$$\mathcal{H}_s = g\beta \mathbf{H} \cdot \mathbf{S} - g_n\beta_n \mathbf{H} \cdot \mathbf{I} + a\mathbf{I} \cdot \mathbf{S}. \tag{1}$$

Here, β and β_n are constants known as the Bohr magneton and nuclear magneton, respectively; g and g_n are the electron and nuclear g factors; a is the hyperfine coupling constant; $\mathbf{H}$ is the external magnetic field; while $\mathbf{I}$ and $\mathbf{S}$ are the nuclear and electron spin operators. The electronic g factor and the hyperfine constant are actually tensors, but for the hydrogen atom they may be treated, to a good approximation, as scalar quantities.

With this spin Hamiltonian and the appropriate wave function it is relatively easy to determine (Appendix B) that the spin interactions give rise to four energy levels which are a function of the external magnetic field:

$$E_1 = \tfrac{1}{2}g\beta H - \tfrac{1}{2}g_n\beta_n H + a/4, \tag{2}$$

$$E_2 = \tfrac{1}{2}g\beta H + \tfrac{1}{2}g_n\beta_n H - a/4, \tag{3}$$

$$E_3 = -\tfrac{1}{2}g\beta H - \tfrac{1}{2}g_n\beta_n H - a/4, \tag{4}$$

$$E_4 = -\tfrac{1}{2}g\beta H + \tfrac{1}{2}g_n\beta_n H + a/4. \tag{5}$$

Equations (3) and (4) are not exact, but they represent a good approximation at a magnetic field of 3000 G or larger. The exact solutions are shown graphically in Fig. 1.

The allowed transitions occur between levels E_1 and E_3, and between levels E_2 and E_4 as indicated by the dashed lines. These transitions are governed by selection rules which require that the electron spin changes by one unit while the nuclear spin remains unchanged. Under certain rather restricted conditions these selection rules no longer apply and "forbidden" transitions occur.

The transitions shown in Fig. 1 occur when microwave radiation of a proper frequency ν interacts with the spin system. In the usual ESR

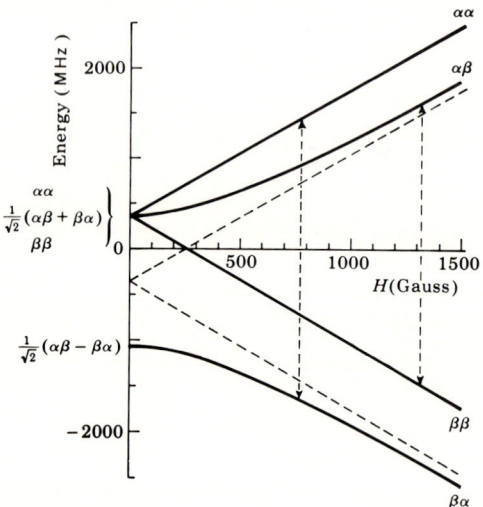

FIG. 1. Exact and approximate energy levels of the hydrogen atom in varying magnetic fields from "Introduction to Magnetic Resonance" by Alan Carrington and Andrew D. McLachlan, p. 20. Harper, New York, 1967. Reprinted by permission of Harper & Row, Publishers, Inc.

experiment the sample is bathed in microwave radiation of a constant frequency; the corresponding energy $h\nu$ is depicted by the length of the dashed lines. Meanwhile, the external magnetic field is increased at a uniform rate. At the point where the appropriate energy level splitting is just equal to $h\nu$, the microwave radiation is absorbed in the sample in the process of promoting electrons from the ground state to the excited state. The absorption of energy is detected, and the first derivative of this quantity is plotted as a function of the magnetic field.

For the hydrogen atom, two such resonance conditions occur, giving rise to two lines separated by 506 G, which is just the value of a for the hydrogen atom [Eq. (1D)]. The spectrum would look the same for a single crystal or for a polycrystalline sample because the g factor and the hyperfine constant are isotropic.

A more general form of the spin Hamiltonian for a paramagnetic species with *one* unpaired electron may be written as

$$\mathcal{H}_s = \beta \mathbf{H} \cdot \mathbf{g} \cdot \mathbf{S} + \sum_{i=1}^{n} (\mathbf{S} \cdot \mathbf{a} \cdot \mathbf{I}_i - g_n \beta_n \mathbf{H} \cdot \mathbf{I}_i), \tag{6}$$

where n is the number of different nuclei with magnetic moments. In this equation it is important to note that both the g factor and the hyperfine coupling constants are tensors. This means that the energy level spacing will depend on the orientation of each molecule with respect to the external field. For polycrystalline samples, all orientations will be present with equal probability. These polycrystalline spectra, however, have unique features which usually can be analyzed to give the principal values of the g and hyperfine tensors as will be described in Section III.C.

If two or three unpaired electrons are present so that the total spin is greater than one-half, additional terms must be added to the spin Hamiltonian of Eq. (6). The new terms may be written as

$$\mathcal{H}_D = D(S_z^2 - \tfrac{1}{3} S^2) + E(S_x^2 - S_y^2). \tag{7}$$

In this equation D and E are known as zero-field splitting terms. As the name implies, they are a measure of the extent of the energy splitting of the spin states which occur even in the absence of an external magnetic field. Organic molecules in the triplet state and most transition metals with d^2 to d^8 electrons have $S > \tfrac{1}{2}$. Although the form of Eq. (7) is identical for both cases, the origins of D and E are quite different. For triplet state organic molecules, the two terms may be related to direct dipolar interactions between two electron spins; but for transition metal ions, the two spins are indirectly coupled through the orbital angular momentum. (The interaction is formally the same as that described in Appendix D.) The manifestation of zero-field splitting in the spectra is referred to as "fine-structure."

The magnitude of D and E is a strong function of the crystal or ligand field of the metal ion: In good octahedral symmetry both D and E are zero; trigonal and tetragonal distortions lead to positive values for D; whereas, more asymmetric distortions result in positive values of D and E. The energy splitting of Cr^{3+} in an octahedral environment and after a tetragonal distortion is shown in Fig. 2. On the surface, of course, these ions are subject to large and not very uniform distortions from octahedral symmetry. Consequently, their polycrystalline spectra often cover several thousand gauss. For this reason it is difficult to obtain resolved spectra for any of the surface transition metal ions except those with an effective $S = \frac{1}{2}$.

Up to this point the paramagnetic species have been considered as isolated entities on the surface or in the bulk phase. It is clear, however, that many supported catalysts are actually clusters of ions. If these ions are paramagnetic, such as Cr_2O_3 at temperatures above about 30°, then another term must be included in the spin Hamiltonian to account for the exchange energy. This term is written as

$$\mathcal{H}_{exch} = J\mathbf{S}_1 \cdot \mathbf{S}_2, \qquad (8)$$

where J is the exchange integral. The exchange process allows spin states on neighboring ions to exchange rapidly. As a result of this process the hyperfine structure is averaged out and the width of the resonance ab-

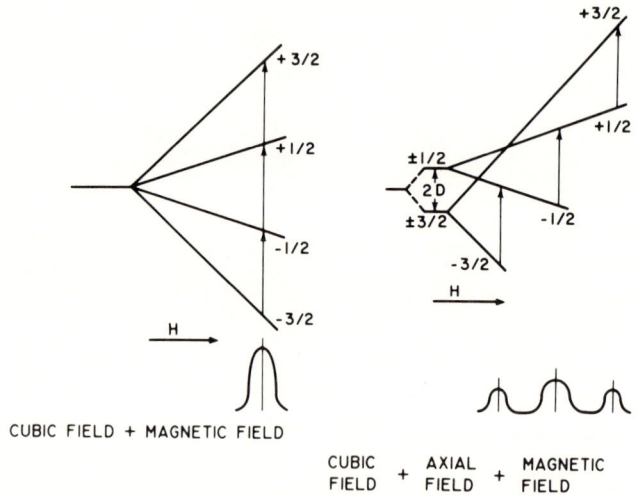

FIG. 2. Energy level splitting of Cr^{3+} in octahedral environment and after tetragonal distortion.

sorption is much less than would be predicted on the basis of dipolar spin–spin interactions, provided the paramagnetic species are identical. The latter phenomenon is known as exchange narrowing.

B. From g and Hyperfine Tensors to Molecular Structure

1. *The g Tensor*

For almost all paramagnetic species g values differ from the free electron g value of 2.0023 because of interactions between the magnetic moments due to the electron spin and orbital motion. The extent of the deviation from g_e, the free electron g value, depends upon the orbital containing the unpaired electron, the presence of a low-lying excited state, and upon the spin–orbit coupling constant. The equations which show this dependence are developed in Appendix C. Since electrons with angular momentum are in directional orbitals (p, d, and f), it is not surprising that g values depend upon the orientation of the paramagnetic species. As a corollary, g values are isotropic and very close to 2.0023 when unpaired electrons are in s orbitals, such as for the hydrogen atom. Of course, much more information is available if one or more of the g-tensor components varies appreciably from g_e. In surface studies the g tensor has been used mainly (1) to aid in the assignment of spectra, (2) to study the effect of the crystal field upon the energy levels of a paramagnetic species, and (3) to determine the motion of an adsorbed radical.

Although a number of factors are usually considered when assigning an ESR spectrum to a particular species, the experimental g tensor *must* be consistent with the theoretical g tensor, provided the latter can be determined with any degree of certainty. Very often certain qualitative features of the g-tensor can be reliably predicted; the nitric oxide molecule and the superoxide ion may be used to illustrate this point. The filling of the molecular orbitals for the NO and O_2^- molecules is shown in Fig. 3. For NO, the $2p\sigma$ orbital may be higher in energy than the $2p\pi$ orbitals; however, this will not affect our discussion. In a crystalline field of orthorhombic or lower symmetry the degenerate $2p\pi^*$ level will be split by the

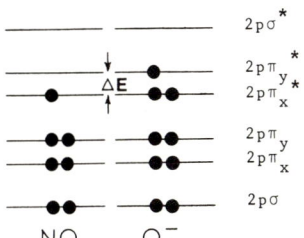

Fig. 3. Filling of the molecular orbital for NO and O_2^-.

amount ΔE. Although the unpaired electron will be mainly in the $2p\pi_x^*$ level for NO and in the $2p\pi_y^*$ level for O_2^-, there will be partial occupancy of the lowest excited states. The unpaired electron will be promoted to the $2p\pi_y^*$ state in NO and one of the $2p\pi_x^*$ electrons will be promoted to the $2p\pi_y^*$ orbital in O_2^-. For reasons outlined in Appendix C, excited states formed by the promotion of an electron from a half-filled orbital to an empty orbital result in at least one principal g value which is less than g_e. The converse is also true; that is, excited states formed by the promotion of an electron from a filled orbital to a half-filled orbital result in at least one principal g value which is greater than g_e. The theory also shows that g_{zz} is the component which is shifted for NO and O_2^-; g_{xx} and g_{yy} are approximately equal to g_e. Both of these molecules have been extensively studied in the adsorbed state and in single crystals (Section IV.A.3). In all cases the experimental spectra agree well with the theory.

Although the qualitative aspects of the g tensor may be easily determined, a calculation of the exact values requires a knowledge of ΔE, which, in the case of NO and O_2^-, is a strong function of the crystal field. The adsorbed NO and O_2^- molecules offer an excellent opportunity to study crystal field

TABLE I

g Values of Several Inorganic Radicals

Radical	Environment	g_1	g_2	g_3	Ref.
NO	MgO	1.89	1.996	1.996	17
NO	ZnO	1.94	1.999	1.999	18
NO	ZnS	1.91	1.997	1.997	18
NO	NaY-Zeolite	1.86	1.989	1.989	19
NO	Decat. Y-Zeolite	1.95	1.996	1.996	19
NO	γ-Al_2O_3	1.96	1.996	1.996	20
O_2^-	MgO	2.077	2.007	2.001	21
O_2^-	ZnO	2.051	2.008	2.002	21
O_2^-	TiO_2-Rutile	2.030	2.008	2.004	22
O_2^-	TiO_2-Rutile	2.020	2.009	2.003	22
O_2^-	TiO_2-Anatase	2.025	2.009	2.003	22
O_2^-	TiO_2-Anatase	2.024	2.009	2.003	22
O_2^-	SnO_2	2.028	2.009	2.002	23
O_2^-	NaO_2 (bulk)	2.175	2.000	2.000	24
O_2^-	KCl (bulk)	2.436	1.951	1.955	25
CO_2^-	MgO	2.0017	2.0029	1.9974	26
CO_2^-	NaCOOH (bulk)	2.0014	2.0032	1.9975	27
CO_2^-	$CaCO_3$ (bulk)	2.0016	2.0032	1.9973	28

interactions at specific adsorption sites. Results of such studies for NO and O_2^- are summarized in Table I. It is, of course, obvious that the spectra of these molecules cannot be characterized by one set of g values. On the other hand, if the energy levels are already widely separated, the environment will have little effect on the g tensor as illustrated by the principal g values of the CO_2^- ion. Here it is possible to use the absolute values to "fingerprint" the species.

To a limited extent the degree of motion of an adsorbed paramagnetic molecule may be deduced from the g tensor. Rapid tumbling of the paramagnetic species relative to a time scale of 10^{-9} sec results in an averaging of the principal g values so that

$$g_{av} = \tfrac{1}{3}(g_{xx} + g_{yy} + g_{zz}). \tag{9}$$

Or if there is rapid rotation around one axis, say the z axis, then

$$g_{\parallel} = g_{zz} \quad \text{and} \quad g_{\perp} = \tfrac{1}{2}(g_{xx} + g_{yy}). \tag{10}$$

The NO_2 molecule offers an example which illustrates this point. The spectrum of NO_2 molecules rigidly held on MgO at $-196°$ is characterized by $g_{xx} = 2.005$, $g_{yy} = 1.991$, and $g_{zz} = 2.002$ (*29*). If this molecule were rapidly tumbling, one would expect a value of $g_{av} = 1.999$. The spectrum of NO_2 absorbed in a 13X molecular sieve indicates an isotropic $g_{av} = 2.003$ (*30*), which is within experimental error of the predicted value for NO_2 on MgO. The hyperfine constants confirm that NO_2 is rapidly tumbling or undergoing a significant libration about some equilibrium position in the molecular sieve (*31*).

2. *The Hyperfine Tensor*

Interactions between one or more nuclei and the unpaired electron yield a wealth of information concerning molecular structure. In addition, they have proven invaluable in the identification of paramagnetic species. As indicated in Table II many of the common elements have isotopes with nuclear magnetic spins which distinguish them from the other elements. If the isotopes of interest are not sufficiently abundant in the natural form, enriched samples may be purchased. The quantity used in surface studies is usually quite small, so relatively expensive isotopes such as ^{17}O can be studied. In fact, it is possible to recover most of the isotope following an experiment, should the cost require it.

Although the g tensor provides evidence for the identification of a particular spectrum, one should never really be certain until hyperfine structure confirms the identification. The spectrum of the superoxide ion affords a beautiful example of the application of hyperfine structure to establish

TABLE II

Electron Spin Resonance Data for Hyperfine Interaction

Isotope	Abundance (%)	Spin (I)	$\|\psi_{ns}(0)\|^2$ (a.u.)	$\langle r^{-3}\rangle_{np}$ (a.u.)	Isotropic coupling (A_{iso}) (gauss)	Anisotropic coupling (B_o) (gauss)
^{1}H	99.9844	1/2	0.314		508	
^{2}H	0.0156	1			78	
^{10}B	18.83	3	1.408	0.775	242	
^{11}B	81.17	3/2			725	38
^{13}C	1.108	1/2	2.767	1.692	1,130	66
^{14}N	99.635	1	4.770	3.101	552	34
^{15}N	0.365	1/2			775	48
^{17}O	0.037	5/2	7.638	4.974	1,660	104
^{19}F	100	1/2	11.966	7.546	17,200	1,084
^{23}Na	100	3/2			317	
^{25}Mg	10.05	5/2				
^{27}Al	100	5/2	2.358	1.055	985	42
^{29}Si	4.70	1/2	3.807	2.041	1,220	62
^{31}P	100	1/2	5.625	3.319	3,640	206
^{33}S	0.74	3/2	7.929	4.814	975	56
^{35}Cl	75.4	3/2	10.643	6.709	1,680	100
^{37}Cl	24.6	3/2			1,395	84
^{67}Zn	4.12	5/2	4.522		376	84

that the paramagnetic species was indeed a diatomic oxygen radical. Oxygen-16 has a zero nuclear spin; hence, the spectrum assigned to O_2^- shows no hyperfine structure. Tench and Holroyd (*32*) used oxygen-17 and observed the spectrum of $^{17}O^{17}O^-$ on magnesium oxide. The nuclear spin of ^{17}O is $\frac{5}{2}$ so a paramagnetic species containing one oxygen-17 atom would yield $2I + 1$ or six lines of approximately the same amplitude for each principal direction; a total of 18 lines for the polycrystalline spectrum. A diatomic $^{17}O^{17}O$ species with equivalent oxygens would yield $2(2I) + 1$ or eleven lines with an amplitude ratio of 1:2:3:4:5:6:5:4:3:2:1 for each principal direction. Should the molecule contain two atoms of oxygen-17 that are not equivalent, the spectrum would consist of 36 lines of the same amplitude for each principal direction. Now the spectrum of O_2^- enriched in oxygen-17 clearly shows two sets of hyperfine lines, one of which contains six components ($^{17}O^{16}O$) and the other eleven components ($^{17}O^{17}O$). The results are shown in Fig. 4. Only the hyperfine splitting along one principal axis is obvious from the spectrum; this corresponds to the hyperfine interaction along the $2p\pi^*$ orbital containing the unpaired electron (Fig. 3). The interaction along the remaining $2p\pi^*$ orbital is quite small and cannot

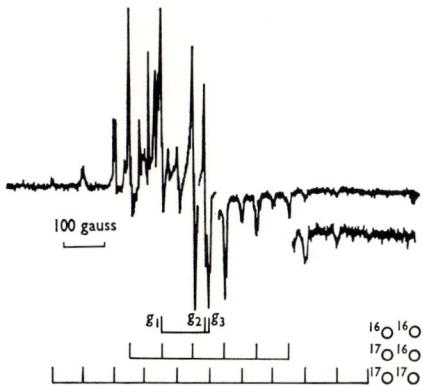

FIG. 4. The ESR spectrum of O_2^- on MgO. The field increases from left to right and the gain has been reduced by five times for the central line; a small portion of the high-field spectrum was overmodulated to show the outermost lines. For clarity, no attempt has been made in this diagram to insert levels for the lines around g_1 (*32*).

be resolved; however, it is possible to detect some interaction in the direction corresponding to the internuclear axis.

Hyperfine interaction has also been used to study adsorption sites on several catalysts. One paramagnetic probe is the same superoxide ion formed from oxygen-16, which has no nuclear magnetic moment. Examination of the spectrum shown in Fig. 5 shows that the adsorbed molecule ion reacts rather strongly with one aluminum atom in a decationated zeolite (*33*). The spectrum can be resolved into three sets of six hyperfine lines. Each set of lines represents the hyperfine interaction with ^{27}Al ($I = \frac{5}{2}$) along one of the three principal axes. The fairly uniform splitting in the three directions indicates that the unpaired electron is mixing with an

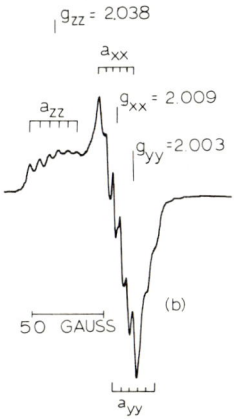

FIG. 5. The ESR spectrum of O_2^- on a decationated Y-type zeolite (*33*).

aluminum s orbital. The superoxide ion has now been used to probe magnetic interactions at adsorption sites on LaY, ScY, and AlHY zeolites as well as on supported V_2O_5. In addition, the NO molecule has been used to a similar advantage.

A fairly detailed treatment of the theory for hyperfine interactions has been given in Appendix D, and it is our intention to show how the results of this development can be used to determine molecular structure. Perhaps the most straightforward way to introduce the subject is to examine the experimental results for the NO_2 molecule adsorbed on MgO (29). This molecule has been extensively studied in the gas, liquid, and solid phase, so that there is ample data for comparison purposes.

It is known that NO_2 is a planar molecule, and the unpaired electron is thought to be mainly in a nonbonding orbital which is described by

$$\psi(4a_1) = c_1 N(s) + c_2 N(p_z) + c_3 O(p_{z1} + p_{z2}) + c_4 O(p_{y1} - p_{y2}). \quad (11)$$

Here, the directions are defined in Fig. 6. In natural NO_2 the ^{17}O content is quite small so the only observable hyperfine structure will be due to ^{14}N, which has a nuclear spin of one. Recent experiments, however, have been carried out using NO_2 enriched in ^{17}O (34). Molecular orbital calculations indicate that c_2 is reasonably large, i.e., the unpaired electron is expected to have considerable nitrogen p_z character.

The polycrystalline spectrum of NO_2 on MgO is somewhat complex, but it yields an unambiguous g and hyperfine tensor which can be checked by comparison with data for NO_2 in single crystals. For NO_2 on MgO, principal values of the hyperfine tensor are $|a_1| = 53.0$, $|a_2| = 49.0$, and $|a_3| = 67.0$ G (29). It should be noted here that neither the signs of the coupling constants nor their directions relative to the molecular coordinates

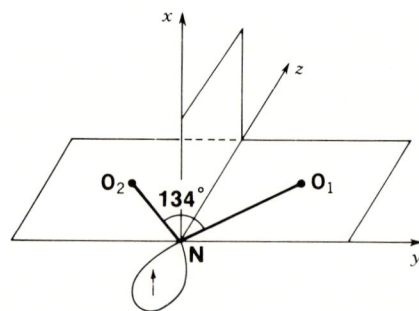

Fig. 6. Coordinate system for the NO_2 molecule (5).

have been determined. In order to determine the signs recourse is made to other data and theoretical considerations. To determine the relative signs the isotropic hyperfine coupling constant as computed from Eq. (14D), using various sign combinations, is compared with the experimental value of A_{iso} for NO_2 molecules which are rapidly tumbling. The value of $|A_{iso}|$ for NO_2 in a 13X molecular sieve is 56.9 G (30). After trying the various combinations of signs it becomes clear that all of the coupling constants must have the *same* sign if the calculated isotropic value ($|A_{iso}| = \frac{1}{3}(a_1 + a_2 + a_3) = 56.5$ G) is to roughly agree with the experimental value for a tumbling molecule.

Before discussing the absolute signs, the experimental hyperfine tensor should be resolved into its isotropic and anisotropic components as in Eq. (14D). The results are

$$\begin{pmatrix} 53.0 & & \\ & 49.0 & \\ & & 67.0 \end{pmatrix} = 56.5 + \begin{pmatrix} -3.5 & & \\ & -7.5 & \\ & & +10.9 \end{pmatrix}, \quad (12)$$

where it has been assumed that the experimental constants are all positive. Now it is apparent that the anisotropic tensor is not of the form

$$\begin{pmatrix} -\alpha & & \\ & -\alpha & \\ & & 2\alpha \end{pmatrix} \quad (13)$$

as it would be if the anisotropic interaction were due only to dipolar interactions between the nitrogen nuclear spin and an unpaired electron in one p orbital on the nitrogen. Hyperfine interaction via spin–orbit coupling is ruled out because the g values are close to g_e; therefore, one concludes that there must be some occupancy of the orthogonal p orbitals. This means that the anisotropic tensor is actually the result of two dipolar interactions [Eq. (10D)] with two radius vectors, each of which is along a coordinate axis of the molecule. Hence, the anisotropic tensor is the sum of two dipolar coupling tensors:

$$\begin{pmatrix} -3.5 & & \\ & -7.5 & \\ & & +10.9 \end{pmatrix} = \begin{pmatrix} -\alpha & & \\ & -\alpha & \\ & & 2\alpha \end{pmatrix} + \begin{pmatrix} 2\beta & & \\ & -\beta & \\ & & -\beta \end{pmatrix}. \quad (14)$$

Solving the simultaneous equations expressed by the tensor summation,

$$-3.5 = -\alpha + 2\beta$$
$$-7.5 = -\alpha - \beta$$
$$+10.9 = 2\alpha - \beta$$

it is found that $\alpha = 6.17$ and $\beta = 1.33$.

Now 2α or 2β for an unpaired electron in a *pure* 2p orbital is equal to $g_n\beta_n\frac{4}{5}\langle 1/r^3\rangle$, where $\langle 1/r^3\rangle$ is defined by Eq. (5D). The symbol B_0 is used to denote the dipolar term for a pure 2p orbital. Theoretical calculations indicate a value of 34 G for ^{14}N as shown in Table II. The ratio of 2α and 2β with respect to B_0 gives the fractional p character of the unpaired electron in the two orthogonal p orbitals. A positive value of B_0 confirms that our choice of signs in Eq. (12) was correct.

The question arises as to the directions of these two orbitals. From Eq. (11) it was anticipated that there would be considerable occupancy of the nitrogen $2p_z$ orbital; hence, the value of $2\alpha/B_0 = 0.361$ is associated with $c_2{}^2$, the fractional occupancy of the p_z orbital. The form of the β tensor requires that the other p orbital be in the x direction or $c_{p_x}^2 = 0.078$. This occupancy of the p_x orbital was not anticipated in Eq. (11).

From the isotropic coupling constant one may calculate $c_1{}^2$, the fractional occupancy of the nitrogen s orbital, which is equal to A_{iso}/A_0. The term A_0 is the Fermi contact interaction for an unpaired electron in a *pure* nitrogen 2s orbital. For NO$_2$ the value of $c_1{}^2 = 56.5/550 = 0.103$. The fraction of the unpaired electron associated with the N nucleus is then

$$c_1{}^2 + c_2{}^2 + c_{p_x}^2 = 0.54. \tag{15}$$

The remainder of the unpaired electron should be shared between the surface and the oxygen p_z orbitals.

Information concerning the geometry of the molecule can be obtained from the hybridization ratio $\gamma = c_2/c_1$. Coulson (7) has shown that for sp^n type hybridization of the atomic orbital on the central atom the bond angle is related to the hybridization ratio. In a molecule with C_{2v} symmetry, such as NO$_2$, the bond angle is given by

$$\theta = 2\cos^{-1}(\gamma^2 + 2)^{-1/2}. \tag{16}$$

This leads to a value of 129° for the ONO bond angle, which may be compared with the accepted value of 134° for free NO$_2$. The smaller bond angle for the adsorbed molecule may be rationalized by assuming that the surface draws the unpaired electron out of the p_z orbital.

This rather lengthy example, using the hyperfine tensor for the adsorbed NO$_2$ molecule, has illustrated the type of information that one can obtain from the hyperfine structure. Such knowledge about the adsorbed species as its identity, stability, motion, electronic and geometric structure, and interaction with surface fields can be obtained under suitable conditions.

C. Relaxation Phenomena

Although relaxation measurements have been widely used in nuclear magnetic resonance studies of solid catalysts and adsorbed molecules, they have not found such favor in similar ESR work. Relaxation phenomena, however, do play a very important role in any magnetic resonance experiment, whether or not this aspect of the problem is studied. In fact, the temperature at which most ESR experiments are conducted is dictated by the relaxation process. Furthermore, even qualitative data on relaxation times can be used as supporting evidence in the identification of a paramagnetic species.

Magnetic resonance is actually a rate process which may be treated in terms of the usual rate equations. Consider a two-level system as shown in Fig. 7, and suppose for the moment that transitions only occur because of interactions between the unpaired electron and the oscillating (microwave) magnetic field. If this is the only interaction, it may be shown that the probability for the downward transition, $P_{\alpha\beta}$, is equal to the probability for the upward transition, $P_{\beta\alpha}$. One can then write the rate equation

$$dN_\beta/dt = N_\alpha P_{\alpha\beta} - N_\beta P_{\beta\alpha} = P(N_\alpha - N_\beta), \qquad (17)$$

where N_α and N_β refer to the population of the $|\alpha\rangle$ and $|\beta\rangle$ states, respectively.

Let us define the population difference as

$$\Delta n \equiv N_\beta - N_\alpha, \qquad (18)$$

and the population sum as

$$N \equiv N_\beta + N_\alpha. \qquad (19)$$

From these two definitions it follows that

$$N_\beta = \tfrac{1}{2}(N + \Delta n) \quad \text{and} \quad N_\alpha = \tfrac{1}{2}(N - \Delta n). \qquad (20)$$

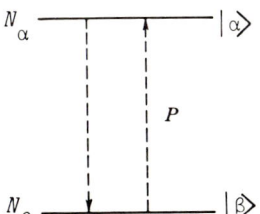

Fig. 7. Stimulated transitions for two spin states.

After substitution, Eq. (17) becomes

$$d(\Delta n)/dt = 2P\,\Delta n, \tag{21}$$

which has the solution

$$\Delta n = \Delta n_0 e^{-2Pt} \tag{22}$$

Here, Δn_0 is the population difference at equilibrium, prior to the time that the microwave power is turned on. Equation (22) shows that the population difference decreases exponentially to zero with increasing time. This is important because the power absorption, which is ultimately measured in the ESR experiment, is related to this population difference by the equation

$$dE/dt = \Delta n\, P\, \Delta E. \tag{23}$$

In effect the sample will no longer absorb microwave power when the levels become equally populated. Such a phenomenon is known as *saturation*.

Let us suppose, however, that transitions may also occur from the upper to the lower level spontaneously. A term to include these net thermal transitions should be added to the term for induced transitions in Eq. (21), which becomes

$$d\Delta n/dt = 2P\,\Delta n - (\Delta n - \Delta n_0)/T_1. \tag{24}$$

The term T_1, which is known as the *spin–lattice relaxation time*, is essentially the reciprocal of a first-order rate constant. At steady state

$$\Delta n = \Delta n_0/(1 + 2PT_1) \tag{25}$$

or

$$dE/dt = \Delta n\, P\, \Delta E = \Delta n_0\, \Delta E P/(1 + 2PT_1). \tag{26}$$

The transition probability P is proportional to the square of the microwave power level. Equation (26) shows that if the product of the microwave power level and the relaxation time are sufficiently small so that $2PT_1 \ll 1$, the rate of energy absorption in the sample (signal amplitude) will be proportional to the population difference and to the power level. If $2PT_1 \gg 1$, saturation occurs and the rate of energy absorption will no longer be proportional to the microwave power level.

Since the rate of energy absorption is proportional to Δn_0, it is perhaps worthwhile to examine this population difference at equilibrium. According to the Boltzmann law

$$\frac{N_\beta}{N_\alpha} = \exp\left(\frac{E_\alpha - E_\beta}{kT}\right) = \exp(g\beta H/kT). \tag{27}$$

After a little algebra it may be shown that

$$\Delta n_0 = \frac{N[\exp(g\beta H/kT) - 1]}{\exp(g\beta H/kT) + 1} \qquad (28)$$

For a temperature of 300°K, $kT \simeq 6 \times 10^6$ MHz and $g\beta H \simeq 9 \times 10^3$ MHz. This means that $g\beta H/kT \ll 1$ or $\exp(g\beta H/kT) \simeq 1$. Using this approximation, it follows that

$$\Delta n_0 \simeq N[\exp(g\beta H/kT) - 1]/2 \qquad (29)$$

or expanding the exponential in a Taylor series we find

$$\Delta n_0 \simeq N g\beta H/2kT. \qquad (30)$$

Provided one does not enter into a region of saturation, Eqs. (26) and (30) show that the signal intensity would be about four times as great at 77°K as at 300°K for a constant microwave power level.

The spin–lattice relaxation time, which was introduced in Eq. (24), is a measure of the time required for magnetic spin energy to be transformed into thermal energy of the lattice. In solids the most important mechanism for this energy transfer occurs via spin–orbit coupling. The electrons in the lattice have a limited degree of orbital motion which gives rise to local magnetic fields. The orbital motion, and hence the local magnetic field, is modulated by lattice vibrations. The magnetic moment of the unpaired electron couples with this oscillating magnetic field just as with the microwave field. As the unpaired electrons flip or return to their ground state the excitation energy is transferred, first into the orbital motion of the lattice electrons and then into quantized lattice vibrations known as phonons.

Thus far we have shown how saturation problems can arise if the relaxation time is too long; an experimentally difficult situation also results when the relaxation time is too short. According to the uncertainty principle

$$\Delta E \, \Delta t \geq \hbar/2, \qquad (31)$$

where Δt is interpreted as the lifetime of the state whose energy is uncertain by ΔE. If the unpaired electron spends a very short time in the excited state, there is a large uncertainty in the resonance magnetic field and the absorption lines become broad. The relaxation time T_1 is increased by lowering the temperature of the sample; thus, it is often possible to obtain good spectra at 77°K; whereas, at 300°K the lines may be broad and poorly resolved.

For systems in which the inhomogeneities in the local magnetic field exceed the linewidth as determined by the uncertainty principle, still another type of relaxation or energy transfer is apparent. Since the local magnetic field is not homogeneous, some spins will be at resonance while others will not. These resonant and nonresonant spins, however, are coupled by dipolar and exchange interactions so that energy can be transferred from one spin system to another. The rate of energy transfer is characterized by T_2, the spin–spin relaxation time. Eventually, however, the spin energy must be transferred to the lattice via a spin–lattice relaxation mechanism.

As Adrian (3) has pointed out, one may have a cross-relaxation phenomenon whereby a readily relaxed paramagnetic molecule, such as molecular oxygen, may relax another paramagnetic species which has a large T_1. This exchange of energy between the different magnetic species usually takes place via magnetic dipole interactions. Dipolar broadening of an ESR spectrum upon adsorption of O_2 is usually a test to determine whether a paramagnetic species is on the surface; yet, it is possible to actually narrow the signal of a saturated spin system by adsorption of a limited amount of oxygen.

III. Experimental Considerations

A. THE SPECTROMETER

The ESR spectrometer is an amazingly sensitive and reliable instrument. The widespread application of the ESR technique can be to a large degree attributed to commercial availability of these research instruments at a moderate cost, since most surface and catalytic chemists would find it a difficult task to construct a spectrometer from the basic hardware. It is not our intention here to give a detailed discussion of each component in the instrument; such information can be found, for example, in the texts by Poole (35) and by Wilmshurst (36). Rather, it appears necessary to give a general description of the instrument, paying particular attention only to those components that the experimentalist can readily alter to his advantage.

As indicated in the previous sections, one is measuring microwave adsorption in a sample; hence, the spectrometer must have a microwave source, a location where the microwaves interact with the sample, a detector, and a means for amplifying and presenting the signal. The microwave generator is known as a klystron, which in the ESR experiment produces about 300 mW of power. The power level to the sample is controlled by introducing an attenuator between the klystron and the sample. Samples which saturate easily are exposed to a power level in the microwatt region. The microwave energy is transmitted to the sample through a metal waveguide.

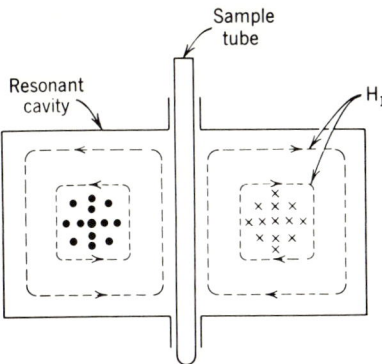

Fig. 8. ESR sample tube in a rectangular TE_{102} resonant cavity. The symbols × and · represent the E_1 field or the current distribution (35).

The sample is exposed to the electromagnetic radiation in a resonant cavity, which is fabricated from a highly conductive metal. It has dimensions comparable to the wavelength. At resonance the cavity is capable of sustaining microwave oscillations which form standing waves of different configurations called modes. The TE_{102} resonant cavity is one of the more popular designs. The magnetic field and the electric field lines of force along with the conventional position of the sample are shown in Fig. 8. One should note that the H_1 field has a maximum along the sample tube; whereas, the E_1 field has a minimum along a plane which passes through the sample and is perpendicular to the plane of the figure. This configuration becomes important in the study of materials such as liquid water, which would interact strongly with the E_1 field. Aqueous, as well as other lossy samples, may be effectively studied in this cavity by using a flat cell placed in the plane where the E_1 field is a minimum.

The cylindrical resonant cavity which operates in the TE_{011} mode is another very useful design. The H_1 and E_1 field lines of force for this cavity are shown in Fig. 9. One commercially available cavity of this type has a bottom which can be screwed in and out, thus changing the resonant microwave frequency over a limited range.

For routine studies with the ESR spectrometer, it is most convenient to work at X-band frequencies (~9.5 MHz or 3 cm). The sample is usually contained in a 4 or 5 mm diameter quartz tube having a sensitive region about 2 cm in length. An alternative frequency is at Q-band (~35,000 MHz or 1 cm). Here, the cavity dimensions are much smaller and the diameter of the sample tube is less than 2 mm. This creates some problems in handling and degassing powder samples. By varying the frequency it is possible to determine which features in a spectrum are due to Zeeman interactions

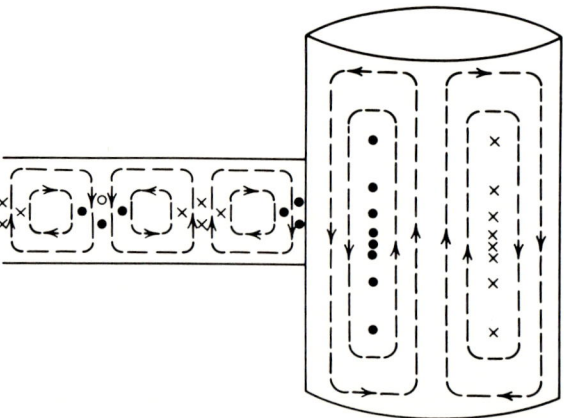

Fig. 9. The TE$_{011}$ cylindrical resonant cavity. The manner in which the waveguide couples to the cavity is also shown. The symbols × and · represent the E$_1$ field or the current distribution (*35*).

and which are due to hyperfine interactions. Zeeman splitting is a function of the magnetic field (and therefore the frequency); whereas, hyperfine splitting is not a function of the field. The resolution is often enchanced at one or the other of the two frequencies. The absolute sensitivity of the Q-band system is more than an order of magnitude greater than at X-band, but the smaller sample size and g-value anisotropy often lead to a decreased signal-to-noise ratio at that frequency.

At either frequency the sensitivity of the instrument is quite remarkable. The exact signal-to-noise ratio depends upon a number of factors including apparent line width (including g and hyperfine anisotropy), ease of saturation, the temperature, and the linear density of the sample in the quartz tube. For a relatively narrow line with peak-to-peak separation of two gauss it is possible to observe the spectrum for concentrations as low as 10^{14} spins per gram of sample. As the spectrum becomes more anisotropic, the sensitivity of course decreases. Lowering the temperature increases the sensitivity since the population difference Δn increases [(Eqs. (26) and (30)].

Two of the primary components responsible for the high sensitivity of the instrument are the modulation and the phase detection systems. Superimposed upon the static magnetic field is a variable field which is modulated typically at a frequency of 100 kHz. This modulation is effected by applying an alternating current to a set of coils on the cavity walls. The modulating field converts the resonance signal into an alternating signal as shown in Fig. 10. By means of phase detection the coherent alternating signal can be separated from noise which is characterized by random phases. Several interesting observations can be made from Fig. 10. First, it is obvious that the signal voltage is the difference between two points on the absorption curve; therefore, in the limit, the signal voltage is the derivative of the absorption curve. In addition, one should also note that the amplitude of the signal voltage increases with increasing modulation amplitude up to a certain point. When the modulation amplitude becomes larger than the resonance linewidth, the signal becomes distorted because the signal voltage is no longer a good approximation to the true derivative of the absorption curve. The spectrometer is usually operated at low enough modulation amplitudes so that these distortions do not occur.

The temperature of the sample in the cavity may be easily altered within the range from 77° to 600°K. A temperature controller is commercially available in which dry nitrogen is passed through a heat exchanger (immersed in liquid nitrogen for cooling), over a heating filament, and then over the sample tube. For cooling to 77°K it is more efficient to immerse the sample in a special dewar having a lower section that fits into the cavity. The boiling nitrogen increases the noise level; but this can be

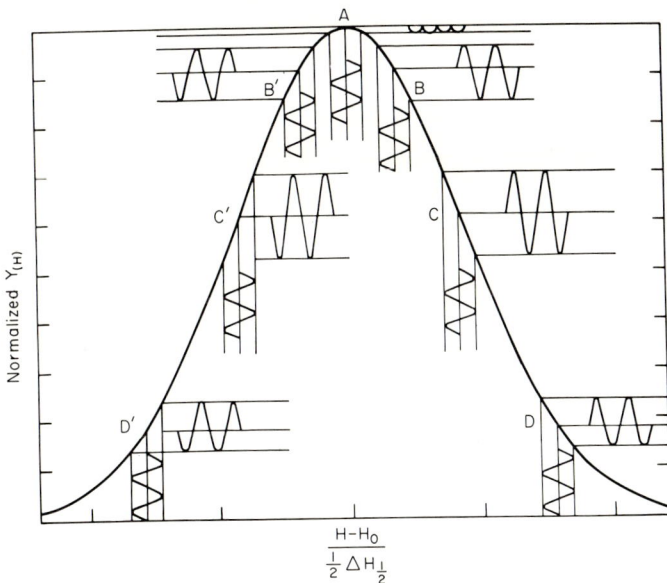

FIG. 10. The ESR signal produced at various points on the resonant line in a magnetic field modulated spectrometer. The vertical magnetic field modulation interacts with the bell-shaped adsorption curve $[Y_{(H)}]$ to produce the horizontal ESR signal. Here ΔH is the half amplitude line width and H_0 is the center of resonance (35).

minimized by attaining uniform boiling, as opposed to bumping, and by making certain adjustments on the instrument. Temperatures down to about 15°K can be achieved by boiling liquid helium and passing the cold gas over the sample. To achieve temperatures of 4°K or less it is necessary to immerse the sample in liquid helium. This is accomplished by using a special cavity which fits inside of a liquid helium dewar or by constructing a dewar which fits inside of a conventional X-band cavity (*35*). In both situations it is difficult to remove the sample for minor modifications and then to replace it in the liquid helium.

B. Quantitative Determination of Spin Concentration

Experiments on catalysts often require the measurement of an absolute spin concentration. Although this is a somewhat tedious task, it can be carried out to a moderate degree of accuracy. The number of spins per gram of unknown n_u is given by

$$n_u = (A_u/A_s)(g_s/g_u)^2 n_s/(\rho_l L), \qquad (32)$$

where A_u and A_s are the areas under the resonance absorption curves (the double integral of the derivative curves) of the unknown and a standard, respectively; g_u and g_s are the g values of the unknown and a standard; n_s is the number of spins in a standard; and ρ_l is the mass per unit length of unknown in the sample tube. The standard and unknown here must have the same total spin; if not, a factor $S_s(S_s + 1)/S_u(S_u + 1)$ should be included as a product on the right-hand side of the equation. The term L is an experimentally determined factor that enables one to use a line unknown and a point standard; it is the length of a line sample with X spins per unit length that one would use to get the same signal intensity as from a point sample with X spins at the center of the cavity. For Eq. (32) to be valid, both the unknown and the standard must be observed at the same modulation amplitude and power level. Overmodulation is no problem, but the power level *must* be sufficiently low so that neither the unknown nor the standard is near a point of saturation.

A variety of standards have been used in different laboratories. The author has found two standards to be suitable: one is a small single crystal of freshly recrystallized $CuSO_4 \cdot 5H_2O$ weighing approximately 50 µg, and the other is powdered phosphorous doped silicon embedded in a polyethylene sheet. The latter standard prepared by E. A. Gere of the Bell Telephone Laboratories, can also be used to determine g values. For maximum accuracy, the standard should be placed in the cavity along with the unknown because each different loading of the cavity will slightly alter the standing wave of the microwave field. The standard is usually glued onto the outside of the sample tube and positioned halfway between the top and the bottom of the cavity. Since it is displaced from the vertical axis by about 2 mm (assuming that the center of the sample is along the axis) there will be some error introduced because of a variation in the H_1 field across the cavity. This error can be minimized in the TE_{102} cavity by turning the sample tube so that the standard is displaced along the short dimension of the cavity.

Perhaps the greatest source of error is introduced by the double integration of the experimental derivative curve. The exact location of the baseline is critical since the outer regions or "wings" of the spectrum are weighted more heavily than the central portion. One necessary requirement is that the areas enclosed by the curve above and below the baseline must be equal. After the baseline and the initial and terminal points on the spectrum have been determined, the integration can be carried out rather easily by numerical techniques.

Although the error in the absolute spin concentration will vary from one type of study to another, a reasonable value is about $\pm 20\%$. The error in relative concentrations between samples having *identical* line shapes is probably closer to $\pm 2\%$. In the latter case the relative amplitude of the derivative spectrum may be used; hence, the troublesome double integration may be avoided.

A word of caution is in order concerning the determination of the absolute spin concentration for very broad lines of somewhat doubtful origin. If such spectra are due to a ferromagnetic phase, the difference in spin population cannot be described by a Boltzmann distribution since the spins are aligned in some systematic manner, even in the absence of an external field. Because of this large difference in spin population for the ferromagnetic phase compared to the paramagnetic standard, much more microwave power would be adsorbed in a given number of molecules of the former compared to the latter. A naive assumption that both were described by a Boltzmann distribution would lead one to calculate an erroneously large (by orders of magnitude) concentration for the ferromagnetic phase.

C. Analysis of Polycrystalline Spectra

Virtually all catalysts studied by the ESR technique are composed of small crystallites or an amorphous material. For such samples, the spectrum is the envelope of the spectra from all possible orientations of the radical with respect to the external magnetic field. In order to obtain meaningful data it must be possible to extract the principal g and hyperfine values from these polycrystalline spectra. A relatively straightforward analysis of the spectra can be made provided the resolution is adequate.

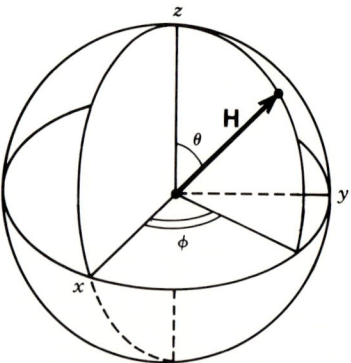

Fig. 11. Coordinate system showing the **H** vector and a sphere of unit area.

Consider a coordinate system in which the orientation of the paramagnetic species is defined by the orthogonal axes and **H** is a vector as described in Fig. 11. The sphere has unit area and every orientation or point on the sphere is equally probable. Now the number of radicals with a magnetic field orientation between θ and $\theta + \Delta\theta$ and between ϕ and $\phi + \Delta\phi$ is given by the solid angle $\Delta\Omega$ where

$$\Delta\Omega = \sin\theta\, \Delta\theta\, \Delta\phi. \tag{33}$$

A fraction of these radicals will contribute to energy absorption at a magnetic field between H and $H + dH$. This fraction depends on the individual line width ΔH as well as on the distance from resonance $H_r - H$.

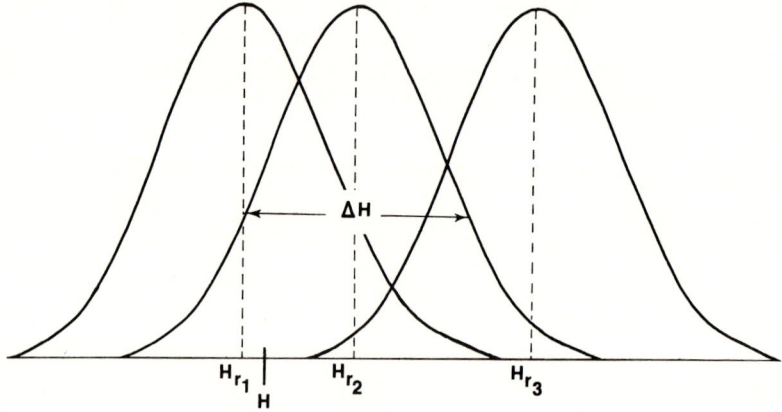

Fig. 12. Illustration of the manner in which three resonance lines from three different crystallites contribute to the adsorption at the magnetic field H.

The fraction may be written as the function $f(H_r - H, \Delta H)$, where H_r is the resonance field, or center of the individual line. It should be pointed out that the resonance field is itself a function of orientation. Perhaps this concept can best be clarified by the illustration in Fig. 12. We are trying to determine the total energy absorption at the magnetic field H. One radical, oriented so that its absorption is centered at H_{r_1}, contributes to the absorption at H. It is clear that the extent of the contribution depends upon the value of $H_r - H$ and upon the line width ΔH. Another orientation, corresponding to resonance at H_{r_2}, contributes less while a third orientation, with resonance at H_{r_3}, contributes a negligible amount at H.

The total absorption I at a particular magnetic field is the product of the function $f(H_r - H, \Delta H)$ and the element of solid angle $\Delta \Omega$ summed over all solid angles, or

$$I = \tfrac{1}{4}\pi \sum f(H_r - H, \Delta H) \, \Delta \Omega_i$$

$$= \tfrac{1}{4}\pi \sum_{j=0}^{2\pi} \sum_{i=0}^{\pi} f(H_r - H, \Delta H) \sin \theta_i \, \Delta \theta_i \, \Delta \phi_j. \qquad (34)$$

The sum includes all solid angles since, in principle, all orientations may

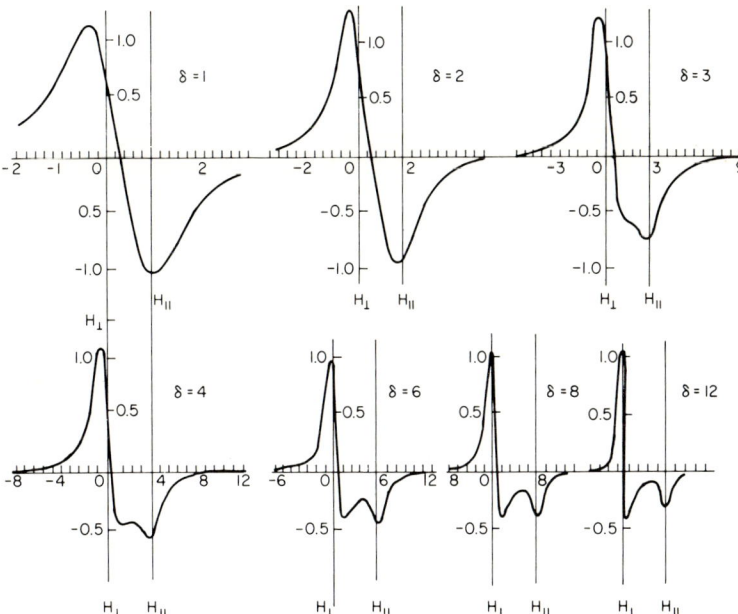

FIG. 13. Theoretically calculated shape of the asymmetric ESR line based upon the Lorentzian form of the individual lines. Reprinted from (37) with permission of Plenum Press.

contribute to the absorption at a particular field. In integral form it thus becomes

$$I = \tfrac{1}{4}\pi \int_0^{2\pi} \int_0^{\pi} f(H_r - H, \Delta H) \sin\theta \, d\theta \, d\phi, \qquad (35)$$

where the term $\tfrac{1}{4}\pi$ is a normalization factor.

As indicated earlier, one is usually concerned with the derivative curve rather than the absorption curve. The transformation can be made easily by taking the derivative of Eq. (35) which yields

$$dI/dH = \tfrac{1}{4}\pi \int_0^{2\pi} \int_0^{\pi} f'(H_r - H, \Delta H) \sin\theta \, d\theta \, d\phi. \qquad (36)$$

This means that the absorption curves in Fig. 12 are replaced by derivative curves.

The explicit form of the function $f'(H_r - H, \Delta H)$ depends on the shape of the individual derivative curves. If the absorption curve can be described by a Lorentzian function, then

$$f'(H_r - H, \Delta H) = \frac{16}{\pi(3^{1/2}\Delta H)^3}(H_r - H)\left[1 + \frac{4}{3}\left(\frac{H_r - H}{\Delta H}\right)^2\right]^{-2}. \qquad (37)$$

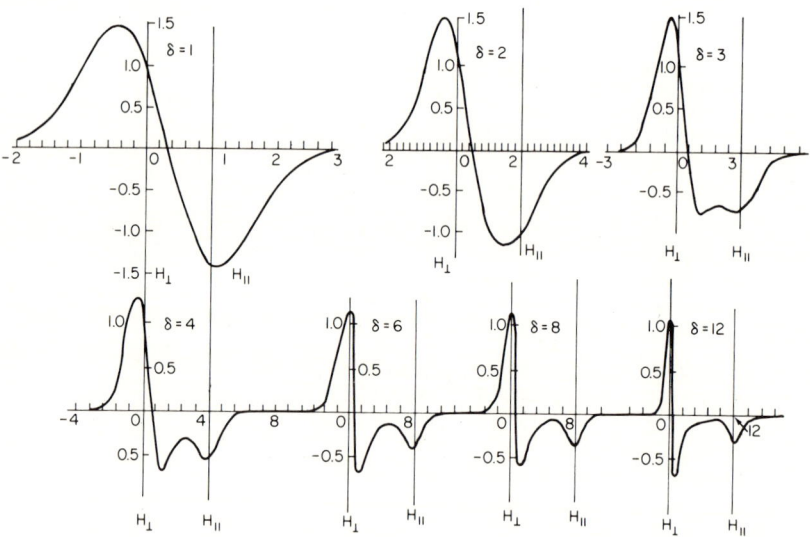

FIG. 14. Theoretically calculated shape of the asymmetric ESR line based upon the Gaussian form of the individual lines. Reprinted from (*37*) with permission of Plenum Press.

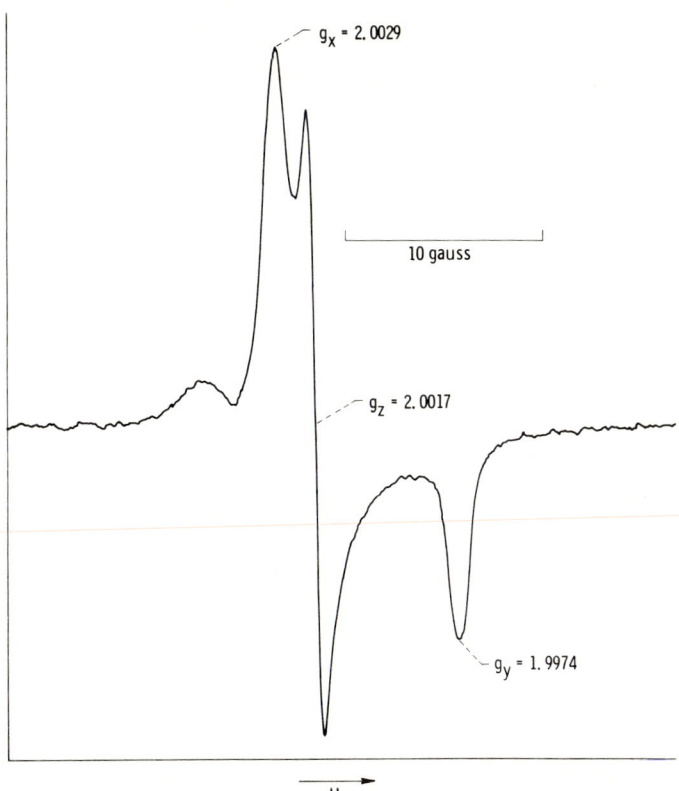

Fig. 15. The ESR spectrum of CO_2^- on the surface of a polycrystalline sample of MgO (26). The g values may be compared with $g_x = 2.0032$, $g_y = 2.0014$, and $g_z = 1.9975$ for CO_2^- in a single crystal of sodium formate (27).

Here, H is the magnetic field variable and ΔH is the magnetic field separation between the maximum and the minimum in the derivative curve. This "linewidth" may be a function of orientation also; however, in most calculations it is assumed to be constant.

The resonance field H_r as a function of orientation is determined from the appropriate spin Hamiltonian. For example, assume that the radical of interest is experiencing only Zeeman interactions and that it has one unique symmetry axis. For this case $g_{xx} = g_{yy} = g_\perp$ and $g_{zz} = g_{\parallel}$. At any orientation the g value is described by

$$g = (g_\perp^2 \sin^2 \theta + g_{\parallel}^2 \cos^2 \theta)^{-1/2}, \tag{38}$$

and therefore,

$$H_r = (h\nu_0/\beta)(g_{\parallel}^2 \cos^2 \theta + g_\perp^2 \sin^2 \theta)^{-1/2}. \tag{39}$$

The term ν_0 represents the actual microwave frequency.

A number of authors have given solutions to Eq. (36) based upon Eqs. (37) and (39). One of the more complete descriptions is reported by Lebedev (37), whose results are shown in Fig. 13 for a Lorentzian line shape. In this figure $H_\perp$ and $H_\parallel$ refer to the solutions of Eq. (39) when θ equals $\pi/2$ and 0, respectively; the origin is taken at $H_\perp$. Also,

$$\delta = \Delta H_{an}/\Delta H_i$$

where

$$\Delta H_{an} = |H_\parallel - H_\perp|$$

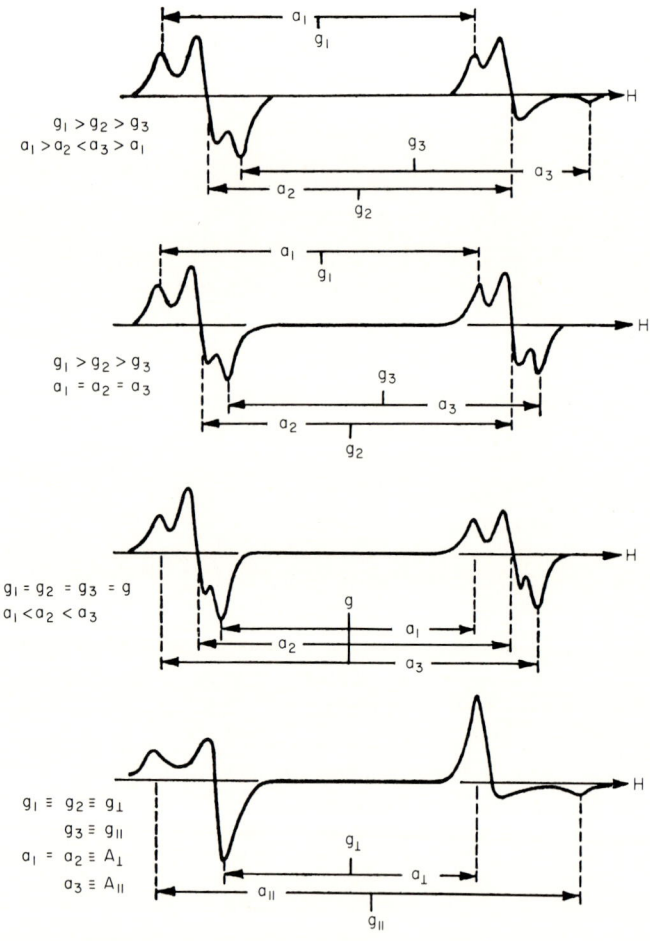

FIG. 16. Typical powder spectra for radicals with one spin $-\tfrac{1}{2}$ nucleus (7).

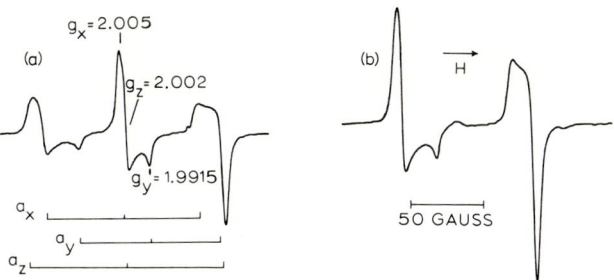

FIG. 17. The ESR spectra of the NO_2 molecule on MgO. Spectra were recorded at $-180°$: (a) $^{14}NO_2$ adsorbed, (b) $^{15}NO_2$ adsorbed (29) with permission Academic Press.

and

$$\Delta H_i = 0.865\, \Delta H. \tag{40}$$

A similar set of curves is shown in Fig. 14 for the Gaussian form of the individual line. By comparing experimental spectra with the theoretical curves described in these two figures it is possible to extract reasonably accurate g value data.

An additional degree of complexity is introduced if $g_{xx} \neq g_{yy} \neq g_{zz}$. Theoretical curves for this case have been obtained by Kneubühl (38). One result is shown in Fig. 15; a clear-cut example of such a curve is found in the spectrum of CO_2^- on MgO (26). It is interesting to note how closely

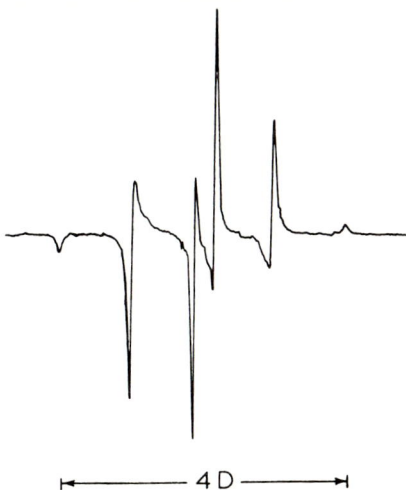

FIG. 18. The polycrystalline spectrum of Cr^{3+} $(S=\frac{3}{2})$ in an axial field (39).

the g values derived from the polycrystalline sample agree with those from CO_2^- in the sodium formate single crystal (27).

Hyperfine interactions likewise produce characteristic inflections in the derivative curve. Anisotropic hyperfine coupling is usually accompanied by anisotropic g values and as a result, the powder spectra are often quite complex. Typical powder spectra for paramagnetic species having one nucleus with $I = \frac{1}{2}$ are shown in Fig. 16. An unambiguous analysis of the more complex experimental spectra often requires the use of two microwave frequencies and a variation in the nuclear isotopes. The latter technique is illustrated by a comparison of the spectra for $^{14}NO_2$ and $^{15}NO_2$ on MgO as shown in Fig. 17.

Although there are no examples of well-resolved fine-structure spectra for surface complexes, polycrystalline spectra for bulk species have been

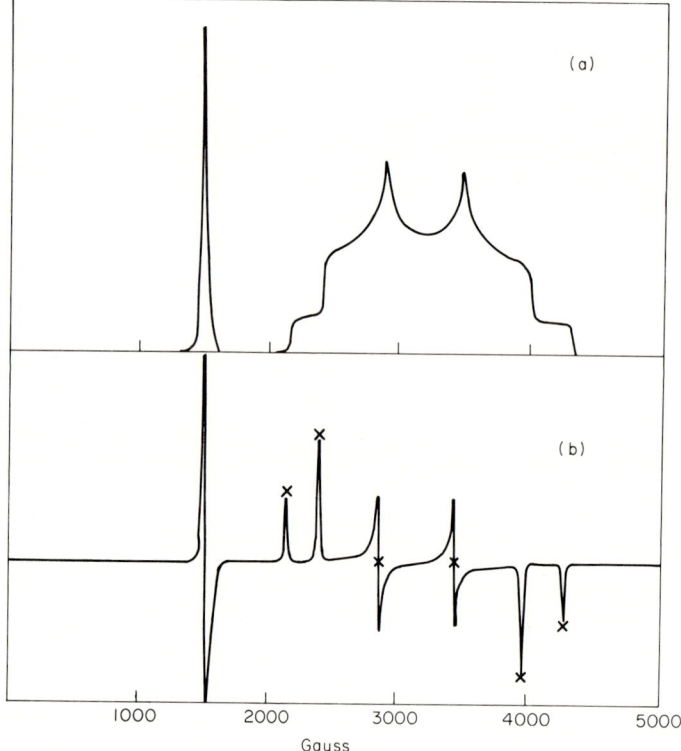

Fig. 19. Computed absorption (a) and derivative (b) spectra for naphthalene in the triplet state with $D = 1.07 \times 10^3$ G and $E = 1.65 \times 10^2$ G, $g = g_e$. The sharp, low-field line is due to $\Delta m = 2$ transitions (40).

reported by a number of investigators. For paramagnetic ions with axial symmetry the spin Hamiltonian formed by combining Eqs. (1) and (7) results in resonance absorption at magnetic fields which are given by the equation

$$H_r(m_s \to m_s - 1) = (h\nu_0/g\beta) - D(m_s - \tfrac{1}{2})(3\cos^2\theta - 1)$$
$$- (D\cos\theta\sin\theta)^2(g\beta/2h\nu_0)$$
$$\times [24m_s(m_s - 1) + 9 - 4S(S+1)]$$
$$- (D\sin^2\theta)^2(g\beta/8h\nu_0)$$
$$\times [2S(S+1) - 6m_s(m_s - 1) - 3] \qquad (41)$$

Here, θ is the angle between the magnetic field vector and the unique symmetry axis. Any anisotropy in the g value is assumed to be small compared to the zero field splitting effects. For Cr^{3+}, which is characterized by $S = \tfrac{3}{2}$, and $m_s = \tfrac{3}{2}, \tfrac{1}{2}, -\tfrac{1}{2}, -\tfrac{3}{2}$, the polycrystalline spectrum has the shape indicated in Fig. 18 (39). An example of the polycrystalline spectrum for the $S = 1$ case in which both D and E are nonzero is shown in Fig. 19 (40). A numerical evaluation of D and E may be made from the structure indicated in the spectrum.

IV. Applications

A. Adsorbed Atoms and Molecules

1. *Monatomic Radicals*

Perhaps the simplest adsorbed species is the hydrogen atom, which has been stabilized on various surfaces including silica gel, silica-alumina, alumina, and zeolites. The atoms are formed by X or γ irradiation of the solid under vacuum at $-196°$. Pariiskii and Kazanskii (41) conclude that the hydrogen atoms are formed by the radiolysis of surface OH groups. The expected two-line spectrum is isotropic and shows a hyperfine splitting of 506 G, which is very close to the theoretical value of 508 G. This confirms that the hydrogen atom is only weakly interacting with the surface. Within experimental error (± 0.2 G) the hyperfine splitting is independent of the silica-to-alumina ratio; however, the line width varies from 0.88 G on pure silica to 4.37 G on pure alumina (42). The broadening on alumina apparently reflects a dipolar interaction between ^{23}Al nuclei and the adsorbed hydrogen atom.

On silica gel the recombination reaction *in vacuo* follows first-order kinetics with a half-life of about 300 sec at $-130°$ (43). The activation

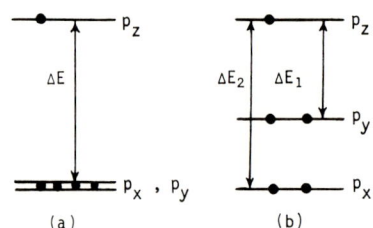

FIG. 20. Energy level schemes for the O⁻ ion in (a) tetragonal symmetry and (b) orthorhombic symmetry.

energy is about 1–2 kcal/mole. The number of hydrogen atoms formed on irradiated alumina gel is much smaller than on silica gel, and their disappearance is stepwise; i.e., at each temperature only a fraction of the atoms disappear, and the remainder are essentially stable at that temperature. It has been shown that the rate of disappearance of hydrogen atoms is increased in the presence of oxygen or ethylene and new radicals are formed. The annealing of hydrogen atoms in ethylene leads to an apparent six-line spectrum which is similar, though not identical, to the spectrum of the ethyl radical prepared by the radiolysis of ethane adsorbed on a silica gel surface. Kazanskii et al. (43) and Kazanskii and Pariiskii (44) have attributed the new spectrum to that of a polymer $CH_3(CH_2)_n\dot{C}H_2$; however, as they point out, it may also be due to a deformed ethyl radical.

Gardner (45) has observed the spectrum of Cl atoms adsorbed on a silica-gel surface at 77°K. The experimental results indicate that the orbital degeneracy of the 3p atomic orbital has been removed as a result of the electrostatic interaction with the surface. From the occupancy of the atomic orbitals one would predict that $g_\perp > g_\parallel \simeq 2.00$ and indeed the experimental g values are $g_\perp = 2.012$ with $g_\parallel = 2.003$. The hyperfine coupling indicates that the unpaired electron is highly localized in the 3p orbitals.

Undoubtedly, the most controversial spectrum in surface studies has been that of the O⁻ species. As indicated in Appendix C, the theoretical spectrum depends upon the energy level configuration for the 2p orbitals. For the energy level scheme shown in Fig. 20a the principal components of the g tensor are approximately given by

$$g_{zz} = g_e \tag{42}$$

and

$$g_{xx} = g_{yy} = g_e + 2\lambda/\Delta E. \tag{43}$$

Mikheikin and co-workers (46) pointed out that the energy levels for O⁻ on TiO₂ would better be described by the scheme shown in Fig. 20b, for which the g components are

$$g_{zz} = g_e \tag{44}$$

$$g_{xx} = g_e + 2\lambda/\Delta E_1 \tag{45}$$

and

$$g_{yy} = g_e + 2\lambda/\Delta E_2. \tag{46}$$

It is apparent from either set of equations that the spectrum of O^- will be fairly anisotropic for reasonable values of λ and ΔE. Furthermore, two principal components of the g tensor will be greater than g_e; whereas, the third component will nearly equal g_e.

A symmetric spectrum with $g = 2.003 \pm 0.001$ and $\Delta H \simeq 3$ G has been observed when oxygen or nitrous oxide is adsorbed on TiO_2, SnO_2, and ZnO. Although this spectrum was assigned to the O^- species by Kwan (47) and later by van Hooff (48), it is inconsistent with the theoretical spectrum, making the assignment doubtful.

A different spectrum having $g_{zz} = 2.002$, $g_{xx} = 2.009$, and $g_{yy} = 2.020$ was observed following the adsorption of oxygen on TiO_2. Mikheikin and co-workers (46) suggested that this spectrum was consistent with that of O^-; however, to achieve the small value for g_{xx} it was necessary to assume a rather large value for ΔE_1. As will be discussed in Section IV.A.3, the spectrum was later proven to be that of O_2^-.

A more reasonable spectrum for O^- has now been observed by Shvets, Vorotyntsev, and Kazanskii (49) using vanadium pentoxide supported on silica gel. The spectrum is complicated because of the ^{51}V hyperfine splitting; yet, the authors are able to deduce a value of $g_\perp = 2.026$ with $g_\perp > g_{||}$. The species is formed by adsorption of oxygen or N_2O on the sample at temperatures greater than $-78°$. Under most conditions the spectrum of O_2^- is also observed, but the spectrum of O^- in a pure form may be obtained by adsorbing the oxygen for several minutes at $300°$, evacuating the excess oxygen, and cooling the sample rapidly to $-196°$. The species reacts rapidly at room temperature with hydrogen, methane, ethylene, and carbon monoxide.

Recently the spectrum of O^- has been observed on MgO by Lunsford and co-workers (50, 51). The species was formed by adsorption of N_2O at low temperatures onto MgO which contained trapped electrons. By using $N_2^{17}O$ it was shown that the species was indeed O^-. The spectrum shown in Fig. 21 is characterized by $g_\perp = 2.042$ and $g_{||} = 2.0013$ with $a_\perp = 19.5$ and $a_{||} = 103$ G. From the hyperfine coupling it may be shown that the unpaired electron is localized mainly in one 2p orbital. Both the g values and the hyperfine coupling constants are consistent with the energy level diagram of Fig. 20a. The results are also consistent with the spectrum of

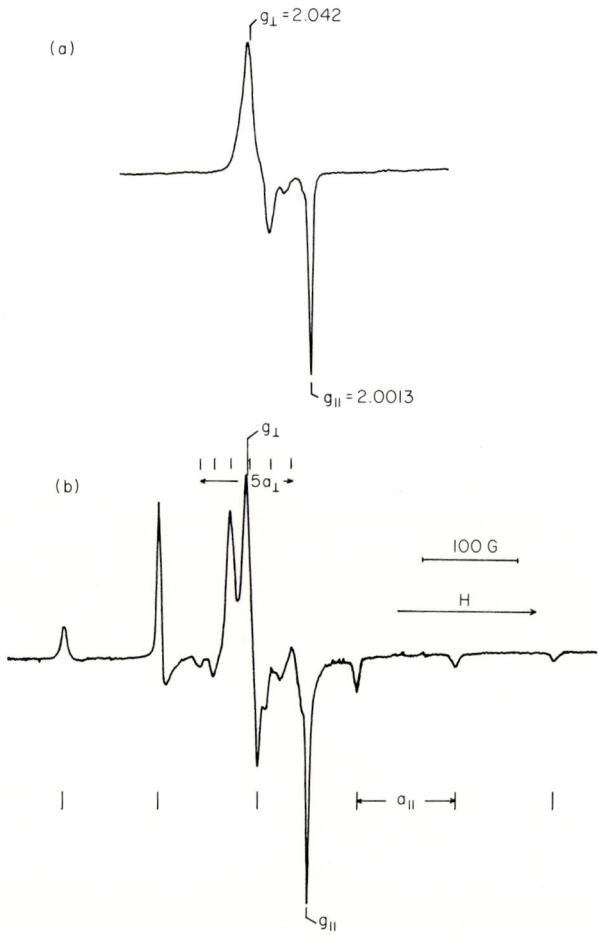

Fig. 21. (a) Spectrum of $^{16}O^-$ on magnesium oxide; (b) spectrum of O^- enriched with 71.9% oxygen-17 (51).

O^- which has now been observed in a number of different crystals. Upon exposure to molecular oxygen the O^- spectrum is replaced by another spectrum which has been attributed to the O_3^- species (52). Over a period of several hours at room temperature the ozonide ion disproportionates to form O_2^-.

2. *Organic Molecules*

In an exceptionally clear-cut experiment methyl radicals have been formed on silica (Vycor glass) by using 2537-Å radiation to decompose adsorbed methyl iodide (*53–55*). The normal methyl radical, the trideuteromethyl radical, and a 54% ^{13}C methyl radical were studied. The intensity ratio and the hyperfine splitting are in excellent agreement with predicted values. The ^{12}CH$_3\cdot$ radical is characterized by a four line spectrum having an intensity ratio of 1.0:3.0:3.0:1.0 with a g value of 2.0024 ± 0.0001 and a hyperfine splitting of 23 G. On the basis of the splitting and the symmetry of the lines it is clear that the CH$_3\cdot$ radical is planar and is free to tumble on the surface or in the pores, even at liquid nitrogen temperatures. The radical is quite stable under vacuum at room temperature, having a half-life of 100 hr; however, it is less stable in the presence of H$_2$, D$_2$, O$_2$, NO, or CH$_3$I. On the same Vycor samples another methyl radical that is weakly bonded to a surface boron impurity has been detected. Each of the four lines is further split into four additional lines as a result of the interaction with ^{11}B ($I = \frac{3}{2}$). The boron coupling constant is only 2.6 G.

Methyl radicals formed on a silica gel surface are apparently less mobile and less stable than on porous glass (*56, 57*). The spectral intensity is noticeably reduced if the samples are heated to −130° for 5 min. The line shape is not symmetric, and the linewidth is a function of the nuclear spin quantum number. Hence, the amplitude of the derivative spectrum does not follow the binomial distribution 1:3:3:1 which would be expected for a rapidly tumbling molecule. A quantitative comparison of the spectrum with that predicted by relaxation theory has indicated a tumbling frequency of 2×10^7 and 1.3×10^7 sec^{-1} for CH$_3\cdot$ and CD$_3\cdot$, respectively (*57*).

The ethyl radical has been produced on the surface of silica gel by γ irradiation of adsorbed ethane (*58*) and hexane (*59*) at −196°. Irradiation of hexane produced other radicals which were not identified. Kazanskii

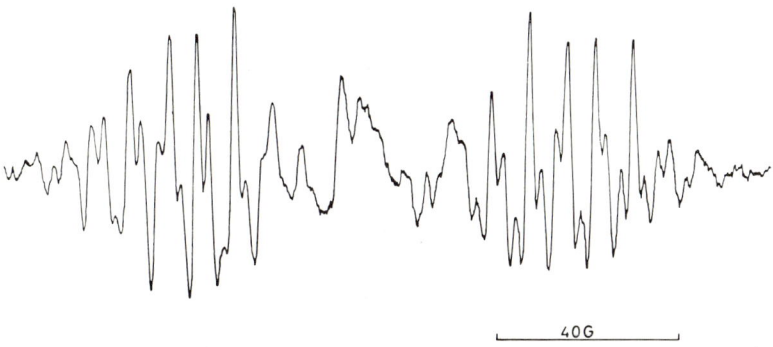

Fig. 22. Spectrum observed during irradiation of mesitylene in a 13× molecular sieve at 0° (*64*).

and Paruskii (58) have suggested that the adsorbed ethyl radicals are connected to the surface at only one end. The spectrum is a quartet having a pronounced sharp central doublet with a separation of about 27 G.

When benzene adsorbed on silica gel was irradiated with γ rays, a rather complex symmetric spectrum was observed. The spectrum consisted of a triplet of about 48 G splitting with each component being further split into a quartet of lines. The latter splitting was about 10.6 G. This spectrum is attributed to the cyclohexadienyl radical (I) which is formed by the

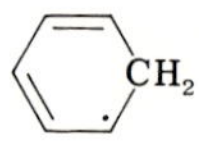

(I)

reaction of a hydrogen atom with adsorbed benzene. Strelko and Suprunenko (60, 61) have used deuterated silica gel to show that the hydrogen atom is split off from the surface hydroxyl groups of the silica gel. By way of contrast, Tanei (62) has reported that UV-irradiated benzene on alumina produced the cyclohexadienyl radical, but on silica gel the spectra were strongly dependent on the amount of benzene adsorbed. Three different spectra were observed: (1) a six-line spectrum with a splitting of 6 G was assigned to the phenyl radical; (2) a seven-line spectrum with a splitting of 4.5 G was assigned to the benzene cation radical; and (3) an apparent nine-line spectrum with a 2.4 G splitting was assigned to a benzene dimer cation radical. No cyclohexadienyl radicals were observed. Upon γ irradiation of benzene adsorbed on silica gel, Edlund and co-workers (63) observed the cyclohexadienyl radical as well as the cation radicals.

A reaction analogous to the formation of the cyclohexadienyl radical apparently occurs when mesitylene is adsorbed in a 13X molecular sieve and then irradiated with high-energy electrons (64). The resulting spectrum shown in Fig. 22 agrees well with the theoretical spectrum predicted for the radical (II) where the numbers indicate the splitting from the

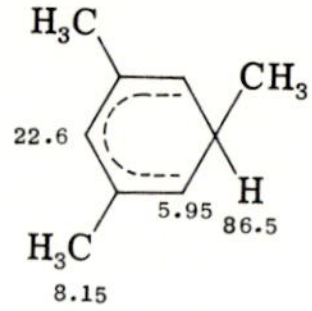

(II)

various protons. The usefulness of molecular sieves for radical trapping during irradiation has been clearly demonstrated by means of this experiment.

Photolysis of alcohols adsorbed on various solids produces relatively stable radicals. Ono and Keii (*65*) have demonstrated that methanol adsorbed on alumina produced a triplet signal when the alcohol was UV irradiated at liquid nitrogen temperatures. The signal was assigned to AlOCH$_2$·, which is thought to be the decomposition product of the surface complex AlOMe. The radical was quite stable to $-90°$ and could be observed at room temperature. Other investigators have proposed that the radical ·CH$_2$OH was formed from methanol on various metal oxides (*66*). Irradiation of ethanol produced a quintet signal which was ascribed to AlOC·HOMe, formed by the decomposition of the surface ethoxide. Similar radicals were observed by Borovskii and Kholmogoov (*67*) upon the photodissociation of ethanol adsorbed on silica gel.

The photodecomposition of isopropyl alcohol on silica gel produces a seven-line spectrum having a hyperfine separation of 20.7 G and an amplitude ratio of 1:6.7:20.2:31:21.1:7.4:1.5 (*68*). This spectrum was attributed to SiOĊMe$_2$ formed from the ether surface groups. In addition to this spectrum the spectrum of the methyl radical was also observed. Irradiation of adsorbed tert-butyl alcohol produced a three-line spectrum which was attributed to SiOMe$_2$OĊH$_2$ (*68*). Apparently the splitting from the methyl protons was too small to be observed.

Although the work discussed thus far has covered primarily neutral organic radicals, there are many types of cation and anion radicals that are stabilized on the surface. Some of these ion radicals are formed through photochemical processes; however, many others are spontaneously generated on a surface. The type of radical ion that is formed depends on the oxidizing or reducing character of particular sites on the surface, as well as on the ionization potential and the electronegativity of the adsorbed molecule.

One of the most commonly studied systems involves the adsorption of polynuclear aromatic compounds on amorphous or certain crystalline silica-alumina catalysts. The aromatic compounds such as anthracene, perylene, and naphthalene are characterized by low ionization potentials, and upon adsorption they form paramagnetic species which are generally attributed to the appropriate cation radical (*69, 70*). An analysis of the well-resolved spectrum of perylene on silica-alumina shows that the proton hyperfine coupling constants are shifted by about four percent from the corresponding values obtained when the radical cation is prepared in H$_2$SO$_4$ (*71*). The linewidth and symmetry require that the motion is appreciable and that the correlation times are comparable to those found in solution.

The mechanism for the formation of these cation radicals remains a somewhat puzzling problem. It is now reasonably clear that the catalyst must be in an oxidized state before the radical cation can be formed (*72, 73*). This state of the catalyst is usually formed when the material is calcined in air or oxygen to remove carbonaceous material from the surface. It remains in this oxidized form, even after evacuation at 500°, since the radicals can be produced following such a pretreatment. Subsequent addition of oxygen to the surface greatly enhances the number of radicals that are formed. Collision broadening may occur when the pressure of oxygen becomes significant, and oxidation of the radical may also take place if the sample is heated in the presence of oxygen (*74*). In a quantitative experiment Dollish and Hall (*75*) showed that the concentration of radical ions which can be formed on decationated zeolites depends on both the amount of available oxygen and the presence of dehydroxylated sites. The oxygen may be either in the gas phase or tenaciously chemisorbed during pretreatment. In the former case a 1:1 relationship apparently exists between the number of perylene radicals which are formed and the number of O_2 molecules added, up to some limiting value. Following a high-temperature evacuation only a fraction of the original oxygen was effective. The state of this oxygen is uncertain; however, it was established that only oxygen which can be reversibly chemisorbed and reduced off as water is effective in radical-ion formation.

According to the results of Ben Taarit and co-workers (*76*) and Neikam (*77*) Ce(III) Y zeolites will not form anthracene cation radicals but upon oxidation to Ce(IV) the radicals are readily formed. This experiment suggests that one role of oxygen during calcination may be to oxidize certain cations. The surface may be oxidized by molecules other than oxygen since the chlorination of γ-alumina by carbon tetrachloride considerably increases the sites responsible for the acceptor character. These sites, which oxidize perylene into the paramagnetic radical ion, have been attributed to biocoordinated positive aluminum atoms (*78*).

For cationic zeolites Richardson (*79*) has demonstrated that the radical concentration is a function of the electron affinity of the exchangeable cation and the ionization potential of the hydrocarbon, provided the size of the molecule does not prevent entrance into the zeolite. In a study made on mixed cationic zeolites, such as MgCuY, Richardson used the ability of zeolites to form radicals as a measure of the polarizing effect of one metal cation upon another. He subsequently developed a theory for the catalytic activity of these materials based upon this polarizing ability of various cations. It should be pointed out that infrared and ESR evidence indicate that this same polarizing ability is effective in hydrolyzing water to form acidic sites in cationic zeolites (*80, 81*).

Hirschler and co-workers (*82*) have reported the formation of a radical when benzene was adsorbed on a calcium exchanged Y-type zeolite. A hyperfine splitting of 5.0 G is in agreement with the benzene cation radical which was formed by irradiation of adsorbed benzene (63). The spectra observed following adsorption of certain olefins on a rare-earth-exchanged Y-type zeolite present a very interesting problem. Pentene-2 and 2-methylbutene-2 produce spectra which arise from six equivalent protons with the unpaired electron delocalized over two carbon atoms. It is clear from these spectra that the paramagnetic species is not a simple cation radical formed by the removal of an electron from the olefin. The data could be explained by the formation of $C_2H_6^+$ cations, but ethane has a very high ionization potential (11.5 eV).

Recently, Muha (83) has found that the concentration of cation radicals is a rather complex function of the half-wave potential; the concentration goes through a maximum at a half-wave potential of about 0.7 V. The results were obtained for an amorphous silica-alumina catalyst where the steric problem would not be significant. To explain the observed dependence, the presence of dipositive ions and carbonium ions along with a distribution in the oxidizing strengths of the surface electrophilic sites must be taken into account. The interaction between the different species present is explained by assuming that a chemical equilibrium exists on the surface.

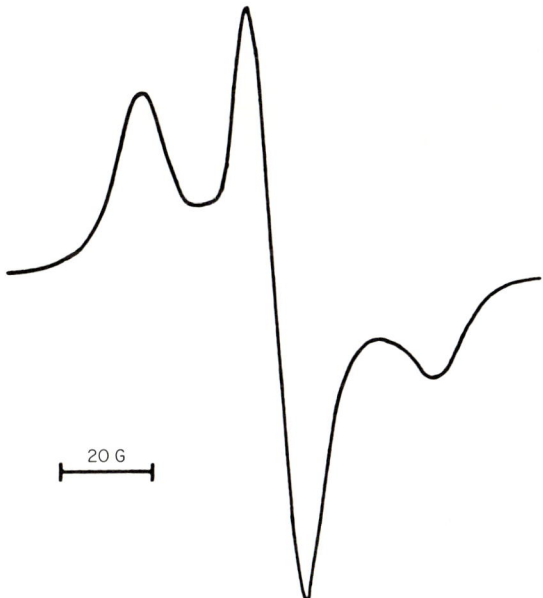

FIG. 23. The ESR spectrum of trinitrobenzene adsorbed from benzene at 20° on alumina dehydrated at 750° (88).

Anion radicals are also formed when hydrocarbons having large electronegativities (1–3 eV) such as tetracyanoethylene and various nitrocompounds are adsorbed on metal oxide surfaces. Tetracyanoethylene readily forms a stable anion radical when the neutral molecule is adsorbed on zinc oxide or alumina (84–86). The radical is characterized by a nine-line ESR spectrum with a hyperfine splitting of about 1.6 G. From the spectrum it may be concluded that the unpaired electron is interacting with four equivalent atoms of the cyano groups. The number of sites for radical formation on Al_2O_3 decreases with increasing activation temperature from 100° to 450°.

The generation of anion radicals of various mono-, di-, and trinitro aromatic compounds by electron transfer at the surface of partially dehydrated catalytic aluminas and silica-aluminas has also been reported (87–90). The ESR spectrum for trinitrobenzene, as shown in Fig. 23, can be analyzed in terms of a fixed molecule in which the unpaired electron is experiencing strong anisotropic interactions with one nitrogen atom. This appears to be a characteristic of all adsorbed nitro radicals. In contrast to the case for tetracyanoethylene, a maximum reducing activity in alumina occurs with samples dehydrated in the 600° to 700° range. Flockhart and co-workers (88) interpret this as evidence for two different reducing sites: an unsolvated hydroxyl ion, which is predominant at the lower dehydration temperatures, and an excess oxide ion associated with a defect center, which is predominant at the higher dehydration temperatures. Typical cracking catalysts of low alumina content, which are powerful electron acceptors, have only feeble reducing activity.

It is also possible to form radical cations and radical anions on the same alumina or silica-alumina surface (88). One of the more interesting observations was that a marked enhancement of the radical anion spectrum for trinitrobenzene results when perylene is adsorbed on an alumina surface, and similarly the radical cation signal is reenforced by adsorption of trinitrobenzene. The linewidths of the spectra confirm that the radical ions are separated by a distance greater than 10 Å. This means that the electron must be transferred through the lattice or that the ions separate after the transfer step, which seems unlikely. Oxygen was still required for the formation of the radical cation.

Radical anions of tetracyanoethylene and trinitrobenzene have been observed on TiO_2 and MgO. For nitro-benzene compounds adsorbed on well-degassed MgO, Tench and Nelson (90) reported spectra similar to those observed for the radical anions on alumina and silica-alumina. After discussing several possibilities for the source of electrons in MgO they suggested that some of the lattice oxide ions may act as electron donors. One of the difficulties with this proposal is the failure to detect the resulting

O⁻ species. Similarly, Che, Naccache, and Imelik (*91*) proposed that the electron donor centers are associated with OH⁻ groups which are present on the surface following activation at temperatures below 350°. Again, it is difficult to devise a mechanism in which one obtains a single paramagnetic ion from two relatively immobile diamagnetic species. Obviously, the details for the formation of the radical cations and anions are not well understood.

A variety of radical ions have been produced by photochemical reactions involving adsorbed species. One of the simplest of these is the radical ion which is formed upon γ irradiation of ethylene adsorbed on silica gel (*92*). The resulting nine-line spectrum has been attributed to $(CH_2=CH_2)^+$; however, it may be due to the corresponding negative ion. The neutral ethyl radical is also formed under these conditions.

The radical cation of benzene has been produced by γ irradiation of benzene adsorbed on silica gel (*62, 63*). The seven-line spectrum shown in Fig. 24 is expected for a molecule having six equivalent protons. An experimental coupling constant of 4.4 G, compared to a value of 3.75 G for the negative ion, gives strong support for attributing the spectrum to a positive ion. The g value is also consistent with this assignment.

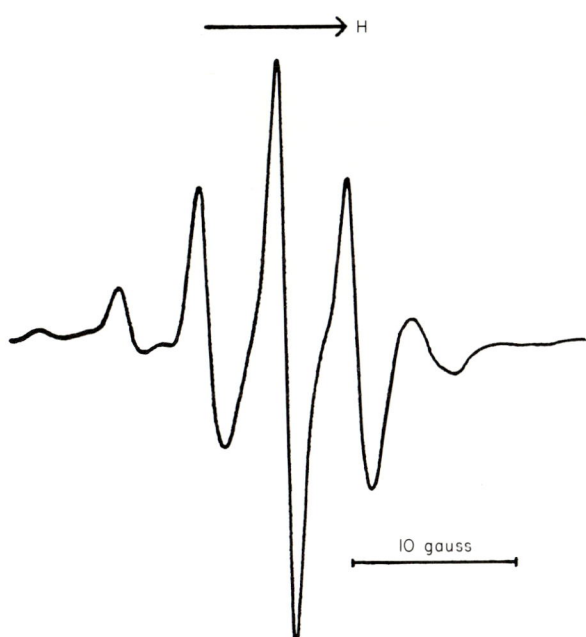

FIG. 24. The ESR spectrum of the benzene cation radical (*63*).

Aromatic radicals obtained from γ irradiation of naphthalene, anthracene, phenanthrene, and biphenyl adsorbed on silica gel have been studied; however, there exists some disagreement as to whether certain spectra should be assigned to the cation or the anion radical. Most notable of these is the spectrum of the biphenyl ion, where the largest coupling constant is 5.7 G according to Wong and Willard (93) or 6.6 G according to Kinell, Lund, and Shimizu (94). The value of 6.6 G is clearly indicative of the cation, as pointed out by Kinell and co-workers; yet, the value of 5.7 G is rather close to the splitting observed for the anion in solution. The assignment of the naphthalene spectrum to the cation radical, as proposed by Kinell and co-workers, is favored because of the good agreement between the theoretical and observed spectrum for naphthacene, in contrast to the disagreement with the observed splitting for the naphthalide ion in solution. These results indeed point out that a degree of uncertainty is involved in

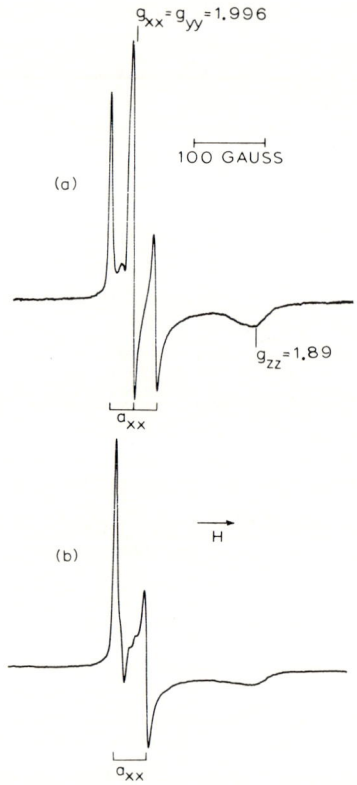

FIG. 25. The ESR spectra recorded at −180° after NO adsorption on MgO that had been degassed at 800°: (a) ^{14}NO adsorbed, (b) ^{15}NO adsorbed (17).

making assignments concerning the sign of the paramagnetic ion. This uncertainty is magnified when one realizes that significant solvent effects have been observed for both cations and anions. It would be of interest to form the radical anions by adsorbing naphthalene or biphenyl on a sodium-silica gel system.

The spectra of both the adsorbed radical cation and radical anion have been reported for phenanzine(III) (*95*) adsorbed on silica-alumina and

(III)

MgO, respectively. Photolysis by an incandescent lamp is necessary to produce the radical on MgO, and the light significantly alters, both in amplitude and shape, the spectrum of the species on silica-alumina. Both spectra are interpreted in terms of very anisotropic hyperfine interactions. This is consistent with work on radical anions such as nitrobenzene on MgO (*90*); however, it is a bit surprising in light of the motional averaging found for most cations on silica-alumina.

3. *Inorganic Molecules*

There exist a number of simple inorganic molecules which have an odd number of electrons and are therefore paramagnetic. The fifteen-electron nitric oxide molecule is an example of such an "odd" molecule. Although the ESR spectrum of nitric oxide had been studied in the gas phase, the expected spectrum of the stationary molecule had not been observed prior to the work of Lunsford (*17*) on NO adsorbed on magnesium oxide. Attempts to observe NO in rare-gas matrices had given negative results, probably because of strong spin–orbit coupling and dimerization of the molecule at low temperatures. The spectra shown in Fig. 25a and b were observed at $-180°$ when ^{14}NO and ^{15}NO, respectively, were adsorbed on MgO that had been degassed at 800°. The ^{14}N hyperfine splitting was 33 G along one principal direction and <10 G along another. The results are consistent with a more recent investigation of NO in a single crystal of NH_3OHCl where it was reported that the comparable hyperfine splittings were 36.4 G and 11.5 G (*96*). The third principal hyperfine component was 5.4 G.

The NO spectrum has now been studied for the molecule adsorbed on ZnO, ZnS (*18*), γ-Al_2O_3, silica-alumina, silica-magnesia (*20*), an A-type zeolite (*97*), H-mordenite (*98*), and a variety of Y-type zeolites including NaY (*19*), MgY, CaY, BaY, SrY (*81*), decationated-Y (*19*), ScY, LaY, and AlHY (*99*). The nitric oxide molecule has mainly been used as a probe to determine crystal and magnetic field interactions at specific adsorption sites. From values of g_{zz} (H parallel to the N–O axis) the splitting of the $2p\pi_x^*$ and $2p\pi_y^*$ levels by the crystal field have been determined.

One type of adsorption site, found on γ-alumina and on all acidic aluminosilicate catalysts, is thought to be an aluminum ion which is exposed to the surface at an oxide ion vacancy (19, 20, 81). The site is characterized by hyperfine interactions between the unpaired electron and the aluminum ion, as well as by very strong crystal field interactions. The aluminum hyperfine interaction splits each of the nitrogen hyperfine lines into six lines. In Fig. 26 the experimental spectrum for ^{15}NO on γ-Al$_2$O$_3$ is compared with a theoretical spectrum which was generated by summing the contribution from six spectra of ^{15}NO on MgO that had been displaced equally with respect to one another. The maximum number of sites, as indicated by the spin concentration, was found to be $4(\pm 1) \times 10^{12}$ sites/cm^2 for γ-alumina and two types of silica-alumina. This compares favorably with the number of α sites for CO$_2$ adsorption as determined by Peri (100).

A similar site has been observed for decationated-Y, AlHY, MgY, CaY and SrY zeolites; but not for BaY and NaY zeolites. The number of sites on MgY, CaY, and SrY is about 50 times less than the number observed

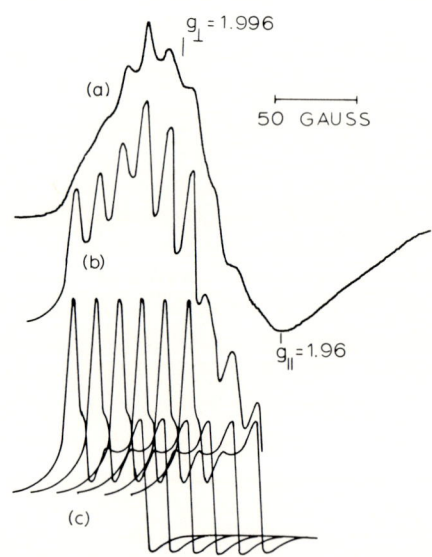

FIG. 26. (a) The ESR spectrum of ^{15}NO on dehydrated γ-alumina; (b) theoretical spectrum that resulted from the summation of the six curves (amplitude reduced) in (c); (c) six equally spaced ESR spectra of ^{15}NO on MgO (20) with permission Academic Press.

for decationated-Y and AlHY. These ESR results are qualitatively in agreement with the infrared data on acidic sites in zeolites (*80*). It should be pointed out, however, that the absolute number of sites determined by the spin concentration is almost two orders of magnitude less than the number of Bronsted or Lewis acid sites as determined by infrared spectroscopy (*101*). This discrepancy implies that either the sites observed by means of the two types of spectroscopy are not the same, or that the ESR spectrum of nitric oxide reflects only a small fraction of the total number of sites. A definitive answer to the latter problem is complicated by the fact that part of the adsorbed nitric oxide undergoes both association and disproportionation reactions on zeolite surfaces (*102*). The quantitative value of the ESR data as a means of determining site concentration is supported by the observation that the integrated NO spectrum and the O_2^- spectrum, which will be discussed later, were in agreement for decationated-Y and AlHY zeolites. It is evident in these two zeolites that the primary adsorption site involves an aluminum ion.

Nitrogen dioxide is another odd molecule that has been studied in the adsorbed state. The first work in the sorbed state was carried out by Colburn et al. (*30*) who studied NO_2 in a 13X molecular sieve. As previously discussed in Section II.B.2, the isotropic spectrum of the sample at room temperature revealed that the NO_2 molecule was in rapid motion. An elegant ESR study of the motion of NO_2 in zeolites has been conducted by Pietrzak and Wood (*31*). By comparing computer simulated spectra with experimental spectra over a wide range of temperatures, it was concluded that there was no preferred axis of rotation for NO_2 in the zeolites and that rotation greater than 35° from the equilibrium position was strongly hindered. This work demonstrates that motional narrowing and an isotropic spectrum can be obtained by restricted motion about an equilibrium position. As the authors point out, a libration of 90° total angle, i.e., a 45° displacement from the equilibrium position, is sufficient to give a completely isotropic spectrum. The spectra of NO_2 on MgO and ZnO indicate that the molecule is rigidly adsorbed with reference to a time scale of 10^{-9} sec (*29, 103*).

The molecules ClO_2 and NF_2, each of which contain 19 valence electrons, have been studied in molecular sieves (*31, 104*). For ClO_2 the paramagnetic molecules were not sufficiently separated by a dimer and strong exchange narrowing effects were observed. These completely washed out any hyperfine structure. By diluting the ClO_2 with ethylene the exchange interactions could be avoided and the ^{35}Cl splitting was observed. The spectrum at 77°K is characterized by $g_{xx} = 2.0020$, $g_{yy} = 2.0187$, and $g_{zz} = 2.0123$ with $a_{xx} = 76.5$ G, $a_{yy} = -9.2$ G, and $a_{zz} = -8.0$ G.

A five-line spectrum attributed to the NH_2 radical has been observed by Sorokin and co-workers (*105*) in γ-irradiated zeolites containing am-

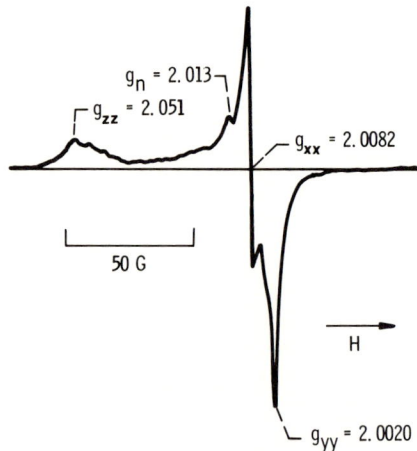

FIG. 27. The ESR spectrum of O_2^- on degassed zinc oxide. The spectrum was recorded with the sample at $-190°$ (*21*).

monia. The assignment of the spectrum is open to question, however, since it is neither consistent with the theoretical nine-line spectrum nor with the experimental spectrum of NH_2 trapped in an inert argon matrix or in an ammonia matrix. It is conceivable that a unique set of nitrogen and hydrogen hyperfine splitting constants would produce a five-line spectrum by means of overlapping lines; but if this were indeed the case, the coupling constants must be appreciably different from those found for NH_2 in solid argon or ammonia (*106*). Obviously the assignment needs to be verified by using nitrogen-15 and deuterium isotopes.

One might expect that molecular oxygen could be readily studied in the adsorbed state since it is a stable paramagnetic molecule; yet, considerable difficulty arises because the molecule contains *two* unpaired electrons in its ground state. This along with a strong spin–orbit coupling would result in a highly anisotropic spectrum for the rigid molecule. These anisotropic interactions are averaged out for the tumbling molecule, but they are replaced by other complications which result from the coupling of angular momenta with molecular rotations. The gas phase spectrum exhibits over 40 lines ranging from 3000 to 11,000 G at X-band frequencies. Keys (*107*) has compared the spectrum of oxygen adsorbed on γ-alumina with the spectrum of oxygen gas at 77°K. The position of the lines is essentially the same, although the intensity and line widths are considerably smaller for oxygen on alumina. Keys suggested that these results could be explained in terms of the stabilization of a diamagnetic O_4 dimer on the alumina.

In addition to the previously described work on adsorbed inorganic molecules that are neutral, a vast amount of effort has gone into studies of molecule ions, the most popular being the superoxide ion. At the time of this writing no less than thirty papers have been published on the subject. A partial spectrum of the O_2^- species on ZnO was first reported by Kokes (108), and thereafter the entire spectrum of O_2^- on ZnO and MgO was analyzed on the basis of the g tensor by Lunsford and Jayne (21). More recently, as described in Section II.B.2, Tench and Holroyd (32) used oxygen-17 to prove that the spectrum is indeed due to a diatomic species and that both oxygen nuclei are equivalent. The radical ion may either be formed by the spontaneous transfer of an electron from the surface of a metal or an n-type semiconductor, or by irradiation of an insulator material. The basic spectrum is illustrated by that of O_2^- on ZnO as shown in Fig. 27, but a review of the literature will quickly reveal that numerous complications arise for certain systems.

One such complication appears when the O_2^- is held at different types of adsorption sites. In such cases two or more overlapping spectra result from the ion being in different environments. The spectrum of O_2^- on TiO_2 (rutile), for example, has two low field maxima; one corresponding to $g_{zz} = 2.030$ and the other corresponding to $g_{zz} = 2.020$. Again, it has been proved using oxygen-17 that both spectra are the result of a diatomic species (22).

The existence of a variety of adsorption sites in the cationic form of Y-type zeolites is also evident from ESR spectra of the superoxide ion as shown in Fig. 28. Here, only the low-field maxima are shown. At least

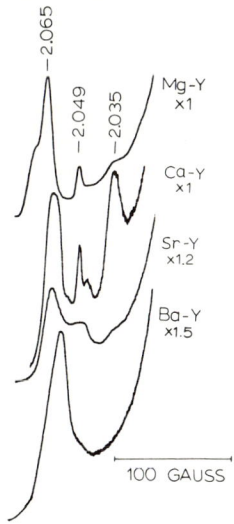

Fig. 28. The ESR spectra of O_2^- on γ-irradiated alkaline earth zeolites. Only the low-field maxima are shown here (110).

three different adsorption sites were observed for each of the alkaline-earth cationic zeolites. The predominance of a particular spectrum depended upon the type of irradiation used. It was observed that γ irradiation favored the formation of O_2^- exposed to the supercage; whereas, UV irradiation favored the formation of the ions in the sodalite units or hexagonal prisms. It is perhaps worthwhile to note here that even for γ-irradiated samples rather large differences have been noted in the position of g_{zz} for the NaY zeolite. For the dominant spectrum Kasai (109) observed a value of $g_{zz} = 2.113$, Wang and Lunsford (110) observed a value of $g_{zz} = 2.074$, and Tupikov et al. (111) observed a value of $g_{zz} = 2.068$. On many other surfaces, however, the spectra are quite reproducible from one laboratory to another.

Because of the complex spectrum and a rather low signal-to-noise ratio, a discrepancy has arisen concerning the oxygen-17 hyperfine splitting. For O_2^- on MgO Tench and Holroyd (32) assigned $a_{xx} = 77 \pm 2$, $a_{yy} = 0 \pm 4$, and $a_{zz} = 15 \pm 2$ G (Fig. 4); similar values of a_{xx} were given for O_2^- on the rutile and anatase forms of TiO_2. On the other hand, Codell and co-workers (112) assigned a value of 34.6 G for the direction corresponding to a_{xx}. It is doubtful that the parameters would vary this much between ZnO and the other two solids. A more reasonable explanation is that the spectrum assigned to $^{17}O^{17}O^-$ on ZnO was actually a mixture of the spectrum of $^{17}O^{16}O^-$ and $^{17}O^{17}O^-$, in which case the value of a_{xx} would be about 70 G for O_2^- on ZnO. Indeed, Tench and Lawson (113) have recently shown that the spectrum of $^{17}O_2^-$ on ZnO can be described by $a_{xx} = 80 \pm 2$, $a_{yy} = 0 \pm 5$, and $a_{zz} = 15 \pm 4$ G.

Another type of hyperfine interaction arises when the superoxide ion is formed on a cation which has a nonzero nuclear spin. The spectrum of O_2^- on a decationated Y-type zeolite reflects hyperfine interaction with the aluminum ions just as the NO spectrum did (33). The spectrum (shown in Fig. 5) of O_2^- on a decationated zeolite after γ irradiation is characterized by $|a_{zz}| = 6.5 \pm 0.5$, $|a_{yy}| = 5.7 \pm 0.5$, and $|a_{xx}| = 4.7 \pm 0.5$ G. The isotropic nature of the hyperfine splitting ($a_{xx} \simeq a_{yy} \simeq a_{zz}$, assuming the same sign) as well as its magnitude indicates that the splitting is mainly the result of a small aluminum s-orbital contribution to the wavefunction of the unpaired electron. Magnetic dipole interaction would have led to a smaller, anisotropic splitting. The simple six-line sets confirm that the unpaired electron is interacting with one ^{27}Al nucleus. Similar hyperfine interactions have been observed between the superoxide ion and Sc^{3+} or La^{3+} ions in Y-type zeolites (114). Shvets and co-workers (49, 115, 116) have observed well-resolved hyperfine structure from ^{51}V using vanadium pentoxide supported on silica gel. The ESR spectrum is characterized by $g_\perp = 2.023$, $g_2 = 2.011$, $g_3 = 2.004$ with $|a_1| = 9.7$, $|a_2| = 6.8$, and $|a_3| = 5.9$ G.

The state of the superoxide ion has been summarized by Naccache et al. (22). It appears probable that an ionic model is most suitable for the adsorbed species since the hyperfine interaction with the adjacent cation is relatively small. Furthermore, the equivalent ^{17}O hyperfine interaction suggests that the ion is adsorbed with its internuclear axis parallel to the plane of the surface and perpendicular to the axis of symmetry of the adsorption site. Hence, the covalent structures suggested by several investigators have not been verified by ESR data.

The role of O_2^- in the oxidation of various reactants, including hydrocarbons, has been briefly investigated but no definitive data has been obtained. In most of the studies the ESR spectra have been observed following addition of a reactant to a surface which has the superoxide ion. The ESR spectrum of the superoxide ion remains essentially unperturbed upon addition of hydrogen, methane, carbon monoxide or ethylene; how-

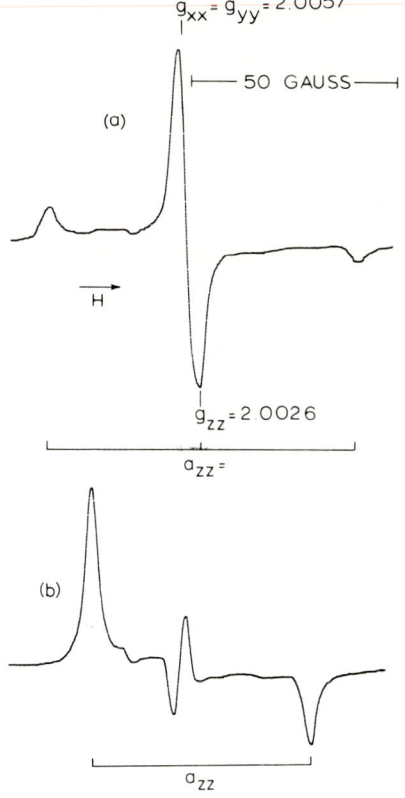

FIG. 29. The ESR spectra of the NO_2^{2-} ion on ZnO. The spectra were recorded at $-196°$ with a concentration of 10^{-3} gm of NO/gm of ZnO: (a) $^{14}NO_2^{2-}$, (b) $^{15}NO_2^{2-}$ (18).

ever, the spectrum disappears over a period of minutes when propylene, isobutene, or butene are added to the surface at room temperature. It is important to note that the reaction did not produce the spectrum of another radical, which is indeed disappointing since this would be an ideal way to follow the reaction. Van Hooff (23) has suggested that the reaction involves the formation of an allyl radical according to the equation

$$CH_2 = CH - CH_2 - CH_3 + 2O_2^- \rightleftarrows \begin{array}{c} CH_2 \cdots CH \cdots CH_3 + H \\ \downarrow \qquad\qquad \downarrow \\ O_2^- \qquad\qquad O_2^- \end{array} \qquad (47)$$

At temperatures near 100° Horiguchi and associates (117) have shown that CO reacts with the superoxide ion on ZnO, but the O_2^- spectrum remained at a fixed level following the addition of a mixture of CO and O_2. Hydrogen did not react with O_2^- at that temperature. Upon addition of nitric oxide at room temperature the O_2^- signal instantaneously disappeared.

Another very interesting molecule ion has been observed following adsorption of NO on MgO and ZnO (17, 18). By using ^{14}N and ^{15}N isotopes it has been demonstrated, as shown in Fig. 29, that the principal hyperfine coupling constants are $|a_{zz}| = 38$ and $|a_{zz}| = 54$ G for the two isotopes, respectively. The central region of the spectrum in Fig. 29 is primarily the result of the $m_I = 0$ nuclear spin state for ^{14}N with some contributions from the states $m_I = 1, -1$ in the x and y directions. For ^{15}N, no $m_I = 0$ state exists and only contributions from $m_I = \frac{1}{2}, -\frac{1}{2}$ are apparent in Fig. 30b.

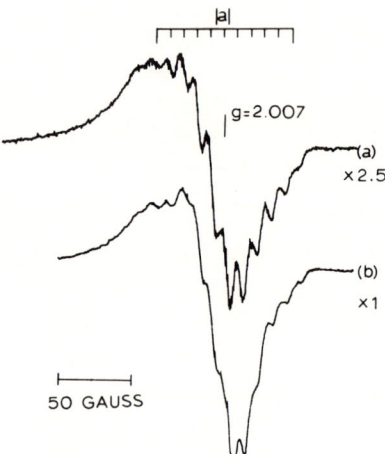

FIG. 30. The ESR spectra of γ-irradiated Y-type zeolites that were dehydroxylated at 600° prior to irradiation: (a) AlHY, (b) NH$_4$Y.

In the figure the central region is the result of overlapping lines in the x and y directions where $a_{xx} \simeq 0$. The species is stable on MgO, but on ZnO it can be removed by brief evacuation at room temperature. The g values and hyperfine splitting are quite comparable to the spectrum of NO_2^{2-}, which has been observed in irradiated single crystals of KNO_3, KNO_3/KCl, and KN_3 (118–120). Furthermore, the spectrum is consistent with that of the isoelectronic ClO_2 molecule. The NO_2^{2-} species is probably formed by the reaction of NO with an oxide ion adjacent to a cation vacancy on the surface.

The appearance of this same spectrum upon addition of oxygen or other electron acceptors to certain samples of TiO_2 has lead to a great deal of confusion concerning its origin. The spectrum was at first assigned by Iyengar and co-workers (121) to the O_2^+ ion, and by Fukuzawa and associates (122) to a bulk defect. Iyengar and Kellerman (123) later showed conclusively that the paramagnetic species contained nitrogen which was introduced when $TiCl_4$ was hydrolyzed with aqueous ammonia. In a similar study it has been shown that NO_2 is a paramagnetic impurity in certain commercial aluminas. This NO_2 can be reduced with H_2 to NO and then reoxidized (124).

The spectrum of SO_2^- radical ions has been observed by Mischenko and co-workers (125) following adsorption of SO_2 on partially reduced TiO_2 and ZnO. The spectra are characterized by $g_{||} = 2.005$ and $g_\perp = 2.001$ for TiO_2 and $g_1 = 2.007$, $g_2 = 2.004$, and $g_3 = 2.002$ for ZnO. Schoonheydt and Lunsford (126) demonstrated that $^{33}SO_2^-$ is adsorbed on MgO at two different sites: one is characterized by $g_1 = 2.010$, $g_2 = 2.005$, and $g_3 = 2.003$ with $A_3 = 59$ G; the other by $g_1 = 2.008$, $g_2 = 2.003$, and $g_3 = 2.001$ with $A_3 = 55$ G. Both of these studies indicate that the molecule ion is stationary on the surface relative to a time scale of 10^{-9} sec.

A detailed study of the CO_2^- species on MgO has been carried out by Lunsford and Jayne (26). Electrons trapped at surface defects during UV irradiation of the sample are transferred to the CO_2 molecule upon adsorption. By using $^{13}CO_2$ the hyperfine structure was obtained. The coupling constants are $a_{xx} = 184$, $a_{yy} = 184$, and $a_{zz} = 230$ G. An analysis of the data, similar to that carried out in Section II.B.2 for NO_2, indicates that the unpaired electron has 18% 2s character and 47% 2p character on the carbon atom. An OCO bond angle of 125° may be compared with an angle of 128° for CO_2^- in sodium formate.

B. Surface Defects

Irradiation of high surface area silica has produced several well-defined paramagnetic centers, one of which appears to be an intrinsic defect in the solid. Kohn (*127*), and later Kazanskii and associates (*128*), showed that γ irradiation of silica gel resulted in a sharp signal at $g = 2.0006 \pm 0.0004$. The admission of oxygen did not widen the signal and its ease of saturation remained practically unchanged, indicating that the species is largely a bulk rather than a surface defect. A similar line was observed by Muha (*129*) in γ-irradiated porous glass, although the line is slightly asymmetric. The values of $g_{\parallel} = 2.0016$ and $g_{\perp} = 2.0003$ are in good agreement with the E_1' center observed in quartz crystals (*130*). This center has been identified as an electron trapped in an oxygen vacancy between two silicon atoms. Muha and Yates (*131*) have shown that this center interacts with methanol but not with oxygen or a number of other gases.

A second type of defect is associated with boron or aluminum impurities that are present in SiO_2. In porous glass Muha (*129*) observed a rather complex spectrum which results from hyperfine interaction with ^{10}B and ^{11}B isotopes. The spectrum is characterized by $g_{\parallel} = 2.0100$, $g_{\perp} = 2.0023$, $a_{\parallel} = 15$ and $a_{\perp} = 13$ G for ^{11}B. The paramagnetic defect is apparently a hole trapped on an oxygen atom which is bonded to a trigonally coordinated boron atom. This center is irreversibly destroyed upon adsorption of hydrogen.

A similar signal has been observed by Mishchenko and Boreskov (*132*) for γ irradiated silica gel containing aluminum impurities. The spectrum is not well resolved; however, the six-line hyperfine structure with $A = 8.5$ G can be clearly seen. It is at least consistent with a spectrum observed in irradiated quartz, which has been attributed to a hole trapped on an oxygen atom that is bonded to an aluminum impurity. Adsorption of hydrogen at room temperature results in the disappearance of the spectrum on silica gel. Perhaps the most significant feature of this defect is the role which it plays as an active site in the hydrogen–deuterium exchange reaction. A good correlation has been observed between the behavior of the center and the catalytic activity with respect to thermal annealing and exposure to hydrogen. As Boudart (*133*) has pointed out in a recent article, this is one of the few examples of an active site that has been characterized with any degree of certainty.

Considerable effort has been devoted to studies of paramagnetic defects that are formed in zeolites following γ irradiation. The particular defects depend upon the cationic form of the zeolite and upon the degree of dehydration. Nozaki and Turkevich (*134*) and Stamires (*135*) along with Kasai (*109*) have reported a thirteen-line spectrum with a hyperfine splitting of 32 to 35 G and a g factor of 1.997 to 1.999. The center is ascribed to an unpaired electron interacting with four equivalent Na^+ ($I = \frac{3}{2}$) ions in the crystal lattice. Although the position of the four Na^+ ions is not certain, it has been proposed that they are tetrahedrally arranged at site II positions. Stamires (*135*) has pointed out that this center is observed in samples degassed at 350° but not at 550°; whereas, Kasai (*109*) reported the spectrum for samples degassed at 540°. On the more extensively dehydrated samples Stamires and Turkevich (*136*) noted another spectrum which consisted of a single broad line with a ΔH of 18 G and a g value of 2.0020. This spectrum is attributed to holes trapped on oxygen atoms which join adjacent $SiO_{4/2}$ units. Hydrogen–deuterium exchange experiments indicate that this center may be the active site for this reaction.

Wang and Lunsford (*110*) showed that three paramagnetic centers were formed upon γ irradiation of a CaY zeolite. The first magnetic center has $g_\perp = 2.046$ and $g_{||} = 2.0011$, the second center has a g value of 1.999, and the third center is rather broad ($\Delta H \simeq 18$ G) with $g = 2.015$. Both the first and third centers are characterized by short relaxation times. Upon exposure to 150 Torr of oxygen the first center increases in amplitude, the second is irreversibly destroyed, and the third is reversibly broadened. The O_2^- spectrum was observed following evacuation of the excess oxygen. Although it is not possible to make a positive identification of the irradiation-induced centers from these results, one can speculate as to the types of defects that might be involved. The anisotropic spectrum is similar to that observed for V-type centers in ionic crystals. Infrared and X-ray evidence has indicated that the predominant dehydroxylation process may lead to an M^+–O–M^+ configuration, where an oxide in the sodalite unit bridges a site I′ cation to a site II′ cation. One can then imagine a cation defect such as Ca^+–O– which would trap a hole to form the V-type center. The spectrum with $g = 2.015$ is also thought to be a V-type center. In this case the hole is localized on a lattice oxygen. The remaining center with $g = 1.999$ is assigned to an electron trapped on the calcium ion.

Results for centers found in decationated zeolites differ between investigators, perhaps because of different degassing conditions. Following partial dehydroxylation at 500° and subsequent γ irradiation, Stamires and Turkevich (*136*) observed a six-line spectrum with a hyperfine splitting of 5 G and a g value of 2.0017. They attributed the spectrum to an electron trapped at an $AlO_{3/2}$ defect, where the Al is part of the zeolite lattice. They reported, however, that there was no interaction between this center and molecular oxygen. This observation is indeed difficult to reconcile with their proposed center, since all of the lattice aluminum ions are on the surface of the supercage and would be accessible to molecular oxygen. On samples that were more extensively dehydroxylated at 600° Wang and Lunsford (*114*) detected the spectrum shown in Fig. 30b, which may be resolved into eleven lines whose relative intensities are approximated by the ratios 1:2:3:4:5:6:5:4:3:2:1. An apparent g value of 2.007 and an $|a|$ of 10 G were obtained. It was difficult to saturate the spectrum and the lines were broadened reversibly by molecular oxygen. The spectrum has been attributed to a hole trapped on an oxygen atom which is between two equivalent aluminum ions. The expected anisotropy of the spectrum has not been clearly analyzed. The same spectrum (Fig. 30a), although somewhat more intense, has been detected in a γ-irradiated AlHY zeolite which was formed by exchanging the zeolite in an aluminum nitrate solution.

Extensive study has been devoted to paramagnetic defects that are formed on high-surface area alkaline earth oxides, particularly magnesium oxide. The work carried out by Wertz et al. (137, 138) and Henderson and Wertz (139) on bulk defects formed in MgO single crystals has been quite valuable in the identification of the surface defects. Both the bulk and surface defects may be divided into two classes: those in which an electron

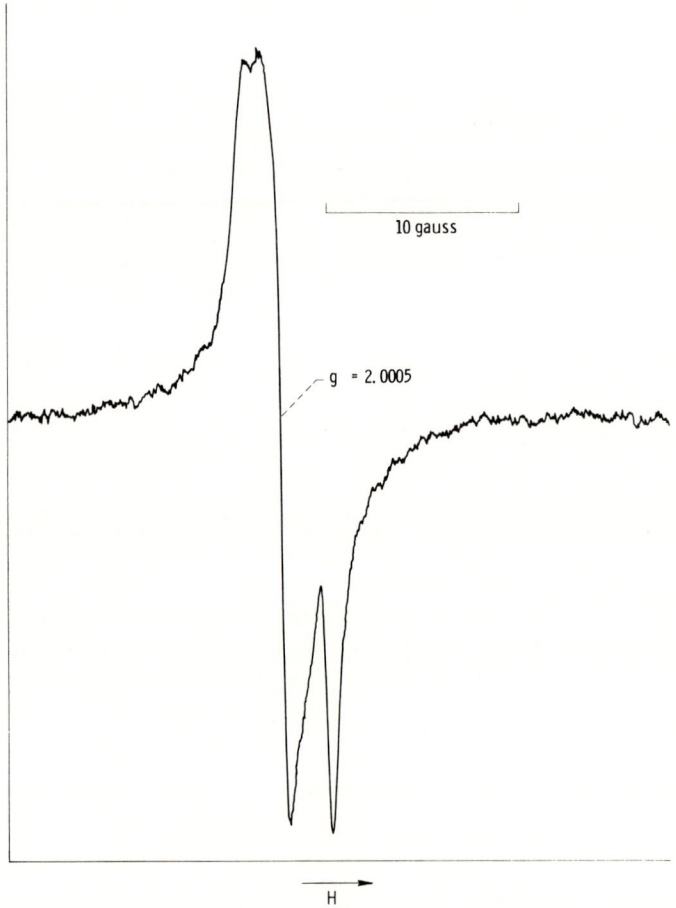

FIG. 31. The ESR spectrum of the S_H center on magnesium oxide (141).

is trapped and those in which a hole is trapped. For samples degassed under vacuum at 850° or higher temperatures, electrons trapped at two distinct sites are observed following γ or UV irradiation. One of these, known as the S center, is thought to be an electron trapped at an oxide ion vacancy on the surface, which makes it a surface analog of an F-type center. It has an apparent g value of 2.0007 and is very easily saturated with microwave power. The other center, known as the S_H or S' center, is clearly associated with a hydrogen atom. The spectrum of this center shown in Fig. 31 was at first interpreted in terms of g value anisotropy by Lunsford and Jayne (140); however, Tench and Nelson (141) conclusively demonstrated that the splitting between the two minima is the result of a weak dipolar interaction between the unpaired electron and a proton. Hyperfine interaction with ^{25}Mg has been observed for both centers. Another F-type center with $g_\perp = 1.9998$ and $g_{||} = 2.0012$ is observed following irradiation of samples degassed at temperatures from 300° to 700°. Here, the hydrogen influences the formation of the center but no hyperfine splitting with either protons or ^{25}Mg can be detected. Perhaps the trapped electron is less localized in this defect. The role of the UV light may be only to form atomic hydrogen since Smith and Tench (142) have shown that the same defects are formed upon exposure of the magnesium oxide to hydrogen atoms produced in a microwave discharge. Trapped electrons in such surface defects have been used to form O_2^-, CO_2^-, and SO_2^- on magnesium oxide.

It is now felt that two types of hole-trapping centers, or V centers, are observed on the surface of magnesium oxide (143). One of these is observed following dehydration of the sample at temperatures greater than 800°; whereas, the other, shown in Fig. 32, is observed following decomposition of the hydroxide at 300°. The former type is thought to be a V_1 center at, or near to the surface. This surface center is destroyed by exposure to H_2 or O_2. The other center, observed on the samples degassed at 300°, is probably the surface counterpart of the V_{OH} center which Kirklin and co-workers (144) have analyzed by a double resonance experiment. The proton hyperfine splitting amounts to only 1.7 G in the perpendicular direction and 0.8 G in the parallel direction; hence, it is probably not resolved in the polycrystalline spectrum. An attempt has been made to compare the spectra formed from $Mg(OD)_2$ and $Mg(OH)_2$, but the results are not conclusive. The surface V_{OH} center is reversibly broadened by molecular oxygen, and it is unaffected by molecular hydrogen.

There is considerable evidence available which indicates that these surface V_{OH} centers are the active sites for the irradiation induced catalytic activity of MgO for the hydrogen–deuterium exchange reaction (143). In particular, a correlation exists between the V_{OH} center concentration and the induced catalytic activity (1) for samples degassed at different tem-

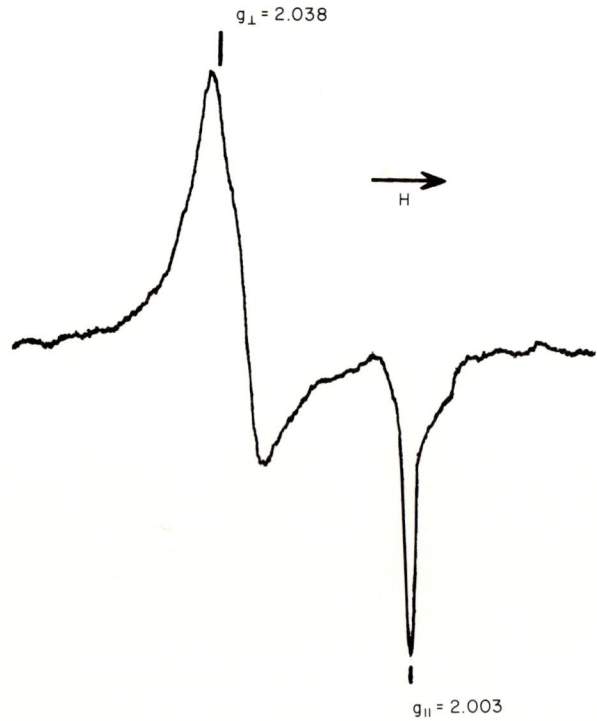

FIG. 32. The ESR spectrum of the V_{OH} center on magnesium oxide, produced by ultraviolet irradiation of a sample degassed at 300° *(143)*.

peratures, (2) for thermal decay at −78°, 0°, and 30°, (3) for response to irradiation at 2537-Å UV light, and (4) for response to irradiation at several wavelengths of UV light. Taylor *(145)* has recently provided a critical review of this topic.

C. Transition Metal Ions

The electron paramagnetic resonance spectrum of transition metal ions has been widely used to interpret the state of these ions in systems of catalytic interest. Major emphasis has been placed on supported chromia because of its catalytic importance in low-pressure ethylene polymerization and other commercial reactions. Earlier work on chromia-alumina catalysts has been reviewed by Poole and MacIver *(146)*. On alumina it appears that the chromium is present in three general forms: the δ phase, which is isolated Cr^{3+} on the surface or in the lattice; the β phase, which is clusters of Cr^{3+}; and the γ phase, which is isolated Cr^{5+} on the surface. The δ and β phases are characterized by apparent linewidths ranging from 200 to 1300 G. Such linewidths are consistent with a large zero-field splitting term which is expected for the d^3 ion in low symmetry. Extensive spin–spin interaction is also expected for the cluster.

The Cr^{5+} ion has only one unpaired electron; hence, no zero-field splitting is expected. Indeed, a well-resolved spectrum has been observed for the ion on alumina (*147, 148*), silica gel (*149–151*), silica-alumina (*152–154*), and magnesium oxide (*155*). The line may be resolved into parallel and perpendicular g values. As van Reijen and Cossee (*151*) have shown, the values of $g_\perp$ range from 1.970 to 1.975; whereas, the values of $g_{||}$ range from 1.898 to 2.002, depending on the treatment of a Cr/SiO_2 sample. These authors have suggested that Cr^{5+} in two different symmetries is present: one has a long relaxation time and can be observed at room temperature, but the other has a very short relaxation time and can be observed only at very low temperatures ($-253°$).

Numerous attempts have been made to correlate the presence of this species with the catalytic activity for ethylene polymerization. Kazanskii and co-workers showed that a correlation did exist between the catalytic activity and the intensity of the Cr^{5+} signal (observed at room temperature) provided the spectrum was obtained before the polymerization reaction (*152*). Ayscough et al. (*153*) found that no direct correlation between signal strength and activity could be established, but a nearly linear relation was found between the activity and the difference in signal strength of ethylene-treated catalysts before and after exposure to water vapor. More recently Kazanskii and Turkevich (*156*) observed a virtual disappearance of the Cr^{5+} signal (observed at room temperature) upon treatment of the catalyst with ethylene at 100°. At the same time the catalyst activity increased by an order of magnitude. These authors pointed out that the disappearance of the signal was due either to a change in the valence of Cr^{5+} or to a change in the symmetry of the ion so that it has a very short relaxation time. This negative correlation, though it can be explained, is not very satisfactory proof that the Cr^{5+} ion is the active site for the polymerization reaction. Furthermore, Karra and Turkevich (*157*) briefly reported that Cr^{5+} signals observed at temperatures as low as 1.6°K did not correlate with the catalytic activity.

As one might expect, Mo^{5+} has a spectrum that is quite similar to Cr^{5+} since both pentavalent ions have one d electron. The spectrum of Mo^{5+} is typically characterized by $g_{||} = 1.86$ and $g_\perp = 1.93$, although the parallel component may vary significantly depending upon the ligands attached to the ion (*158*). Cornaz and co-workers (*159*) were able to detect the signal only at low temperatures, but Sancier and co-workers (*160*) have observed the spectrum at 390° during the catalytic oxidation of propylene. Both

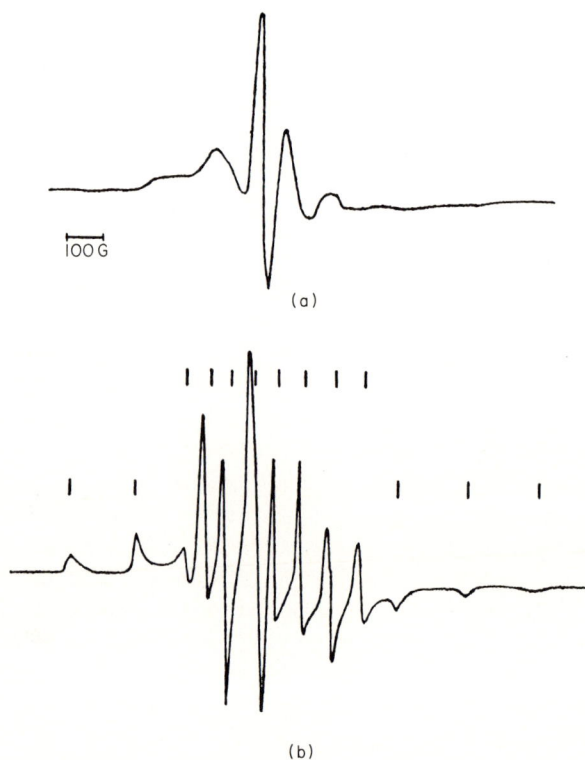

FIG. 33. The ESR spectra of V/SiO$_2$ after vacuum treatment at 500°; spectra at −253°: (a) without further treatment, (b) after contact with H$_2$O vapor (151).

supported MoO$_3$ and Bi$_2$O$_3$–MoO$_3$ catalysts have been studied because of their importance in this oxidation reaction. An inverse correlation was observed between the Mo^{5+} concentration and the catalytic activity; hence, it was suggested that Mo^{6+} was the active oxidation state (160). On the other hand, Boreskov et al. (161) have reported a direct correlation between the intensity of the Mo^{5+} signal and the activity of a catalyst for ethylene polymerization.

Studies of V^{4+}, which is also a 3d^1 ion, have been carried out by several investigators (151, 162, 163). The spectra are often complex due to a nuclear spin of $\frac{7}{2}$; however, good resolution has been achieved for the ion in a particular symmetry. After a vacuum treatment of vanadium on silica gel at 500° van Reijen and Cossee (151) observed the spectra shown in Fig. 33. Both spectra were obtained with the sample at 20°K. The authors attributed the first spectrum to the (VO$_4$)$^{4-}$ ion and the second to the

$(VO)^{2+}$ ion in a square pyramidal coordination. On alumina only the latter configuration is stable. Tanaka and Matsumoto (163) attributed a poorly resolved spectrum on silica gel to the formation of small crystallites of a VO^{2+} salt. Upon dilution with an alkali salt the resolution increased. They also observed much better resolution on alumina, indicating that the vanadyl ion is more uniformly dispersed on this surface.

Divalent copper has only one unpaired electron (hole) since it is a $3d^9$ ion. The spectrum also does not exhibit zero-field splitting; however, it is complicated by anisotropic g values and hyperfine coupling constants in a low-symmetry crystalline field. The best-resolved spectra have been obtained by using zeolites or other ion exchange materials so that one can observe the spectrum of isolated ions. Nicula, Stamires, and Turkevich (164) carried out an extensive study of Cu^{2+} in X and Y zeolites. They observed that the spectrum was a function of both the dehydration conditions and the concentration. The effect of dehydration on the ESR spectrum of a CuY zeolite containing four and sixteen Cu^{2+} ions per unit cell is shown in Fig. 34a and b, respectively. The spectrum of the dehydrated zeolite containing four copper ions per unit cell is characterized by $g_\perp = 2.0662$ and $g_{||} = 2.3301$ with $a_\perp = 17.7$ G and $a_{||} = 178$ G. In the hydrated copper zeolite the copper ion is thought to be encased in a hydration cell of water molecules. This tumbles sufficiently rapidly so as to average any effects of the nonuniform crystalline field of the ligands or the hyperfine structure due to the nuclear magnetic moment. The result is the single line

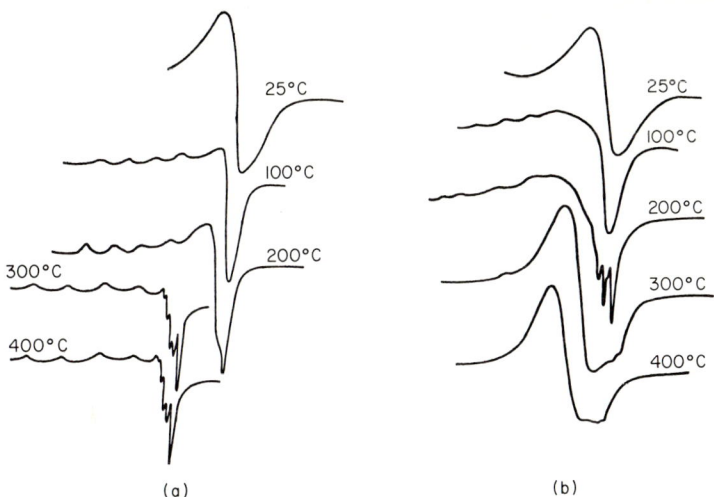

FIG. 34. The ESR spectra of Cu^{2+} in a CuY zeolite (a) containing four Cu^{2+} ions per unit cell and (b) containing 16 Cu^{2+} ions per unit cell (164).

in the hydrated state. By way of contrast Richardson *(165)* has reported an anisotropic signal, even for a hydrated Cu–Mg zeolite, which is similar in shape to that shown in Fig. 34a (100°).

One interesting feature of the spectrum that was not discussed by Nicula *et al.* is the broad, rather symmetric line which appears in dehydrated zeolites containing sixteen Cu^{2+} ions per unit cell. Tikhomirova and Nikolaveva *(166)* have reported that a similar line is observed for Cu^{2+} in A-type zeolites. They attribute this to a spin exchange phenomenon. Such exchange interactions have been studied in detail by Fujiwara *et al. (167)* using Cu^{2+} in polyvinyl alcohol gels. The spectra are essentially identical to those found in zeolites. Since the spin exchange interaction requires that two copper ions be within about 8 Å of one another, one is tempted to take this as evidence for a $Cu^+–O–Cu^+$ linkage, similar to that which has been proposed for other divalent zeolites *(168)*.

Tikhomirova and Nikolaeva *(166)*, as well as Slot and Verbeek *(169)*, have noted that two different spectra may be resolved in the zeolites. It appears that this reflects the presence of Cu^{2+} in two different sites. In addition, the relative intensities of the two sites may be varied by altering the other cation in the zeolite, suggesting that the different cations compete for the same site.

The spectrum of Mn^{2+} in zeolites has been used to study the bonding and cation sites in these crystalline materials. This is a $3d^5$ ion; hence, one would expect a zero-field splitting effect. A detailed analysis of this system was carried out by Nicula *et al. (170)*. When the symmetry of the environment is less than cubic, the resonance field for transitions other than those between the $+\frac{1}{2}$ and $-\frac{1}{2}$ electron spin states varies rapidly with orientation, and that portion of the spectrum is spread over several hundred gauss. The energies of the $\pm\frac{1}{2}$ levels are given by the equation

$$E_{\pm 1|2} = \pm \tfrac{1}{2} g\beta H \pm \tfrac{1}{2} Am \pm (A^2/4g\beta H)\{I(I+1) - m^2\}$$
$$- 17mA^2/4g\beta H \pm D^2 \sin^2 \theta / g\beta H$$
$$\pm D^2 \cos^2 \theta \sin^2 \theta (72Am/g\beta H - 8)/g\beta H. \qquad (48)$$

Here, A is the nearly isotropic nuclear coupling constant, I is the nuclear spin $(I_{Mn} = \tfrac{5}{2})$, and m is the particular nuclear spin state. It may be observed that the zero field splitting term D has a second-order effect which must be considered at magnetic fields near 3,000 G (X-band). In addition to this complication nuclear transitions for which $\Delta m = 1$ and 2 must also be considered. The analysis by Barry and Lay *(171)* of the Mn^{2+} spectrum in a CsX zeolite is shown in Fig. 35. From such spectra these authors have proposed that manganese is found in five different sites, depending upon the type of zeolite, the primary cation, and the extent of dehydration.

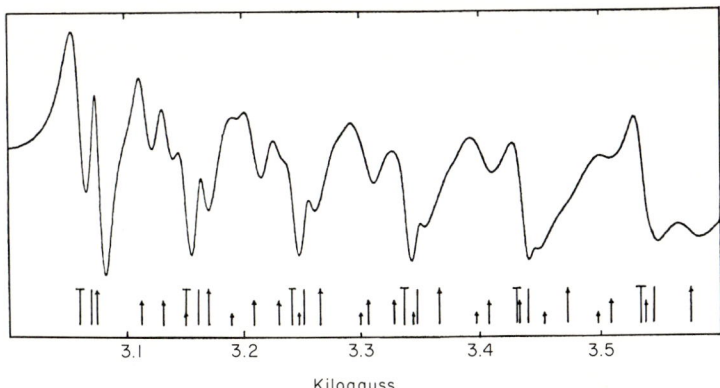

FIG. 35. The ESR spectrum of Mn^{2+} in a CsMnX zeolite treated at 300°. The longest vertical lines indicate transitions with $\Delta m = 0$, for the medium lines $\Delta m = 1$, for the shortest lines $\Delta m = 2$. $\perp$ indicates $\theta = 90°$, $\uparrow$ indicates $\theta = 42°$, $|$ indicates $\theta = 0°$ (171).

By going to fields of 12,000 G, Dzhaskiashvili and co-workers (172) have eliminated most of the anisotropy which arises from the latter two terms in Eq. (48). The spectra of Mn^{2+} in the zeolite appear as symmetric lines except where two species are present. Although the spectra can be much more easily interpreted under these conditions, all information concerning the zero-field splitting term is lost. The hyperfine splitting constants varied from 87 G in a partially dehydrated NaY zeolite to 99 G in a fully dehydrated zeolite. The smaller hyperfine constant is attributed to manganese ions surrounded by water molecules; whereas, the larger hyperfine constant is attributed to a manganese ion in an electric field of high symmetry. However, if the assignments of Barry and Lay (171) are corrected, a rather large zero-field splitting effect (indicating a strong axial field component) exists for the Mn^{2+} ion in a similar dehydrated NaY zeolite. They further demonstrated that the hyperfine splitting constant and the zero-field splitting parameter varied independently.

One of the most promising techniques for studying transition metal ions involves the use of zeolite single crystals. Such crystals offer a unique opportunity to carry out single crystal measurements on a large "surface area" material. Suitable crystals of the natural large pore zeolites are available, and fairly small crystals of the synthetic zeolites can be obtained. The spectra in the faujasite-type crystals will not be simple because of the magnetically inequivalent sites; however, the lines should be sharp and symmetric. Work on Mn^{2+} in hydrated chabazite has indicated that there is only one symmetry axis in that material (173), and a current study in the author's laboratory on Cu^{2+} in partially dehydrated chabazite tends to confirm this observation.

Appendix A. Quantum Mechanics for Spin Systems

I. PROPERTIES OF WAVEFUNCTIONS

A. Normalized

$$\int_{\text{all space}} \psi_i^* \psi_i \, d\tau = \langle \psi_i | \psi_i \rangle = 1 \tag{1A}$$

B. Orthogonal, normalized

$$\int_{\text{all space}} \psi_i^* \psi_j \, d\tau = \langle \psi_i | \psi_j \rangle = 0, \quad i \neq j$$
$$= 1, \quad i = j \tag{2A}$$

C. Construction of a wavefunction out of known functions.

$$|\phi_n\rangle = c_{n1}|\psi_1\rangle + c_{n2}|\psi_2\rangle + c_{n3}|\psi_3\rangle + \cdots$$
$$= \sum c_{nj}|\psi_j\rangle \tag{3A}$$

c_{n1}, c_{n2}, etc. are constants. Because of normalization, $c_{n1}^2 + c_{n2}^2 + \cdots = 1$.

D. Spin functions for electrons are denoted by $|\alpha_e\rangle$ for $m_s = +\frac{1}{2}$ and $|\beta_e\rangle$ for $m_s = -\frac{1}{2}$.

II. SPIN OPERATORS

A. For the z direction $(H\|z)$: $S_z|\alpha_e\rangle = +\frac{1}{2}|\alpha_e\rangle$, $S_z|\beta_e\rangle = -\frac{1}{2}|\beta_e\rangle$, and in general $S_z|m_s\rangle = m_s|m_s\rangle$

B. For the x and y directions it is more convenient to use shift operators,

1. defined by
$$S^+ = S_x + iS_y$$
$$S^- = S_x - iS_y$$

2. from this definition $S_x = \frac{1}{2}(S^+ + S^-)$
$$iS_y = \frac{1}{2}(S^+ - S^-)$$

3. operations
$$S^+|\alpha_e\rangle = 0, \quad S^-|\beta_e\rangle = 0$$
$$S^+|\beta_e\rangle = |\alpha_e\rangle, \quad S^-|\alpha_e\rangle = |\beta_e\rangle$$

in general for spin S and eigenfunction m_s

$$\langle m_s + 1 | S^+ | m_s \rangle = \langle m_s | S^- | m_s + 1 \rangle = [S(S+1) - m_s(m_s+1)]^{1/2} \tag{4A}$$

III. ORBITAL ANGULAR MOMENTUM OPERATORS

A. Operate on the m_L part of the wavefunction, otherwise the same as spin operators.

B. Operations

$$L_z \,|\, m_L\rangle = m_L \,|\, m_L\rangle \qquad (5A)$$

$$\langle m_L + 1 \,|\, L^+ \,|\, m_L\rangle = \langle m_L \,|\, L^- \,|\, m_L + 1\rangle$$
$$= [L(L+1) - m_L(m_L + 1)]^{1/2} \qquad (6A)$$

C. For example, consider the operation $L^- \,|\, m_L = 0\rangle$ for a p orbital ($L = 1$):

$$L^- \,|\, m_L = 0\rangle = 2^{1/2} \,|\, m_L = -1\rangle$$

or simply

$$L^- \,|\, 0\rangle = 2^{1/2} \,|\, -1\rangle$$

IV. Matrix Formulation of the Wave Equation

A. Express the wavefunction (eigenfunction) as the sum of orthogonal, normalized wavefunctions; typically the latter would be spin functions denoted by ψ_j

$$|\phi_n\rangle = \sum_j c_{nj} \,|\, \psi_j\rangle \qquad (7A)$$

B. Now operate on the wavefunctions with an operator P, which may be a spin operator $p_n \,|\, \phi_n\rangle = \sum_j c_{nj} P \,|\, \psi_j\rangle$, here p_n is the corresponding eigenvalue. If the operator is an energy (Hamiltonian) operator, then the eigenvalue is energy.

C. Multiply both sides of the equation by $\langle \psi_i \,|$

$$\langle \psi_i \,|\, p_n \,|\, \phi_n\rangle = \sum_j c_{nj} \langle \psi_i \,|\, P \,|\, \psi_j\rangle, \qquad (8A)$$

and

$$\langle \psi_i \,|\, p_n \,|\, \sum_j c_{nj} \,|\, \psi_j\rangle = \sum_j c_{nj} p_n \langle \psi_i \,|\, \psi_j\rangle \qquad (9A)$$

D. Substitute (9A) into (8A)

$$\sum_j c_{nj} p_n \langle \psi_i \,|\, \psi_j\rangle = \sum_j c_{nj} \langle \psi_i \,|\, P \,|\, \psi_j\rangle \qquad (10A)$$

or

$$\sum_j c_{nj} [\langle \psi_i \,|\, P \,|\, \psi_j\rangle - p_n \langle \psi_i \,|\, \psi_j\rangle] = 0 \qquad (11A)$$

E. As an example let $|\phi\rangle = c_1 \,|\, 1\rangle + c_2 \,|\, 2\rangle$; then from (11A)

$$c_1(\langle 1 \,|\, P \,|\, 1\rangle - p) + c_2 \langle 1 \,|\, P \,|\, 2\rangle = 0 \qquad (12A)$$

$$c_1 \langle 2 \,|\, P \,|\, 1\rangle + c_2(\langle 2 \,|\, P \,|\, 2\rangle - p) = 0 \qquad (13A)$$

F. The array of numbers in Eqs. (12A) and (13A), or the set of equations in (11A), may be expressed as the product of two matrices. Matrix multiplication will show that this is true.

$$\begin{pmatrix} \langle 1 \mid P \mid 1 \rangle - p & \langle 1 \mid P \mid 2 \rangle \\ \langle 2 \mid P \mid 1 \rangle & \langle 2 \mid P \mid 2 \rangle - p \end{pmatrix} \begin{pmatrix} c_1 \\ c_2 \end{pmatrix} = \begin{pmatrix} 0 \\ 0 \end{pmatrix} \quad (14A)$$

A shorthand notation for this is

$$\begin{matrix} & |1\rangle & |2\rangle \\ \langle 1| & \\ \langle 2| & \end{matrix} \begin{pmatrix} P_{11} - p & P_{12} \\ P_{21} & P_{22} - p \end{pmatrix} \begin{pmatrix} c_1 \\ c_2 \end{pmatrix} = 0 \quad (15A)$$

G. Now the eigenvalues p (energy levels of the system, for example) may be found by making the left-most matrix a determinant, and setting it equal to zero.

$$\begin{vmatrix} P_{11} - p & P_{12} \\ P_{21} & P_{22} - p \end{vmatrix} = 0 \quad (16A)$$

This determinant (known as the secular determinant) may then be solved for values of p by standard techniques.

H. To find values of c_1 and c_2 corresponding to each p, combine Eq. (12A) or (13A) with the normalization restriction that $c_1^2 + c_2^2 = 1$.

Appendix B. Determination of Energy Levels from the Spin Hamiltonian

The spin Hamiltonian for the hydrogen atom will be used to determine the energy levels in the presence of an external magnetic field. As indicated in Section II.A, the treatment may be simplified if it is recognized that the g factor and the hyperfine constant are essentially scalar quantities in this particular example. An additional simplification results if the z direction is defined as the direction of the magnetic field. For this case $H = H_z$ and $H_x = H_y = 0$; hence,

$$\mathbf{H} \cdot \mathbf{S} = (0, 0, H_z) \begin{pmatrix} S_x \\ S_y \\ S_z \end{pmatrix} = H_z S_z. \quad (1B)$$

After expanding the dot product $\mathbf{I} \cdot \mathbf{S}$ Eq. (1) becomes

$$\mathcal{H}_s = g\beta H S_z - g_n\beta_n H I_z + a[I_x S_x + I_y S_y + I_z S_z], \quad (2B)$$

or in terms of shift operators (Appendix A),

$$\mathcal{H}_s = g\beta H S_z - g_n\beta_n H I_z + a[\tfrac{1}{2}(I^+S^- + I^-S^+) + I_z S_z]. \quad (3B)$$

Before going on to calculate the energy levels it is necessary to digress and briefly describe the wavefunction. The spin Hamiltonian only operates on the spin part of the wavefunction. Every unpaired electron has a spin vector $S = \tfrac{1}{2}$ with spin quantum numbers $m_s = +\tfrac{1}{2}$ and $m_s = -\tfrac{1}{2}$. The wavefunctions for these two spin states are denoted by $|\alpha_e\rangle$ and $|\beta_e\rangle$, respectively. The proton likewise has $\mathbf{I} = \tfrac{1}{2}$ with spin wavefunctions $|\alpha_n\rangle$ and $|\beta_n\rangle$. In the present example these will be used as the basis functions in our calculation of energy levels, although it is sometimes convenient to use a linear combination of these spin states.

As shown in Appendix A it is possible to formulate the wave equation in terms of a product of two matrices:

$$\begin{array}{c} \\ \langle \alpha_e\alpha_n| \\ \langle \alpha_e\beta_n| \\ \langle \beta_e\alpha_n| \\ \langle \beta_e\beta_n| \end{array} \begin{pmatrix} |\alpha_e\alpha_n\rangle & |\alpha_e\beta_n\rangle & |\beta_e\alpha_n\rangle & |\beta_e\beta_n\rangle \\ \mathcal{H}_{11}-E & \mathcal{H}_{12} & \mathcal{H}_{13} & \mathcal{H}_{14} \\ \mathcal{H}_{21} & \mathcal{H}_{22}-E & \mathcal{H}_{23} & \mathcal{H}_{24} \\ \mathcal{H}_{31} & \mathcal{H}_{32} & \mathcal{H}_{33}-E & \mathcal{H}_{34} \\ \mathcal{H}_{41} & \mathcal{H}_{42} & \mathcal{H}_{43} & \mathcal{H}_{44}-E \end{pmatrix} \begin{pmatrix} c_1 \\ c_2 \\ c_3 \\ c_4 \end{pmatrix} = 0 \quad (4B)$$

where $\mathcal{H}_{11} = \langle \alpha_e\alpha_n | \mathcal{H}_s | \alpha_e\alpha_n\rangle$, $\mathcal{H}_{12} = \langle \alpha_e\alpha_n | \mathcal{H}_s | \alpha_e\beta_n\rangle$, etc. These matrix elements may be evaluated by allowing $\mathcal{H}_s$ to operate on the wavefunction. Expanding $\mathcal{H}_{11}$ and operating on the wavefunction gives

$$\langle \alpha_e\alpha_n | g\beta H S_z - g_n\beta_n H I_z + a[\tfrac{1}{2}(I^+S^- + I^-S^+) + I_z S_z]| \alpha_e\alpha_n\rangle \quad (5B)$$

$$= \tfrac{1}{2}g\beta H \langle \alpha_e\alpha_n | \alpha_e\alpha_n\rangle - \tfrac{1}{2}g_n\beta_n H \langle \alpha_e\alpha_n | \alpha_e\alpha_n\rangle$$
$$+ \tfrac{1}{4}a \langle \alpha_e\alpha_n | \alpha_e\alpha_n\rangle \quad (6B)$$

$$= \tfrac{1}{2}g\beta H - \tfrac{1}{2}g_n\beta_n H + \tfrac{1}{4}a, \quad (7B)$$

since $\langle \alpha_e\alpha_n | \alpha_e\alpha_n\rangle = 1$ and $I^+S^- | \alpha_e\alpha_n\rangle = I^-S^+ | \alpha_e\alpha_n\rangle = 0$. The matrix element $\mathcal{H}_{12}$ leads to

$$\langle \alpha_e\alpha_n | g\beta H S_z - g_n\beta_n H I_z + a[\tfrac{1}{2}(I^+S^- + I^-S^+) + I_z S_z]| \alpha_e\beta_n\rangle \quad (8B)$$

$$= \tfrac{1}{2}g\beta H \langle \alpha_e\alpha_n | \alpha_e\beta_n\rangle - \tfrac{1}{2}g_n\beta_n H \langle \alpha_e\alpha_n | \alpha_e\beta_n\rangle + \tfrac{1}{2}a \langle \alpha_e\alpha_n | \beta_e\beta_n\rangle$$
$$- \tfrac{1}{4}a \langle \alpha_e\alpha_n | \alpha_e\beta_n\rangle = 0, \quad (9B)$$

since $\langle \alpha_e\alpha_n | \alpha_e\beta_n\rangle = \langle \alpha_e\alpha_n | \beta_e\alpha_n\rangle = 0$.

In order to evaluate the energy levels of the system the left-most matrix is expressed as a determinant and is set equal to zero. Actually, the secular determinant is a systematic means of solving the set of equations in Eq. (11A). The 4 × 4 determinant becomes

$$\begin{vmatrix} \frac{1}{2}(\Delta_e - \Delta_n) + \frac{1}{4}a - E & 0 & 0 & 0 \\ 0 & \frac{1}{2}(\Delta_e + \Delta_n) - \frac{1}{4}a - E & \frac{1}{2}a & 0 \\ 0 & \frac{1}{2}a & -\frac{1}{2}(\Delta_e + \Delta_n) - \frac{1}{4}a - E & 0 \\ 0 & 0 & 0 & -\frac{1}{2}(\Delta_e + \Delta_n) + \frac{1}{4}a - E \end{vmatrix} = 0.$$

(10B)

Here, $\Delta_e = g\beta H$ and $\Delta_n = g_n\beta_n H$. Because of symmetry the determinant can be resolved into two 1 × 1 determinants and one 2 × 2 determinant. Solving for E gives

$$E_1 = \tfrac{1}{2}(\Delta_e - \Delta_n) + \tfrac{1}{4}a \qquad (11B)$$

$$E_2 = [\tfrac{1}{4}(\Delta_e + \Delta_n)^2 + \tfrac{1}{4}a^2]^{1/2} - \tfrac{1}{4}a \qquad (12B)$$

$$E_3 = -[\tfrac{1}{4}(\Delta_e + \Delta_n)^2 + \tfrac{1}{4}a^2]^{1/2} - \tfrac{1}{4}a \qquad (13B)$$

$$E_4 = -\tfrac{1}{2}(\Delta_e - \Delta_n) + \tfrac{1}{4}a. \qquad (14B)$$

The variation of these energy levels as a function of external magnetic field is shown in Fig. 1. It should be noted that the hyperfine term, a, is not a function of field; whereas, Δ_e and Δ_n are linear functions of the field. With an X-band spectrometer the magnetic field is on the order of 3000 G, which means that Δ_e and Δ_n are about 9,000 and 15 MHz, respectively. By comparison a is about 1,400 MHz. To a good approximation at high fields,

$$E_2 \simeq \tfrac{1}{2}(\Delta_e + \Delta_n) - \tfrac{1}{4}a \qquad (15B)$$

and

$$E_3 \simeq -\tfrac{1}{2}(\Delta_e + \Delta_n) - \tfrac{1}{4}a. \qquad (16B)$$

Now in order to establish which of the transitions are allowed, it is necessary to determine the wavefunctions that correspond to each energy level. These wavefunctions will have the form

$$|\theta_i\rangle = c_{i1}|\alpha_e\alpha_n\rangle + c_{i2}|\alpha_e\beta_n\rangle + c_{i3}|\beta_e\alpha_n\rangle + c_{i4}|\beta_e\beta_n\rangle, \qquad (17B)$$

where the c's are those in Eq. (7A) and i refers to the particular energy level for which the wavefunction is to be evaluated. Expansion of Eq. (4B) by matrix multiplication leads to the four linear equations,

$$c_{i1}(\mathcal{H}_{11} - E_i) + c_{i2}(0) + c_{i3}(0) + c_{i4}(0) = 0, \qquad (18B)$$

$$c_{i1}(0) + c_{i2}(\mathcal{H}_{22} - E_i) + c_{i3}(\tfrac{1}{2}a) + c_{i4}(0) = 0, \qquad (19B)$$

$$c_{i1}(0) + c_{i2}(\tfrac{1}{2}a) + c_{i3}(\mathcal{H}_{33} - E_i) + c_{i4}(0) = 0, \qquad (20B)$$

$$c_{i1}(0) + c_{i2}(0) + c_{i3}(0) + c_{i4}(\mathcal{H}_{44} - E_i) = 0, \qquad (21B)$$

and a fifth equation is furnished by the normalization restriction

$$c_{i1}^2 + c_{i2}^2 + c_{i3}^2 + c_{i4}^2 = 1. \tag{22B}$$

It may be shown that a solution is $c_{11} = 1$, and $c_{12} = c_{13} = c_{14} = 0$ when $E = E_1$; or a suitable wavefunction is $|\theta_1\rangle = |\alpha_e\alpha_n\rangle$. Likewise, $|\theta_4\rangle = |\beta_e\beta_n\rangle$. For $i = 2$ or 3, $c_{i1} = c_{i4} = 0$ and

$$c_{i2} = \mp \tfrac{1}{2}a\{[\tfrac{1}{2}(\Delta_e + \Delta_n) - \tfrac{1}{4}a - E_i]^2 + (\tfrac{1}{2}a)^2\}^{1/2}, \tag{23B}$$

$$c_{i3} = \pm[\tfrac{1}{2}(\Delta_e + \Delta_n) - \tfrac{1}{4}a - E_i]\{[\tfrac{1}{2}(\Delta_e + \Delta_n) - \tfrac{1}{4}a - E_i]^2 + (\tfrac{1}{2}a)^2\}^{1/2}. \tag{24B}$$

When the magnetic field is large such that $\tfrac{1}{4}(\Delta_e + \Delta_n)^2 \gg \tfrac{1}{4}a^2$, then $c_{22} \simeq 1$ and $c_{23} \simeq 0$ or $|\theta_2\rangle \simeq |\alpha_e\beta_n\rangle$; similarly, $c_{32} \simeq 0$ and $c_{33} \simeq 1$ or $|\theta_3\rangle \simeq |\beta_e\alpha_n\rangle$. These high-field approximations to the wavefunctions are shown along with their respective energy levels in Fig. 1.

With these first-order wavefunctions it is now possible to determine the transition probabilities between each of the four energy levels for the hydrogen atom in an external magnetic field. Transitions from one spin state to another are brought about by the interaction of an oscillating magnetic field, $2H_1 \cos \omega t$, which is the magnetic component of the microwave field. In the conventional TE$_{102}$ mode cavity this oscillating field is perpendicular to the static field, and, therefore, in our frame of reference it is in the x or y direction. This weak field introduces another Zeeman interaction which becomes a time-dependent perturbation

$$\mathcal{H}_1(t) = 2g\beta \mathbf{H}_1 \cdot \mathbf{S} \cos \omega t. \tag{25B}$$

Considering only the x direction, this becomes

$$\mathcal{H}_1(t) = 2g\beta H_1 S_x \cos \omega t. \tag{26B}$$

The time dependent perturbation results in a transition probability from state n to state m which is equal to

$$P_{nm} = (2\pi/\hbar^2)|\langle n|g\beta H_1 S_x|m\rangle|^2 \delta(\omega_{mn} - \omega). \tag{27B}$$

The term $\delta(\omega_{mn} - \omega)$ is a function which requires that the resonance condition be satisfied; that is, the microwave frequency must equal the resonance frequency.

It is now necessary to evaluate the term $|\langle n|g\beta H_1 S_x|m\rangle|^2$ or $g^2\beta^2 H_1^2\langle n|S_x|m\rangle$ for the various states n and m. For example, let $n = |\alpha_e\beta_n\rangle$ and $m = |\alpha_e\alpha_n\rangle$. The integral becomes

$$\langle \alpha_e\beta_n|S_x|\alpha_e\alpha_n\rangle = \langle \alpha_e\beta_n|\tfrac{1}{2}(S^+ + S^-)|\alpha_e\alpha_n\rangle$$
$$= \tfrac{1}{2}\langle \alpha_e\beta_n|\beta_e\alpha_n\rangle = 0 \tag{28B}$$

because of the orthogonality relation; hence, $P_{nm} = 0$ and the transition between these two states is not allowed. On the other hand, evaluation of the integral for $n = |\alpha_e \beta_n\rangle$ and $m = |\beta_e \beta_n\rangle$ leads to a value of $\frac{1}{2}$. These results lead to the more general selection rules

$$\Delta m_s = \pm 1 \quad \text{and} \quad \Delta m_I = 0.$$

The ESR spectrum of the hydrogen atom can be interpreted as transitions between E_1 and E_3 levels and between E_2 and E_4 levels. Resonances occur at a frequency

$$h\nu_{13} = E_1 - E_3$$
$$= \tfrac{1}{2}(\Delta_e - \Delta_n) + \tfrac{1}{4}a + \tfrac{1}{2}(\Delta_e + \Delta_n) + \tfrac{1}{4}a = g\beta H + \tfrac{1}{2}a \qquad (29\text{B})$$

and

$$h\nu_{24} = E_2 - E_4$$
$$= \tfrac{1}{2}(\Delta_e + \Delta_n) - \tfrac{1}{4}a + \tfrac{1}{2}(\Delta_e - \Delta_n) - \tfrac{1}{4}a = g\beta H - \tfrac{1}{2}a \qquad (30\text{B})$$

The hyperfine coupling constant is simply given by

$$a = h\nu_{13} - h\nu_{24}. \qquad (31\text{B})$$

Often the resonance condition and the coupling constant are reported in terms of gauss rather than in energy units. In this case Eqs. (29B), (30B), and (31B) are divided through by $g\beta$ and

$$a' = \frac{a}{g\beta} = h\frac{\nu_{13} - \nu_{24}}{g\beta}. \qquad (32\text{B})$$

The method presented here for evaluating energy levels from the spin Hamiltonian and then determining the allowed transitions is quite general and can be applied to more complex systems by using the appropriate spin Hamiltonian. Of particular interest in surface studies are molecules for which the g values, as well as the hyperfine coupling constants, are not isotropic. These cases will be discussed in the next two sections.

Appendix C. The g Tensor

In the previous section the g value was considered as a scalar quantity, which was indeed a good approximation since the unpaired electron on the hydrogen atom occupies a spherically symmetric s orbital. If the unpaired electron exhibits p or d character the electron possesses both spin and orbital angular momentum. As a result the spin is not quantized exactly along the direction of the external field and the g value becomes a tensor in the Zeeman term of the spin Hamiltonian:

$$\mathcal{H} = \beta \mathbf{H} \cdot \mathbf{g} \cdot \mathbf{S}$$

$$= (H_x, H_y, H_z) \begin{pmatrix} g_{xx} & g_{xy} & g_{xz} \\ g_{yx} & g_{yy} & g_{yz} \\ g_{zx} & g_{zy} & g_{zz} \end{pmatrix} \begin{pmatrix} S_x \\ S_y \\ S_z \end{pmatrix}. \tag{1C}$$

The significance of the terms in the g tensor can best be illustrated by means of a simple example. Let us redefine our coordinates such that the spin is quantized along the z axis. The magnetic field vector is now at some different position in space as shown in Fig. 11. Expanding the Hamiltonian in Eq. (1C) gives,

$$\mathcal{H} = \beta(H_x g_{xz} S_z + H_y g_{yz} S_z + H_z g_{zz} S_z). \tag{2C}$$

Using the spin functions $|\alpha\rangle$ and $|\beta\rangle$, the energy levels are

$$E_1 = (\beta/2)(H_x g_{xz} + H_y g_{yz} + H_z g_{zz}) \tag{3C}$$

and

$$E_2 = -(\beta/2)(H_x g_{xz} + H_y g_{yz} + H_z g_{zz}) \tag{4C}$$

or

$$\Delta E = (\beta/2)(H_x g_{xz} + H_y g_{yz} + H_z g_{zz}). \tag{5C}$$

One may observe from Eq. (5C) how the resonance condition is a function of the off diagonal components of the g tensor.

The form of the g tensor depends upon the choice of orthogonal axes. If the axis system is chosen along the molecular symmetry axes, the tensor contains only diagonal terms with the off-diagonal terms being zero. Conveniently, the principal g values derived from the polycrystalline spectrum are just those of a diagonalized matrix, i.e.,

$$\mathbf{g} = \begin{pmatrix} g_{xx} & 0 & 0 \\ 0 & g_{yy} & 0 \\ 0 & 0 & g_{zz} \end{pmatrix} \tag{6C}$$

One should realize, however, that for polycrystalline samples it is not possible to assign one of the principal g values to a particular molecular axis on the basis of experimental data alone. Such assignments usually are rationalized from theoretical considerations.

The g tensor can be easily related to more fundamental properties of the spin system by comparing the energy from the Zeeman term in the spin

Hamiltonian with the energy from appropriate terms in the true Hamiltonian. The latter terms include the interaction between the external field and the magnetic moment produced by the orbiting electron, the interaction between the external field and the magnetic moment due to electron spin, and the interaction between the orbital magnetic moment and the spin magnetic moment. These interactions may be expressed as a perturbation to the total Hamiltonian for the system where

$$\mathcal{H}_{\text{pert}} = \beta \mathbf{H} \cdot \mathbf{L} + g_e \beta \mathbf{H} \cdot \mathbf{S} + \lambda \mathbf{L} \cdot \mathbf{S} \tag{7C}$$

Here, $\mathbf{L}$ is the angular momentum operator, g_e is the g value for the free electron ($g_e = 2.0023$), and λ is the spin–orbit coupling constant.

Considering only the z direction, this equation becomes

$$\mathcal{H}_{\text{pert}} = \beta H_z L_z + g_e \beta H_z S_z + \lambda L_z S_z \tag{8C}$$

From perturbation theory the energy shift caused by these three terms is given by

$$E = E_0 + \langle \psi_0 | \mathcal{H}_{\text{pert}} | \psi_0 \rangle - \sum_{n \neq 0} \frac{\langle \psi_0 | \mathcal{H}_{\text{pert}} | \psi_n \rangle \langle \psi_n | \mathcal{H}_{\text{pert}} | \psi_0 \rangle}{E_n - E_0} \tag{9C}$$

where E_0 is the unperturbed energy and E_n is the energy of the nth excited electronic state. It is commonly found that only the lowest excited states are important in the summation and that these excited states are often closely spaced compared to the distance from E_0. In this case $E_n - E_0 \simeq \Delta E$. Using the closure relationship ($\sum | \psi_n \rangle \langle \psi_n | = 1$) the energy becomes

$$E = E_0 + \langle \psi_0 | \mathcal{H}_{\text{pert}} | \psi_0 \rangle - (1/\Delta E) \langle \psi_0 | \mathcal{H}_{\text{pert}}^2 | \psi_0 \rangle. \tag{10C}$$

The wavefunction ψ_0 includes both a spin and a space term which are denoted by $| \alpha \rangle$ and $| \theta_0 \rangle$, respectively.

It is now necessary to evaluate the integrals in Eq. (10C) using the $\mathcal{H}_{\text{pert}}$ operator. Terms to first order in S_z will be retained. For the first integral,

$$\langle \psi_0 | \beta H_z L_z + g_e \beta H_z S_z + L_z S_z | \psi_0 \rangle = g_e \beta H_z \langle \alpha | S_z | \alpha \rangle \tag{11C}$$

since

$$\langle \theta_0 | L_z | \theta_0 \rangle = 0. \tag{12C}$$

The latter equation may be shown to be valid for a particular case by considering θ_0 as a p_x wavefunction. The operator L_z may be expressed as $-i(x\partial/\partial y - y\partial/\partial x)$ and $p_x = xf(r)$. Now

$$\langle p_x | - i[x(\partial/\partial y) - y(\partial/\partial x)] | xf(r) \rangle = \langle p_x | iyf(r) \rangle = \langle p_x | ip_y \rangle = 0. \tag{13C}$$

For the second integral,

$$\mathcal{H}_{\text{pert}}^2 = 2g_e\beta^2 H_z^2 S_z L_z + 2\lambda\beta H_z L_z^2 S_z + \text{terms other than first-order in } S_z \quad (14C)$$

Again the term $2g_e\beta^2 H_z^2 S_z L_z$ may be dropped because it leads to an integral like that in Eq. (12C). The second integral in Eq. (10C) therefore gives;

$$\langle\psi_0|\mathcal{H}_{\text{pert}}^2|\psi_0\rangle = 2\lambda\beta H_z\langle\theta_0|L_z^2|\theta_0\rangle\langle\alpha|S_z|\alpha\rangle \quad (15C)$$

and for the energy,

$$E = E_0 + g_e\beta H_z\langle\alpha|S_z|\alpha\rangle - \frac{2\lambda\beta H_z}{\Delta E}\langle\theta_0|L_z^2|\theta_0\rangle\langle\alpha|S_z|\alpha\rangle \quad (16C)$$

The rational is to equate terms derived from the spin Hamiltonian and terms from the true Hamiltonian which are first-order in S_z, i.e.,

$$g_{zz}\beta H_z\langle\alpha|S_z|\alpha\rangle = g_e\beta H_z\langle\alpha|S_z|\alpha\rangle - \frac{2\lambda\beta H_z}{\Delta E}\langle\theta_0|L_z^2|\theta_0\rangle\langle\alpha|S_z|\alpha\rangle. \quad (17C)$$

After dividing through by $\beta H_z\langle\alpha|S_z|\alpha\rangle$, one obtains a first-order expression for g_{zz}:

$$g_{zz} = g_e - 2\lambda\langle\theta_0|L_z^2|\theta_0\rangle/\Delta E. \quad (18C)$$

From this final equation it may be observed that large deviations from the free electron g value occur when (a) the integral $\langle\theta_0|L_z^2|\theta_0\rangle$ is nonzero and (b) excited electronic states are rather close in energy to the ground state. The other principal values of the g tensor have the same form as shown in Eq. (18C).

The application of these equations will be demonstrated by calculating the g values for O^-. Experimental evidence has shown that for O^- trapped on magnesium oxide the configuration is $p_x^2 p_y^2 p_z^1$ where p_x and p_y are degenerate as shown in Fig. 20. This system appears complex because the ground and excited states must be described by four-electron wavefunctions; however, it may be simplified by considering the excited state to be formed by the transfer of a hole from the p_z level to the degenerate p_x, p_y level. Our system now involves only one electron-hole, and the energy levels become inverted in this new system. If ΔE continues to be defined by $E_{p_z} - E_{p_x p_y}$ then the term $E_n - E_0$, which is now equal to $E_{p_x p_y} - E_{p_z}$, must change sign in Eq. (9C) and

$$g_{zz} = g_e + 2\lambda\langle\theta_0|L_z^2|\theta_0\rangle/\Delta E. \quad (19C)$$

The same result also may be reached by letting λ change sign or by considering different spin functions.

The second term in Eq. (19C) may be evaluated after carrying out the operation

$$L_z^2 | \theta_0 \rangle = L_z^2 | p_z \rangle = 0. \tag{20C}$$

since $i(x\partial/\partial y - y\partial/\partial x)zf(r) = 0$. This means that g_{zz} equals the free electron value, g_e. For the x direction;

$$g_{xx} = g_e + 2\lambda \langle \theta_0 | L_x^2 | \theta_0 \rangle / \Delta E. \tag{21C}$$

Again evaluating the operation in the second term:

$$L_x^2 | \theta_0 \rangle = L_x^2 | p_z \rangle = | \theta_0 \rangle, \tag{22C}$$

since

$$i\left(y\frac{\partial}{\partial z} - z\frac{\partial}{\partial y}\right) zf(r) = -iyf(r) \tag{23C}$$

and

$$-i\left(y\frac{\partial}{\partial z} - z\frac{\partial}{\partial y}\right)[-iyf(r)] = zf(r). \tag{24C}$$

It follows then that

$$g_{xx} = g_e + 2\lambda/\Delta E. \tag{25C}$$

In a like manner it may be shown that $g_{yy} = g_{xx}$.

Appendix D. The Hyperfine Tensor

The hyperfine constant a in Eq. (1) was also taken to be a scalar quantity for the hydrogen atom; however, it is in general a tensor because of the various directional interactions in a paramagnetic species. The hyperfine term in the spin Hamiltonian is more correctly written as **S**·**a**·**I**, where **a** is the hyperfine coupling tensor.

Before formally developing the tensor it is perhaps worthwhile to discuss the various types of interactions which contribute to it. The coupling between nuclear and electron magnetic moments are conveniently divided into those which are isotropic and those which depend on orientation. The former is the result of the unpaired electron having a finite probability of being at the nucleus. This type of interaction is termed the *contact interaction*, and is described by the constant,

$$A_{iso} = (8\pi/3)g\beta g_n\beta_n | \psi(0) |^2. \tag{1D}$$

The term $|\psi(0)|^2$ is the square of the absolute value of the wavefunction for the unpaired electron, evaluated at the nucleus ($r = 0$). Now it should be recalled that only s orbitals have a finite probability density at the nucleus; whereas, p, d, or higher orbitals have nodes at the nucleus. This hyperfine term is isotropic because the s wavefunctions are spherically symmetric, and the interaction is evaluated at a point in space.

Although the isotropic interaction is a measure of the s-orbital character of the wavefunction for the unpaired electron, there are certain factors that must be considered when one attempts to write simple wavefunctions for the unpaired electron. Such a case is the benzene radical anion in which the odd electron occupies a molecular π orbital formed by the overlap of carbon $2p_z$ atomic orbitals. If the unpaired electron were truly localized in this π orbital, one would expect no hyperfine interaction due to the protons. Experimental evidence, however, shows considerable hyperfine splitting from the protons, and this can only be accounted for by including the C–H bonding electrons in the molecular orbital of the unpaired electron. In addition, excited electronic states will almost always mix with the ground state wavefunction to introduce some s character. As a rule such indirect contributions account for less than 10% of the hyperfine coupling that would be observed for a total s-orbital wavefunction on the atom concerned. For example, the maximum indirect coupling via the C–H bond as described above, leads to a hyperfine splitting of about 23 G; whereas, a pure s orbital on a hydrogen atom produces a splitting of 506 G.

Aniosotropic hyperfine coupling results primarily from dipolar interactions between a magnetic nucleus and an unpaired electron in a p, d, or f orbital. Such interactions give rise to a Hamiltonian

$$\mathcal{H}_D = g\beta g_n\beta_n \left\{ \frac{\mathbf{S}\cdot\mathbf{I}}{r^3} - \frac{3(\mathbf{S}\cdot\mathbf{r})(\mathbf{I}\cdot\mathbf{r})}{r^5} \right\} \tag{2D}$$

where $\mathbf{r}$ is the radius vector between the two magnetic dipoles. Expanding Eq. (2D) and writing in matrix notation gives

$$\mathcal{H}_D/g\beta g_n\beta_n = -(I_x, I_y, I_z)$$

$$\times \begin{pmatrix} \langle(r^2-3x^2)/r^5\rangle & \langle -3xy/r^5\rangle & \langle -3xz/r^5\rangle \\ \langle -3xy/r^5\rangle & \langle(r^2-3y^2)/r^5\rangle & \langle -3yz/r^5\rangle \\ \langle -3xz/r^5\rangle & \langle -3yz/r^5\rangle & \langle(r^2-3z^2)/r^5\rangle \end{pmatrix} \begin{pmatrix} S_x \\ S_y \\ S_z \end{pmatrix}. \tag{3D}$$

Here the tensor elements are actually expectation values since the electron is distributed over a region of space which is described by a particular atomic orbital. Terms such as $\langle r^2 - 3z^2/r^5 \rangle$ denote integration over all space using the normalized wavefunction. If the unpaired electron is localized in a $2p_z$ atomic orbital, then

$$\left\langle \psi_{2p_z} \left| \frac{r^2 - 3z^2}{r^5} \right| \psi_{2p_z} \right\rangle = -\frac{4}{5} \left\langle \frac{1}{r^3} \right\rangle, \tag{4D}$$

where

$$\left\langle \frac{1}{r^3} \right\rangle = \int_0^\infty f^2(r) \frac{1}{r^3} dr. \tag{5D}$$

The latter integral has been evaluated for a number of atoms of interest and the results are presented in Table II in the form $g_n \beta_n (\frac{4}{5}) \langle 1/r^3 \rangle$.

If **r** points along one of the coordinate axes, the tensor may be diagonalized and rearranged such that

$$\mathcal{H}_D = (I_x, I_y, I_z) \begin{pmatrix} D_{xx} & & \\ & D_{yy} & \\ & & D_{zz} \end{pmatrix} \begin{pmatrix} S_x \\ S_y \\ S_z \end{pmatrix} \tag{6D}$$

where

$$D_{xx} = -g\beta g_n \beta_n \left\langle \frac{r^2 - 3x^2}{r^5} \right\rangle \tag{7D}$$

$$D_{yy} = -g\beta g_n \beta_n \left\langle \frac{r^2 - 3y^2}{r^5} \right\rangle \tag{8D}$$

$$D_{zz} = -g\beta g_n \beta_n \left\langle \frac{r^2 - 3z^2}{r^5} \right\rangle. \tag{9D}$$

The dipolar coupling tensor **D** is defined by Eq. (6D).

Again taking the case of an unpaired electron localized in a $2p_z$ atomic orbital, the dipolar coupling tensor becomes

$$g\beta g_n \beta_n \begin{pmatrix} -\frac{2}{5}\langle 1/r^3 \rangle & & \\ & -\frac{2}{5}\langle 1/r^3 \rangle & \\ & & \frac{4}{5}\langle 1/r^3 \rangle \end{pmatrix}. \tag{10D}$$

This may be compared with an experimental anisotropic tensor which is often resolved into the form

$$\begin{pmatrix} -\alpha & & \\ & -\alpha & \\ & & 2\alpha \end{pmatrix} \quad (11D)$$

and the p character of the wavefunction may be determined in the manner described in Section II.B.2. The absolute signs of the elements of the diagonalized dipolar coupling tensor may be evaluated by comparing the elements in Eqs. (10D) and (11D). Clearly, the sign of α must be positive if g_n is positive and negative if g_n is negative.

An indirect mode of anisotropic hyperfine interaction arises as a result of strong spin–orbit interaction (174). Nuclear and electron spin magnetic moments are coupled to each other because both are coupled to the orbital magnetic moment. The Hamiltonian is

$$\mathcal{H}_{\text{pert}} = a_l \mathbf{I} \cdot \mathbf{L} + \lambda \mathbf{S} \cdot \mathbf{L}. \quad (12D)$$

The formal treatment is quite similar to the derivation of the principal g values as developed in Eqs. (7C) through (18C). The second-order energy term is set equal to the hyperfine term from the spin Hamiltonian, and for the z direction

$$l_{zz} = -\frac{2a_l \lambda}{\Delta E} \langle \theta_0 | L_z^2 | \theta_0 \rangle. \quad (13D)$$

This mode of hyperfine interaction will become important only when the unpaired electron is able to partially occupy a low-lying excited state (ΔE small), and the ground state has orbital angular momentum ($L \neq 0$). The adsorbed nitric oxide molecule and the superoxide ion with ^{17}O are typical examples where hyperfine coupling via spin–orbit interaction may be observed.

It should now be evident that the experimental tensor may be expressed as the sum of an isotropic term resulting from the contact interaction, and a tensor resulting from dipolar interactions and any indirect coupling via the orbital angular momentum. This may be written in the form of an equation:

$$\mathbf{a} = A_{\text{iso}} + \begin{pmatrix} A_{xx} & & \\ & A_{yy} & \\ & & A_{zz} \end{pmatrix}. \quad (14D)$$

Often the tensor on the right-hand side of Eq. (14D) will be in the form of Eq. (11D). In such cases it is clear that the unpaired electron is localized in one p orbital. Alternatively, as shown in Eq. (14) for the NO_2 molecule, the tensor can be resolved into two dipolar coupling tensors with each tensor representing occupancy of orthogonal p orbitals. Such results could also be explained in terms of hyperfine coupling through the orbital angular momentum. The importance of this latter effect can be estimated by evaluating Eq. (13D).

Acknowledgment

The author wishes to acknowledge support for the preparation of this manuscript by the National Science Foundation under Grant GP-19875.

References

1. O'Reilly, D. E., *Advan. Catal.* **12**, 31 (1960).
2. Kokes, R. J., *in* "Experimental Methods in Catalytic Research" (R. B. Anderson, ed.) pp. 436–473. Academic Press, New York, 1968.
3. Adrian, F. J., *J. Colloid Interface Sci.* **26**, 317 (1968).
4. Bersohn, M., and Baird, J. C., "An Introduction to Electron Paramagnetic Resonance." Benjamin, New York, 1966.
5. Carrington, A., and McLachlan, A. D., "Introduction to Magnetic Resonance." Harper, New York, 1967.
6. Bolton, J. R., and Wertz, J. E., "Electron Spin Resonance; Elementary Theory and Applications." McGraw-Hill, New York, 1971.
7. Atkins, P. W., and Symons, M. C. R., "The Structure of Inorganic Radicals." Elsevier, Amsterdam, 1967.
8. Symons, M. C. R., *Advan. Chem.* **36**, 76 (1962).
9. Symons, M. C. R., *Advan. Phys. Org. Chem.* **1**, 283 (1963).
10. Morton, J. P., *Chem. Rev.*, **64**, 453 (1964).
11. Poole, C. P., Dicarlo, E. N., Noble, C. S., Itzel, J. E., and Tobin, H. H. *J. Catal.* **4**, 518 (1965).
12. Ernst, I. T., Fisher, B., Garnett, J. L., and Sollich-Baumgartner, W. A., *Aust. J. Chem.* **19**, 877 (1966).
13. Ernst, I. T., Garnett, J. L. and Sollich-Baumgartner, W. A., *J. Catal.* **3**, 568 (1964).
14. Nicolau, C. S., Thom, H. G. and Pobitschka, E., *Trans. Faraday Soc.* **55**, 1430 (1959).
15. Nicolau, C. and Schnabel, K. H., *Nature (London)* **193**, 871 (1962).
16. Hollis, D. P. and Selwood, P. W., *J. Chem. Phys.* **35**, 378 (1961).
17. Lunsford, J. H., *J. Chem. Phys.* **46**, 4347 (1967).
18. Lunsford, J. H., *J. Phys. Chem.* **72**, 2141 (1968).
19. Lunsford, J. H., *J. Phys. Chem.* **72**, 4163 (1968).
20. Lunsford, J. H., *J. Catal.* **14**, 379 (1969).
21. Lunsford, J. H., and Jayne, J. P., *J. Chem. Phys.* **44**, 1487 (1966).
22. Naccache, C., Meriaudeau, P., Che, M., and Tench, A. J., *Trans. Faraday Soc.* **67**, 506 (1971).
23. van Hooff, J. H. C., Thesis, Univ. of Eindhoven, The Netherlands, 1968.

24. Bennett, J. E., Ingram, D. J. E., and Schonland, D., *Proc. Phys. Soc. Ser. A* **69**, 556 (1956).
25. Känzig, W., and Cohen, M. H., *Phys. Rev. Lett.* **3**, 509 (1959).
26. Lunsford, J. H., and Jayne, J. P., *J. Phys. Chem.* **69**, 2182 (1965).
27. Ovenall, D. W., and Whiffen, D. H., *Mol. Phys.* **4**, 135 (1961).
28. Marshall, S. A., Reinberg, A. R., Serway, R. A., and Hodges, J. A., *Mol. Phys.* **8**, 223 (1964).
29. Lunsford, J. H., *J. Colloid Interface Sci.* **26**, 355 (1968).
30. Colburn, C. B., Ettinger, R., and Johnson, F. A., *Inorg. Chem.* **2**, 1305 (1963).
31. Pietrzak, T. M., and Wood, D. E., *J. Chem. Phys.* **53**, 2454 (1970).
32. Tench, A. J., and Holroyd, P., *Chem. Commun.* 471, (1968).
33. Wang, K. M., and Lunsford, J. H., *J. Phys. Chem.* **73**, 2069 (1969).
34. Luz, Z., Reuveni, A., Holmberg, R. W., and Silver, B. L., *J. Chem. Phys.* **51**, 4017 (1969).
35. Poole, C. P., "Electron Spin Resonance." Wiley (Interscience), New York, 1967.
36. Wilmshurst, T. H., "Electron Spin Resonance Spectrometers." Hilger, London, 1967.
37. Lebedev, Ya. S., *J. Struct. Chem. (USSR)* **4**, 22 (1963).
38. Kneubühl, F. K., *J. Chem. Phys.* **33**, 1074 (1960).
39. Burns, G., *J. Appl. Phys.* **32**, 2048 (1961).
40. Wasserman, E., *J. Chem. Phys.* **41**, 1763 (1964).
41. Pariiskii, G. B., and Kazanskii, V. B., *Kinet. Catal. (USSR)* **5**, 96 (1964).
42. Golubev, V. B., *Zh. Fiz. Khim.* **39**, 2606 (1965).
43. Kazanskii, V. B., Pariiskii, G. B. and Voevodskii, V. V., *Disc. Faraday Soc.* **31**, 203 (1961).
44. Kazanskii, V. B. and Pariiskii, G. B., *Proc. Int. Congr. Catal., 3rd.* **1**, 367 (1964).
45. Gardner, C. L., *J. Chem. Phys.* **46**, 2991 (1967).
46. Mikheikin, I. D., Mashchenko, A. I. and Kazanskii, V. D., *Kinet. Catal. (USSR)* **8**, 1363 (1967).
47. Kwan, T., *Proc. Int. Congr. Catal., 3rd.* **1**, 493 (1964).
48. van Hooff, J. H. C., *J. Catal.* **11**, 277 (1968).
49. Shvets, V. A., Vorotyntsev, V. M., and Kazanskii, V. B., *Kinet. Catal. (USSR)* **10**, 356 (1969).
50. Williamson, W. B., Lunsford, J. H., and Naccache, C., *Chem. Phys. Lett.* **9**, 33 (1971).
51. Wong, N. B., and Lunsford, J. H., *J. Chem. Phys.* **55**, 3007 (1971).
52. Tench, A. J., and Lawson, T., *Chem. Phys. Lett.* **7**, 459 (1970).
53. Fujimoto, M., Gesser, H. D., Garbutt, B., and Cohen, A., *Science* **154**, 381 (1966).
54. Turkevich, J., and Fujita, Y., *Science* **152**, 1619 (1966).
55. Fujimoto, M., Gesser, H. D., Garbutt, B., and Shimizu, M., *Science* **156**, 1105 (1967).
56. Pariiskii, G. B., Zhidomirov, G. M., and Kazanskii, V. B., *Zh. Strukt. Khim.* **4**, 364 (1963).
57. Gardner, C. L., and Casey, E. J., *Can. J. Chem.* **46**, 207 (1968).
58. Kazanskii, V. B., and Pariiskii, G. B., *Kinet. Catal. (USSR)* **2**, 507 (1961).
59. Kinell, P. O., Lund, A., and Vanngard, T., *Acta Chem. Scand.* **19**, 2113 (1965).
60. Strelko, V. V., and Suprunenko, K. A., *Teor. Eksp. Khim.* **2**, 694 (1966).
61. Strelko, V. V., and Suprunenko, K. A., *Khim. Vys. Energ.* **2**, 258 (1968).
62. Tanei, T., *Bull. Chem. Soc. Jap.* **41**, 833 (1968).
63. Edlund, D., Kinell, P. O., Lund, A., and Shimizu, A., *J. Chem. Phys.* **46**, 3679 (1967).

64. Helcke, G. A., and Fantechi, R., *Mol. Phys.* **17,** 65 (1969).
65. Ono, Y., and Keii, T., *J. Phys. Chem.* **72,** 2851 (1968).
66. Vladinirova, V. I., Zhabrova, G. M., Kadenatsi, B. M., Kazanskii, V. B., and Pariiskii, G. B., *Dokl. Akad. Nauk, SSSR* **164,** 361 (1965).
67. Bobrovskii, A. P., and Kholmogorov, V. E., *Khim. Vys. Energ.* **2,** 147 (1968).
68. Bobrovskii, A. P., and Kholmogorov, V. E., *Kinet. Catal. (USSR)* **9,** 909 (1968).
69. Brouwer, D. W., *Chem. Ind.* p. 177 (1961).
70. Rooney, J. J. and Pink, R. C., *Proc. Chem. Soc.* 70 (1961).
71. Muha, G. M., *J. Phys. Chem.* **71,** 633 (1967).
72. Hall, W. K., *J. Catal.* **1,** 53 (1962).
73. Fogo, J. K., *J. Phys. Chem.* **65,** 1919 (1961).
74. Rooney, J. J., and Pink, R. C., *Trans. Faraday Soc.* **58,** 1632 (1962).
75. Dollish, F. R., and Hall, W. K., *J. Phys. Chem.* **69,** 4402 (1965).
76. Ben Taarit, Y., Mathieu, M., and Naccache, C., *Proc. Int. Conf. Mol. Sieve Zeol., 2nd,* 1970.
77. Neikam, W. C., *J. Catal.* **21,** 102 (1971).
78. Bassett, J., Naccache, C., Mathieu, M. V., and Prettre, M., *J. Chim. Phys. Physiochim. Biol.* **66,** 1522 (1969).
79. Richardson, J. T., *J. Catal.* **9,** 172 (1967).
80. Ward, J. W., *J. Phys. Chem.* **72,** 4211 (1968).
81. Lunsford, J. H., *J. Phys. Chem.* **74,** 1518 (1970).
82. Hirschler, A. E., Neikam, W. C., Barmby, D. S., and James, R. L., *J. Catal.* **4,** 628 (1965).
83. Muha, G. M., *J. Phys. Chem.* **74,** 2939 (1970).
84. Hirota, K., Kuwata, K., and Akagi, Y., *Bull. Chem. Soc. Jap.* **38,** 2209 (1965).
85. Flockhart, B. D., Naccache, C., Scott, J. A. N. and Pink, R. C., *Chem. Commun.* p. 238 (1965).
86. Pimenov, Yu. D., Kholmogorov, V. E., and Terenin, A. N., *Dokl. Akad. Nauk, (SSSR)* **163,** 935 (1965).
87. Flockhart, B. D., Leith, I. R., and Pink, R. C., *Chem. Commun.* p. 885 (1966).
88. Flockhart, B. D., Leith, I. R., and Pink, R. C., *Trans. Faraday Soc.* **66,** 469 (1970).
89. Flockhart, B. D., Leith, I. R., and Pink, R. C., *Trans. Faraday Soc.* **65,** 542 (1969).
90. Tench, A. J., and Nelson, R. L., *Trans. Faraday Soc.* **63,** 2254 (1967).
91. Che, M., Naccache, C., and Imelik, B., to be published.
92. Edlund, O., Kinell, P. O., Lund, A., and Shimizu, A., *J. Polym. Sci. Part B* **6,** 133 (1968).
93. Wong, P. K., and Willard, J. E., *J. Phys. Chem.* **72,** 2623 (1968).
94. Kinell, P. O., Lund, A., and Shimizu, A., *J. Phys. Chem.* **73,** 4175 (1969).
95. Seshadri, K. S., and Petrakis, L., *J. Phys. Chem.* **74,** 1317 (1970).
96. Ohigashi, H., and Kurita, Y., *J. Phys. Soc. Jap.* **24,** 564 (1968).
97. Hoffman, B. M., and Nelson, N. J., *J. Chem. Phys.* **50,** 2598 (1969).
98. Gardner, C. L., and Weinberger, M. A., *Can. J. Chem.* **48,** 1317 (1970).
99. Wang, K. M., and Lunsford, J. H., *J. Phys. Chem.* **74,** 1512 (1970).
100. Peri, J. B., *J. Phys. Chem.* **70,** 3168 (1966).
101. Hughes, T. R., and White, H. M., *J. Phys. Chem.* **71,** 2192 (1967).
102. Chao, C., and Lunsford, J. H., *J. Amer. Chem. Soc.* **93,** 71 (1971).
103. Iyengar, R. D., and Rao, V. V., *J. Amer. Chem. Soc.* **90,** 3267 (1968).
104. Colburn, C. B., Ettinger, R., and Johnson, F. A., *Inorg. Chem.* **3,** 455 (1964).
105. Sorokin, Y. A., Kotov, A. G., and Pshezhetskii, S. Y., *Dokl. Akad. Nauk, (SSSR)* **159,** 1385 (1964).

106. Adrian, F. J., Edward, L. C., and Bowers, V. A., *Advan. Chem.* **36**, 50 (1962).
107. Keys, L. K., Ph.D. Dissertation, Pennsylvania State Univ., University Park, Pennsylvania, 1965.
108. Kokes, R. J., *Proc. Int. Congr. Catal., 3rd.*, 1964.
109. Kasai, P. H., *J. Chem. Phys.* **43**, 3322 (1965).
110. Wang, K. M., and Lunsford, J. H., *J. Phys. Chem.* **74**, 1512 (1970).
111. Tupikov, V. I., Malkova, A. I., Sorokin, Y. A., and Pshezhetskii, S. Ya., *Khim. Vys. Energ.* **2**, 352 (1968).
112. Codell, M., Weisberg, J., Gisser, H., and Iyengar, R. D., *J. Amer. Chem. Soc.* **91**, 7762 (1969).
113. Tench, A. J., and Lawson, T., *Chem. Phys. Lett.* **8**, 177 (1971).
114. Wang, K. M., and Lunsford, J. H., *J. Phys. Chem.* **75**, 1165 (1971).
115. Shvets, V. A., Sarichev, M. E., and Kazanskii, V. B., *J. Catal.* **11**, 378 (1968).
116. Shvets, V. A., Vorotyntsev, V. M., and Kazanskii, V. B., *J. Catal.* **15**, 214 (1969).
117. Horiguchi, H., Setaka, M., Sancier, K. M., and Kwan, T., *Proc. Int. Congr. Catal., 4th*, 1968.
118. Cunningham, J., *J. Phys. Chem.* **66**, 779 (1962)
119. Jaccard, C., *Phys. Rev.* **124**, 60 (1961).
120. Mergerian, D., and Marshall, S. A., *Phys. Rev.* **127**, 2015 (1962).
121. Iyengar, R. D., Codell, M., Karra, J. S., and Turkevich, J., *J. Amer. Chem. Soc.* **88**, 5055 (1966).
122. Fukuzawa, S., Sancier, K. M., and Kwan, T., *J. Catal.* **11**, 364 (1968).
123. Iyengar, R. D., and Kellermann, R., *Z. Phys. Chem.* **64**, 345 (1969).
124. Montagna, A. A., Johnson, D. P., Hare, C. R., and Weller, S. W., *J. Catal.* **16**, 391 (1970).
125. Mishchenko, A. I., Pariiskii, G. B., and Kazanskii, V. B., *Kinet. Catal. (USSR)* **9**, 151 (1968).
126. Schoonheydt, R. and Lunsford, J. H., *J. Phys. Chem.* (in press).
127. Kohn, H. W., *J. Chem. Phys.* **33**, 1588 (1960).
128. Kazanskii, V. B., Pariiskii, G. B., and Voevodskii, V. V., *Disc. Faraday Soc.* **31**, 203 (1961).
129. Muha, G. M., *J. Phys. Chem.* **70**, 1390 (1966).
130. Weeks, R. A., and Nelson, C. M., *J. Amer. Ceram. Soc.* **43**, 399 (1960).
131. Muha, G. M., and Yates, D. J. C., *J. Phys. Chem.* **70**, 1399 (1966).
132. Mishchenko, Y. A., and Boreskov, G. K., *Kinet. Catal. (USSR)* **6**, 842 (1965).
133. Boudart, M., *Amer. Sci.* **57**, 97 (1969).
134. Nozaki, F., and Turkevich, *J. Catal. Jap.* **7**, 328 (1965).
135. Stamires, D. N., *in* "Molecular Sieves," pp. 328–332. Soc. of Chem. Ind., London, 1968.
136. Stamires, D. N., and Turkevich, J., *J. Amer. Chem. Soc.* **86**, 757 (1964).
137. Wertz, J. E., Orton, J. W., and Auzins, P. W., *Disc. Faraday Soc.* **31** 140 (1961).
138. Wertz, J. E., Auzins, P., Griffiths, J. H. E., and Orton, J. W., *Disc. Faraday Soc.* **28**, 136 (1959).
139. Henderson, B., and Wertz, J. E., *Advan. Phys.* **17**, 749 (1968).
140. Lunsford, J. H., and Jayne, J. P., *J. Phys. Chem.* **70**, 3464 (1966).
141. Tench, A. J., and Nelson, R. L., *J. Colloid Interface Sci.* **26**, 364 (1968).
142. Smith, D. R., and Tench, A. J., *Chem. Commun.* p. 1113 (1968).
143. Lunsford, J. H., *J. Phys. Chem.* **68**, 2312 (1964).
144. Kirklin, P. W., Auzins, P., and Wertz, J. E., *J. Phys. Chem. Solids* **26**, 1067 (1965).
145. Taylor, E. H., *Advan. Catal.* **18**, 111 (1968).

146. Poole, C. P. and MacIver, D. S., *Advan. Catal.* **17,** 223 (1967).
147. Poole, C. P., Kehl, W. L., and MacIver, D. S., *J. Catal.* **1,** 407 (1962).
148. Kazanskii, V. B., and Pecherskaya, Yu. I., *Kinet. Catal. (USSR)* **2,** 454 (1961).
149. Boreskov, G. K., Bukanaeva, F. M., Dzis'ko, V. A., Kazanskii, V. B., and Pecherskaya, Yu. I., *Kinet. Catal. (USSR)* **5,** 434 (1964).
150. Aleksandrov, I. V., Kazanskii, V. B., and Mikheikin, I. D., *Kinet. Catal. (USSR)* **6,** 439 (1965).
151. van Reijen, L. L., and Cossee, P., *Disc. Faraday Soc.* **41,** 277 (1966).
152. Kazanskii, V. B., and Pecherskaya, Yu. I., *Kinet. Catal. (USSR)* **4,** 244 (1963).
153. Ayscough, P. B., Eden, C., and Steiner, H., *J. Catal.* **4,** 278 (1965).
154. Matsumoto, A., Tanaka, H., and Goto, H., *Bull. Chem. Soc. Jap.* **38,** 1857 (1965).
155. Lunsford, J. H., unpublished work.
156. Kazanski, V. B., and Turkevich, J., *J. Catal.* **8,** 231 (1967).
157. Karra, J., and Turkevich, J., *Disc. Faraday Soc.* **41,** 310 (1966).
158. Peacock, J. M., Sharp, M. J., Parker, A. J., Ashmore, G. P., and Hockey, J. A., *J. Catal.* **15,** 379 (1969).
159. Cornaz, P. F., van Hooff, J. H. C., Pluijm, F. J., and Schuit, G. A. S., *Disc. Faraday Soc.* **41,** 290 (1966).
160. Sancier, K. M., Dozono, T., and Wise, H., *Proc. No. Amer. Meet. Catal. Soc., 2nd,* 1971.
161. Boreskov, G. K., Dzis'ko, V. A., Emel'yanova, V. M., Pecherskaya, Yu. I., and Kazanskii, V. B., *Dokl. Akad. Nauk (SSSR)* **150,** 829 (1963).
162. Boreskov, G. K., Davydova, L. P., Mastikhin, V. M., and Polyakova, G. M. *Dokl. Akad. Nauk (SSSR)* **171,** 648 (1966).
163. Tanaka, H. and Matsumoto, A., *Bull. Chem. Soc. Jap.* **39,** 874 (1966).
164. Nicula, A., Stamires, D., and Turkevich, J., *J. Chem. Phys.* **42,** 3686 (1965).
165. Richardson, J. T., *J. Catal.* **9,** 178 (1967).
166. Tikhomirova, N. N., and Nikolaeva, I. V., *Zh. Strukt. Khim.* **10,** 547 (1969).
167. Fujiwara, S., Katsumata, S., and Seki, T., *J. Phys. Chem.* **71,** 115 (1967).
168. Uytterhoeven, J. B., Schoonheydt, R., Liengme, B. V., and Hall, W. K., *J. Catal.* **13,** 425 (1969).
169. Slot, H. B., and Verbeek, J. L., *J. Catal.* **12,** 216 (1968).
170. Nicula, A., Ursu, I., and Nistor, S., *Rev. Roum. Phys.* **10,** 229 (1965).
171. Barry, T. I., and Lay, L. A., *J. Phys. Chem. Solids* **27,** 1821 (1966); **29,** 1395 (1968).
172. Dzhashiashvili, L. G., Tikhomirova, N. N., and Tsitsishvili, G. V., *Zh. Strukt. Khim.* **10,** 443 (1969).
173. Michoulier, J., and Ducros, P., *Proc. Colloq. AMPERE (At. Mol. Etud. Radio Elec.)* **12,** 215 (1963).
174. Hayes, R. G., *J. Chem. Phys.* **48,** 4806 (1968).

Author Index

Numbers in parentheses are reference numbers and indicate that an author's work is referred to although his name is not cited in the text. Numbers in italics show the page on which the complete reference is listed.

A

Abassin, J. J., 227(69), *263*
Aben, P. C., 103, *113*
Adams, C. M., 125(37, 38), *186*
Adams, C. R., 163, 164(136), *188*
Adrian, F. J., 266, 282, 310(106), *340*, *343*
Aigueperse, J., 4(22, 23), 9(22, 23), 11, *49*
Akagi, Y., 304(84), *342*
Alcock, C. B., 119(14), *185*
Aldag, A., 76(7), *112*
Aldridge, C. L., 41(59), *49*
Aleksandrov, I. V., 321(150), *344*
Alessandrini, E. I., 144(88), *187*
Alexander, E. G., 127(48), 128(48), 135(48), 138(48), *186*
Alferieff, M. E., 65(59), *73*
Alpress, J. G., 78, *113*
Anderson, J. R., 66(69), *73*, 101, 103, 109, *113*
Anderson, P. D., 118(13), 119(13), 182(13), *185*
Arghiropoulos, B., 4(24), *49*
Ashmore, G. P., 321(158), *344*
Atkins, L. T., 164(141), *188*
Atkins, P. W., 266, 292(7), *340*
Auzins, P., 318(137, 138), 319(144), *343*
Averbach, B. L., 183(164), *189*
Avery, N. R., 66(69), *73*
Ayscough, P. B., 321, *344*

B

Baird, J. C., 266, *340*
Bakulin, E. A., 65(55), *73*
Baltz, A., 130(57), *186*
Bank, S., 3(15), *48*
Barberi, P., 227(69), *263*
Barmby, D. S., 302(82), *342*
Barry, T. I., 324, 325, *344*
Bartek, J., 2(5), *48*
Barton, J. C., 172(152), 173(152), *189*
Basset, J., 204(36), *262*, 302(78), *342*
Bastick, M., 227(68), *263*
Bastl, Z., 57(33, 34), 61(33, 34), 62(33, 34), *72*
Bates, P. A., 132(65), *186*
Becker, J. A., 54(7), *71*
Beeck, O., 192, *261*
Beeskow, H., 172(151), *189*
Behrndt, K. H., 125(36), *186*
Bell, A. E., 70(82), *73*
Bellamy, L. J., 44(67), *50*
Belous, M. V., 120, 155(27), *185*
Belser, R. B., 117(9), *185*
Benesi, H. A., 75(5), *112*
Bennett, J. E., 272(24), *341*
Benson, J. E., 76(7), *112*
Benson, S. W., 44(66), *50*
Ben Taarit, Y., 302(76), *342*
Benzinger, T. H., 193(17), 196, *261*
Berger, C., 204(35), *262*
Berger, R. L., 233(72), *263*
Bernard, R., 202(33), 224(33), *262*
Bersohn, M., 266, *340*
Bertrand, A. J., 41(60), *50*
Best, R. J., 149, *188*
Bidwell, L. R., 118(19, 20), *185*
Block, J., 56(17), *72*
Blyholder, G., 65(62), *73*, 90, 92, *113*
Bobrovskii, A. P., 301, *342*
Bogdanovic, B., 42(63), *50*
Boiko, B. T., 129(53), *186*
Bolton, J. R., 266, *340*
Bond, G. C., 1(1, 2), 2(1, 2), 15(1), *48*, 66(67), *73*, 79, 102, 104, *113*, 116(4, 5), 148, 156(125), 174(155), 177(5), *185*, *187*, *188*, *189*

Boreskov, G. K., 144(84), 154(84), 157(84), 158(84), 171(84), 183(165), *187*, *189*, 192(7), *261*, 316, 321(149), 322, *343*, *344*

Boronin, V. S., 75(3, 4), 77, 81, 111(3), *112*

Bortner, M. H., 161(131), 165(131), 168(131), *188*

Bosáček, V., 103(40), *113*

Boudart, M., 4, 5(27), 9(27), *49*, 76, 87(23), *112*, *113*, 316, *343*

Bouwman, R., 126(41), 143(41), 144(83a), 154(41), 171, 180, 181, 182(83a), *186*, *187*

Bowers, V. A., 310(106), *343*

Bozon-Verduraz, F., 4(24, 25), *49*

Bradshaw, A. M., 90, *113*

Brandon, D. G., 79, *113*

Brennan, D., 168(145), *189*, 192(11), *261*

Brenner, S. S., 154(122), *188*

Brie, C., 201(31), 202(31), 211(39), 214(39, 47), 217(52), 225(62), 226(52), 233(39, 47), 237(52), *262*, *263*

Brooks, H., 65(54), *73*

Bros, J. P., 216(49), *262*

Brouwer, D. W., 301(69), *342*

Brown, H. D., 216(51), *262*

Budworth, D. W., 176(156), *189*

Bukanaeva, F. M., 321(149), *344*

Burns, G., 293(39), 295(39), *341*

Burwell, R. L., Jr., 1(3), 2(6), 3, 12, 16, *48*, *49*, 100, 104, *113*

Byrne, J. J., 141(79), 152, 153(117), 182(117), *187*, *188*

C

Calvet, E., 193(14, 16), 196, 197, 198, 200, 201, 205(16), 206(16), 210(16), 214(16), 215(16), 219, 220(16), 222(56), 223, 232(16), *261*, *262*

Camia, F. M., 196(22), 212, 213(46), 214(48), 215(48), 219(56), 221(46), 222(46, 56), 223(61), *261*, *262*

Campbell, J. S., 121(29), 154, 182, 183(29), *185*

Carr, P. F., 152(118, 119, 120), *188*

Carrington, A., 266, 268, 276(5), *340*

Carter, J. L., 22(45), *49*

Casey, E. J., 299(57), *341*

Černý, S., 51(1), 57(22, 25, 26, 27, 28, 31, 35, 35a), 60(27), 61(22, 25, 27, 31, 32), 62(42), *71*, *72*, 192(11, 13), 195(13), 220(13), *261*

Chabert, J., 196(22), 205(38), *261*, *262*

Chalk, A. J., 42(61), *50*

Chan, J. P., 118(15), *185*

Chang, C. C., 44(65), 46(65), 47(65), 48(65), *50*

Chao, C., 309(102), *342*

Charles, S. W., 132(65), *186*

Chaston, J. C., 178(162), *189*

Che, M., 272(22), 305, 311(22), 313(22), *340*, *342*

Cheselske, F. J., 149(111), 152(111), *188*

Chessick, J. J., 196(24), *261*

Chevillot, J. P., 211(41), 226(65), *262*, *263*

Chuikov, B. A., 66(70), *73*

Clarke, J. K. A., 135(69), 141(69, 79), 152, 153(117, 121a), 155, 159, 160(69), 182(117), *186*, *187*, *188*

Codell, M., 312, 315(121), *343*

Cohen, A., 299(53), *341*

Cohen, M., 144(89, 90), 183(164), *187*, *189*

Cohen, M. H., 272(25), *341*

Cohn, G., 116(3), *185*

Colburn, C. B., 273(30), 277(30), 309, *341*, *342*

Coles, B. R., 130, 140(77), 141(77), 148(99), *187*

Conley, R. T., 44(68), *50*

Conner, W. C., 3(8, 9), 13, *48*

Cornaz, P. F., 321, *344*

Cossee, P., 321, 322, *344*

Coten, M., 223(61), *262*

Cotton, F. A., 35(53), *49*

Coué, J., 251(82), 252(82), *263*

Coulson, C. A., 65(62), *73*

Couper, A., 158, 161(127), 162(127), 168(148), 169(127, 148), 171(148), *188*, *189*

Courtois, M., 241(75), *263*

Cram, D. J., 3(16), 42(16), *48*

Crawford, E., 138(72), *187*

Cripps, H. N., 34(51), *49*

Crowell, A. D., 192(3), 226(66), *261*, *263*

Cukr, M., 62(46), 63(46), *72*

Cunningham, J., 315(118), *343*

Curtis, R. M., 75(5), *112*

Cutler, P. H., 65(57), *73*

AUTHOR INDEX

D

Daglish, A. G., 158(129),162(129), 163, 168, *188*
Daneš, V., 103(40, 41), *113*
Darby, J. B., 118(16, 17, 21), *185*
Darken, L. S., 119, *185*
Darling, A. S., 178(162), *189*
Davis, B. J., 148(105), *187*
Davydova, L. P., 322(162), *344*
De Boer, J. H., 192(1), *261*
Decius, J. C., 35(57), 36(57), *49*
Delchar, T. A., 54(10), 70(84), *72, 73*
Dell, R. M., 239(74), 240, *263*
Della Gatta, G., 227(68), *263*
Dent, A. L., 3(10, 11, 12, 13, 14), 4(10, 11, 12, 13, 14), 5(10), 6(10), 7(10), 9(10), 13(13), 14(10), 15(10, 13), 16(10), 17(10, 13), 19(10, 13), 20(13), 23(11, 13), 28(13), 29(12), 36, 38(13), 42(14), 45(14), *48*
Deo, A. V., 48(73), *50*
Dewar, J., 192, *261*
d'Heurle, F. M., 132(67), *186*
Dicarlo, E. N., 266(11), *340*
Dobson, P. J., 147(95c), *187*
Dolejšek, Z., 62(42), *72*
Dollish, F. R., 302, *342*
Dooley, G. J., 145(93), 146(93), *187*
Dorf, R. C., 225(63), *263*
Dorgelo, G. J. H., 121(28), 142, 143(28), 150(28, 114), 154(28), *185, 188*
Dougharty, N. A., 76(7), *112*
Dowden, D. A., 66(68), *73*, 148, 156, *187, 188*
Dozono, T., 321(160), 322(160), *344*
Drechsler, M., 79, *113*
Dubejko, M., 143(83), 162(83), *187*
Ducros, P., 325(173), *344*
Duell, M. J., 140(74), *187*
Dugdale, J. S., 148(100), 171(100), *187*
Duke, C. B., 65(59), *73*
Dunken, M. H., 65(65), *73*
Dupupet, G., 227(68), *263*
Dushman, S., 126, *186*
Duwez, P., 132(66), *186*
Dzhashiashvili, L. G., 325, *344*
Dzis'ko, V. A., 321(149), 322(161),*344*

E

Edel'man, I. S., 130(56), *186*
Eden, C., 321(153), *344*
Edlund, D., 300, 303(63), 305(63), *341*
Edlund, O., 305(92), *342*
Edward, L. C., 310(106), *343*
Eggers, D. F., Jr., 28(48), *49*
Eggers, K. W., 44(66), *50*
Eggs, J., 148(101), *187*
Ehrenreich, H., 65(49), *72*
Ehrlich, G., 54(8, 10), *71, 72*, 192(2), *261*
Eischens, R. P., 3(19), 4(32), 6, 16, *49*, 76, 90, 91, 92, 93, 110, *112, 113*
Eley, D. D., 115(1), 158, 159, 161(127), 162(127, 129), 163, 168, 169(127), *185, 188*
E'liot, J. J., 22(45), *49*
Elliott, R. P., 117(11), *185*
El Shobaky, G., 192(8), 219(53, 54), 241(77), 242(8), 246(53), 247(53, 80), 248(53), 250(8, 54), 251(54), 255(53), 256(53), 257(8), *261, 262, 263*
Elston, J., 204(35), *262*
Emel'yanova, V. M., 322(161), *344*
Emmett, P. H., 121(29), 149, 150, 152(110), 154, 182, 183(29), *185, 188*
Engels, S., 168(149), 170(149), *189*
Erfl, G., 146(95a), *187*
Ernst, I. T., 266(12, 13), *340*
Ettinger, R., 273(30), 277(30), 309(30, 104), *341, 342*
Evans, E. Ll., 119(14), *185*
Eyraud, L., 203(34), 204(35, 37), 216(37), *262*
Eyraud, M., 204(36), *262*

F

Faller, J. W., 35(53), *49*
Fantechi, R., 299(64), 300(64), *342*
Feinstein, L. G., 56(15), *72*
Fetiveau, M., 204(35), *262*
Fetiveau, Y., 204(35), *262*
Fischer, E. O., 35(54), *49*
Fisher, B., 121(32), *185*, 266(12), *340*
Flank, W. H., 116(8), 164(8), 183(8), *185*
Flockhart, B. D., 303(88), 304, *342*
Fogo, J. K., 302(73), *342*
Fonash, S. J., 65(85), *73*

Francis, S. A., 90(25), 91(25), 92(25), 93(25), *113*
Friedel, J., 65(51), *72*
Fritz, H. P., 35, *49*
Fubini, B., 227(68), *263*
Fujiki, Y., 120, *185*
Fujimoto, M., 299(53, 55), *341*
Fujita, Y., 299(54), *341*
Fujiwara, S., 324, *344*
Fukano, Y., 78, *113*, 142(80), *187*
Fukuzawa, S., 315, *343*

G

Gadzuk, J. W., 65(56, 58), *73*
Gale, R. L., 243(78), *263*
Garbutt, B., 299(53, 55), *341*
Gardner, C. L., 296, 299(57), 307(98), *341*, *342*
Garland, C. W., 91, 92, 111, *113*
Garnett, J. L., 266(12, 13), *340*
Garrigues, J. C., 219(57), 221(57), 222(57), *262*
Gerber, R., 125(35), *186*
Gery, A., 202(33), 223(60), 224(33, 60), *262*
Gesser, H. D., 299(53, 55), *341*
Geus, J. W., 75(1, 2), 111, *112*
Gharpurey, M. K., 149, *188*
Gibbens, H. R., 123, 131(60), 132(60), 137(60), 139(73), 142(60), 172(60), 173(60, 73), 174(73), 175(73), 176(73), 177(34), 178(34), *186*, *187*
Gibson, M. J., 147(95c), *187*
Girvin Harkin, C., 76(7), *112*
Gisser, H., 312(112), *343*
Glemza, R., 9(33), *49*
Glev, D. N., 22(44), *49*
Golden, D. M., 44(66), *50*
Goldwaser, D., 211(41), 216(41), 226(65), *262*, *263*
Golubev, V. B., 295(42), *341*
Gonzalez, O. D., 168(147), *189*
Gossner, K., 161(132), 163(132), *188*
Goto, H., 321(154), *344*
Grant, J. T., 145(93), 146(93), *187*
Gravelle, P. C., 192(8), 193(15, 19), 196(22), 199(28), 200(28), 205(38), 211(39), 214(39, 47), 218(19), 219(53, 54, 55), 227(15), 231(15, 55), 232(28, 55), 233(39, 47, 55), 234(55), 235, 237(19), 238(19, 73), 239(19), 240(19, 55, 73), 241(55, 76, 77), 242(8), 244(73), 245(19, 73, 79), 246(19, 53, 73, 79), 247(53, 80), 248, 250(8, 54), 251, 252(82), 255(53), 256(53), 257(8), *261*, *262*, *263*
Grebennik, I. P., 129(51), *186*
Green, J. A. S., 172(152), 173(152), *189*
Green, J. H. S., 28, *49*
Gregg, S. J., 11(36), *49*
Gribko, V. F., 129(51), *186*
Griffiths, J. H. E., 318(138), *343*
Grimley, T. B., 53(4), 66(66), *71*, *73*, 148, 171(150), *188*, *189*
Gruber, H. L., 87(22), *113*
Guénault, A. M., 148(100), 171(100), *187*
Guillaud, C., 201, *262*
Guillouet, Y., 227(69), *263*
Guivarch, M., 225(62), *263*
Gurry, R. W., 119, *185*

H

Haag, W. O., 3(17), 42(17), 46(17), *49*
Haas, T. W., 145(93), 146(93), *187*
Haber, J., 243(78), *263*
Hagstrum, H. D., 53(3), 70(3), *71*
Hall, T. A., 144(87), *187*
Hall, W. K., 45(69), 48(69), *50*, 149, 150, 152(110, 111, 116), *188*, 302, 324(168), *342*, *344*
Haller, G. L., 1(3), 3(3), 16(3), *48*
Hamilton, W. M., 2(6), *48*
Hannan, R. B., 87, *113*
Hansen, M., 117, *185*
Hardt, P., 42(63), *50*
Hardy, W. A., 144(85), 156(85), *187*
Hare, C. R., 315(124), *343*
Harris, L., 128(50), *186*
Harris, L. A., 145, 146(91), 147(91), *187*
Harrison, D. L., 1(4), *48*
Harrison, W. A., 65(50), *72*
Harrod, J. F., 42(61), *50*
Hartman, C. D., 54(7), *71*
Hartmanshenn, O., 227(69), *263*
Hartog, F., 77, 78(13), 79, 80(13), 85(13), 87(24), 100, *112*, *113*
Hathaway, B. J., 111, *113*
Hayes, R. G., 339(174), *344*

Hayward, D. O., 168(145), *189*, 192(4, 6), *261*
Heckingbottom, G., 56(14), *72*
Heimbach, P., 42(63), *50*
Helcke, G. A., 299(64), 300(64), *342*
Henderson, B., 318, *343*
Henglein, A., 53(5), *71*
Herman, Z., 53(6), 62(42), *71*, *72*
Herzberg, G., 20(40), 22, 37, *49*
Hickmott, T. W., 54(8), *71*
Hightower, J. W., 45(69), 48(69), *50*
Hinnen, C., 211(40, 41), 213(40), 216(41), 217(40), 221(40), 222(40), 223(40), 226(65), *262*, *263*
Hirota, K., 304(84), *342*
Hirschler, A. E., 302, *342*
Hládek, L., 57(35, 36), 61(35, 36), *72*
Hoare, F. E., 148(97), 165(142), 176(156), *187*, *188*, *189*
Hoare, J. P., 173(154), *189*
Hockey, J. A., 321(158), *344*
Hodges, J. A., 272(28), *341*
Hoehn, H. H., 34(51), *49*
Hoffman, B. M., 307(97), *342*
Holland, L., 125(39), 126, 127, *186*
Hollis, D. P., 266(16), *340*
Holmberg, R. W., 276(34), *341*
Holroyd, P., 274, 275(32), 311, 312, *341*
Holscher, A. A., 65(61), *73*
Horiguchi, H., 314, *343*
Horiuti, J., 65(60), *73*
Howk, B. W., 34(51), *49*
Hughes, T. R., 309(101), *342*
Hultgren, R., 118(13, 15), 119(13), 182, *185*, 195, *261*
Hunter, D. H., 3(16), 42(16), *48*
Hutchins, G. A., 144(86), *187*

I

Ignat'yev, O. M., 126(42), *186*
Ignat'yeva, L. K., 126(42), *186*
Imelik, B., 305, *342*
Ingram, D. J. E., 272(24), *341*
Innes, R. A., 3(8), *48*
Itzel, J. E., 266(11), *340*
Iwai, I., 46(71), *50*
Iyengar, R. D., 309(103), 312(112), 315, *342*, *343*

J

Jaccard, C., 315(119), *343*
Jacknow, J., 76, 110, *112*
Jacobs, G., 53(5), *71*
Jaeger, H., 77, 106, *113*
James, R. L., 302(82), *342*
Jayne, J. P., 272(21, 26), 291(26), 310(21), 311, 315, 319, *340*, *341*, *343*
Johnson, A. L., 146(95), *187*
Johnson, D. P., 315(124), *343*
Johnson, F. A., 273(30), 277(30), 309(30, 104), *341*, *342*
Jonassen, H. B., 41(59, 60), *49*, *50*
Jongepier, R., 122(33), 150(33, 114), 152(33), 153(121), 175(33), *186*, *188*
Joyner, T. B., 41(60), *50*
Juillet, F., 257(84), *263*

K

Kadenatsi, B. M., 301(66), *342*
Kadlec, V., 64(47), *72*
Kadlecová, H., 64(47), *72*
Känzig, W., 272(25), *341*
Kaesz, H. D., 27, 28, *49*
Karra, J. S., 315(121), 321, *343*, *344*
Kasai, P. H., 312, 316, 317, *343*
Katsumata, S., 324(167), *344*
Kazanskii, V. B., 295, 296, 297, 299(56, 57), 300(57), 301(66), 312(49, 115, 116), 315(125), 316, 321, 322(161), *341*, *342*, *343*, *344*
Kehl, W. L., 321(147), *344*
Keii, T., 301, *342*
Keim, W., 42(63), *50*
Kellermann, R., 315, *343*
Kelley, K. K., 118(13), 119(13), 182(13), *185*
Kemball, C., 101, 103, 109, *113*, 138(71, 72), 167(143), 168(143), 168(144), 177(161), *186*, *187*, *188*, *189*
Kenson, R. E., 164(139), *188*
Kerstter, J., 53(6), *71*
Kesavulu, V., 4(29), *49*
Keys, L. K., 310, *343*
Khasin, A. V., 183(165), *189*
Kholmogorov, V. E., 301, 304(86), *342*

Kinell, P. O., 299(59), 300(63), 303(63), 305(63, 92), 306, *341*, *342*
King, D. A., 192(11), *261*
Kirenskii, L. V., 130(56), *186*
Kirklin, P. W., 319, *343*
Kitzinger, C., 193(17), 196(17), *261*
Klemperer, D. F., 195(21), *261*
Kneller, E., 132(61), *186*
Kneller, E. F., 132(62), *186*
Kneubühl, F. K., 293, *341*
Knights, C. F., 158(130), *188*
Knor, Z., 52(2), 53(2), 57(18, 19, 20, 21, 22, 23, 24, 25, 26, 27, 28, 29, 30, 31, 32, 37, 38, 39), 60(27), 61(22, 25, 27, 31, 32, 40), 62(19, 42), 64(47), 66(2), 68(77), 69(77, 78), 70(77), *71*, *72*, *73*, 228(70), *263*
Koehler, C., 211(41), 216(41), 226(65), *262*, *263*
Kohn, H. W., 316, *343*
Kokes, R. J., 3(8, 9, 10, 11, 12, 13, 14), 4(10, 11, 12, 13, 14), 5(10), 6(10), 7(10), 9(10, 33, 35), 13, 14(10), 15(10, 13), 16(10), 17(10, 13), 19(10, 13), 20(13), 23(11, 13), 28(13), 29(12), 36, 38(13), 42(14), 44(65), 45(14), 46(65), 47(65), *48*, *49*, *50*, 266, 311(108), *340*, *343*
Kosower, E. M., 46(70), *50*
Kotov, A. G., 309(105), *342*
Kovorkian, V., 22(45), *49*
Kowaka, M., 162(133), 168, *188*
Krivanek, M., 103(41), *113*
Kröner, M., 42(63), *50*
Kubaschewski, O., 119(14), *185*, 195, *261*
Kul'kova, N. V., 168(146), *189*
Kulokawa, Y., 4(31), 9(31), *49*
Kummer, J. T., 164, *188*
Kuppers, J., 146(95a), *187*
Kuptsis, J. D., 144(88), *187*
Kurek, T., 199(27), *261*
Kurita, Y., 307(96), *342*
Kuwata, K., 304(84), *342*
Kwan, T., 297, 314(117), 315(122), *341*, *343*

L

Lacker, J. P., Jr., 21(43), 22(43), *49*
Lacman, K., 53(5), *71*
Laffitte, M., 196(22), *261*
Lafitte, M., 223(61), *262*
Lam, V. T., 222(59), *262*
Lambard, J., 227(69), *263*
Lapkin, M., 164(139), *188*
Laurent, M., 202(33), 223(60), 224(33, 60), *262*
Laville, G., 212, *262*
Lawson, T., 298(52), 312, *341*, *343*
Lay, L. A., 324, 325, *344*
Lebedev, Ya. S., 289(37), 290(37), 292, *341*
Lebedeva, M. V., 129(53), *186*
Leith, I. R., 303(88), 304(87, 88, 89), *342*
Lewis, C. E., 111, *113*
Lewis, F. A., 172(152), 173(152), *189*
Liengme, B. V., 324(168), *344*
Light, T. B., 142(82), *187*
Linnett, J. W., 144(85), 156(85), *187*
Ljubarskii, C. D., 103, *113*
Long, J., 257(84), *263*
Lord, R. C., 30(50), 32(50), *49*, 91(29), 92(29), *113*
Love, R. W., 125(36), *186*
Low, M. J. D., 4(30, 32), 6(32), 16(32), *49*
Lucchesi, P. J., 22(45), *49*
Luetic, P., 158(128), 159, *188*
Lund, A., 299(59), 300(63), 303(63), 305(63, 92), 306, *341*, *342*
Lunsford, J. H., 272(17, 18, 19, 20, 21, 26), 273(29), 275(33), 276(29), 291(26), 293(29), 297, 298(51), 302(81), 306(17), 307, 308(20, 81), 309(29, 102), 310(21), 311, 312(33, 114), 313(18), 314(17, 18), 315, 317, 319, 320(143), 321(155), *340*, *341*, *342*, *343*, *344*
Lutinski, F. E., 149(111), 152(111), *188*
Luz, Z., 276(34), *341*

M

McCann, W. H., 142(81), *187*
McClellan, W. R., 34(51), *49*
Machefer, J., 227(69), *263*
MacIver, D. S., 320, 321(147), *344*
McKee, D. W., 3(18), *49*, 177, *189*
McKinney, J. T., 154(122), *188*
McLachlan, A. D., 266, 268, 276(5), *340*
McLane, S. B., 56(16), 70(16), *72*
MacLennan, D. A., 70(84), *73*

McMahon, E., 152(120), *188*
Macqueron, J. L., 196(22), 202(33), 223(60), 224(33, 60), *261*, *262*
MacRae, A. U., 14, *49*
Macrakis, M. S., 56(15), *72*
Mader, S., 120(26), 132, 133(26, 64), *185*, *186*
Malkova, A. I., 312(111), *343*
Manuel, A. J., 176(157), *189*
Marshall, D. J., 144(87), *187*
Marshall, S. A., 272(28), 315(120), *341*, *343*
Marty, G., 219(55), 229(71), 230(71), 231(55), 232(55), 233(55), 234(55), 235(55), 238(55), 240(55, 73), 241(55, 77), 244(73), 245(73), 246(73), 254(71), *262*, *263*
Mashchenko, A. I., 296(46), 297(46), *341*
Mason, R., 34(52), *49*
Mastikhin, V. M., 322(162), *344*
Mathieu, M. V., 204(36), *262*, 302(76, 78), *342*
Matsumoto, A., 321(154), 322(163), 323, *344*
Matthews, J. C., 165(142), *188*
Melmed, A. I., 55(12), *72*
Melsa, J. L., 225(64), *263*
Menzel, D., 172(151), *189*
Mergerian, D., 315(120), *343*
Meriaudeau, P., 272(22), 311(22), 313(22), *340*
Merta, R., 62(44, 45), 64(44, 45), *72*
Metcalfe, A., 168(148), 169(148), 171(148), *189*
Meyer, E. F., 12, *49*
Michel, J., 227(69), *263*
Michel, P., 120, 128(24), *185*
Michoulier, J., 325(173), *344*
Mikheikin, I. D., 296, 297, 321(150), *341*, *344*
Mileshkina, N. V., 70(83), *73*
Mintern, R. A., 178(162), *189*
Mishchenko, A. I., 315, *343*
Mishchenko, Y. A., 316, *343*
Misner, C., 65(52), *73*
Mitchell, D. F., 144(90), *187*
Molinari, E., 4(26), 9(26), *49*
Montagna, A. A., 315(124), *343*
Montgomery, H., 147(96), 148(96), *187*
Moore, D. W., 41(60), *50*
Morabito, J. M., 55(13), *72*

Morgan, A. E., 66(71), *73*
Morton, J. P., 266, *340*
Moss, R. L., 121(30), 122(30), 123, 126(40), 129(30, 54, 55), 130(55), 131(60), 132(60), 134(40), 135(40), 136(40), 137(60), 139(30, 54, 73), 140(30, 74), 141(30, 40), 142(60), 148(105), 158(126), 159, 162(30, 40, 126, 135), 165(40, 135), 167(40), 170(54, 126), 171(126), 172(60), 173(73, 60), 174(73), 175(73), 176(73), 177(34), 178(34), 182(135), *185*, *186*, *187*, *188*
Muller, E. W., 56(16), 68(77), 69(77, 79, 80), 70(16, 77, 80, 81), *72*, *73*
Müller, H., 162(134), *188*
Münster, A., 183(163), *189*
Muetterties, E. L., 34(51), *49*
Muha, G. M., 301(71), 316, *342*, *343*
Muller, R. H., 55(13), *72*
Musco, A., 35(53), *49*
Myles, K. M., 118(17, 18), *185*

N

Naccache, C., 272(22), 297(50), 302(76, 78), 304(85), 305, 311(22), 313, *340*, *341*, *342*
Nagasawa, A., 130(58), *186*
Nagy, D., 65(57), *73*
Nakayama, K., 146(95b), *187*
Neikam, W. C., 302, *342*
Nekipelov, V. N., 144(84), 154(84), 157(84), 158(84), 171(84), *187*
Nelson, C. M., 316(130), *343*
Nelson, N. J., 307(97), *342*
Nelson, R. L., 304, 307(90), 318(141), 319, *342*, *343*
Neugebauer, C. A., 131(59), *186*
Nicholas, J. F., 77, 79, *112*, *113*
Nicholls, D., 1(4), *48*
Nicolau, C., 266(14, 15), *340*
Nicula, A., 323, 324, *344*
Nikolaeva, I. V., 324, *344*
Nikolajenko, V., 103, *113*
Nishikawa, O., 69(79), *73*
Nistor, S., 324(170), *344*
Noble, C. S., 266(11), *340*
Norton, F. J., 177, *189*

Nowick, A. S., 120(26), 132(63, 67, 68), 133(26), *185*, *186*
Nozaki, F., 316, *343*

O

Oberkirch, W., 42(63), *50*
Ohigashi, H., 307(96), *342*
Oldfield, R. C., 128(49), *186*
Olsen, G. H., 222(58), *262*
Olszewski, S., 143(83), 162(83), *187*
O'Neill, C. E., 91, 92, *113*
Ono, M., 146(95b), *187*
Ono, Y., 301, *342*
Opitz, G., 65(65), *73*
O'Reilly, D. E., 266, *340*
Oron, M., 125(37, 38), *186*
Orr, R. L., 118(13), 119(13), 182(13), *185*
Orton, J. W., 318(137, 138), *343*
Ostrovskii, V. E., 168(146), *189*
Ovenall, D. W., 272(27), 291(27), 294(27), *341*

P

Palatnik, L. S., 126(42), 129(53), *186*
Palmberg, P. W., 67(74, 75), *73*, 145(94), 146, *187*
Panitz, J. A., 56(16), 70(16), *72*
Papoulis, A., 211(42), *262*
Pariiskii, G. B., 295, 296, 299(56), 301(66), 315(125), 316(128), *341*, *342*, *343*
Parker, A. J., 321(158), *344*
Parravano, G., 4, 5(27), 9(26, 27), *49*, 161(131), 165(131), 168, *188*, *189*
Patterson, W. R., 168(144), *189*
Peacock, J. M., 321(158), *344*
Pearson, W. B., 141, *187*
Pecherskaya, Yu. I., 321(148, 149, 152), 322(161), *344*
Pells, G. P., 147(96), 148(96), *187*
Pennington, S. N., 216(51), *262*
Peri, J. B., 48(72, 74), *50*, 87, *113*, 308, *342*
Peria, W. T., 145(92), 146(92), *187*
Perry, A. J., 79, *113*
Persoz, M., 200, *261*
Petit, J. L., 196(22), 201, 202, 203(34), 205(38), 211(39), 214(39, 47), 225(62), 233(39, 47), *261*, *262*, *263*
Petrakis, L., 307(95), *342*

Pettifor, D. G., 65(53), *73*
Phillips, J. C., 65(48), *72*
Pietrzak, T. M., 273(31), 309, *341*
Pimenov, Yu. D., 304(86), *342*
Pines, B. Y., 129(51), *186*
Pines, H., 3(17), 42(17), 46(17), *49*
Pink, R. C., 301(70), 302(74), 303(88), 304(85, 87, 88, 89), *342*
Platteeuw, J. C., 103(42), *113*
Pliskin, W. A., 3(19), 4(32), 6(32), 16(32), *49*, 90(25), 91(25), 92, 93, *113*
Pluijm, F. J., 321(159), *344*
Plummer, E. W., 67(74, 76), *73*
Plunkett, T. J., 153(121a), *188*
Plyasova, L. M., 183(165), *189*
Pobitschka, E., 266(14), *340*
Poltorak, O. M., 75(3, 4), 77, 81, 111(3), *112*
Polyakova, G. M., 322(162), *344*
Ponec, V., 51(1), 57(18, 19, 20, 21, 22, 23, 24, 25, 26, 27, 28, 29, 30, 31, 32, 35), 60(27), 61(22, 25, 27, 31, 32, 35, 40), 62(19, 42, 43, 44), 63(43), 64(43, 44), *71*, *72*, 192(13), 195(13), 220(13), *261*
Poole, C. P., 266(11), 282, 283(35), 284(35), 285(35), 286(35), 320, 321(147), *340*, *341*, *344*
Popovskii, V. V., 192(7), *261*
Popplewell, J., 132(65), *186*
Powell, D. B., 21, *49*
Praat, A. P., 42(62), *50*
Pradilla-Sorzana, J., 21(43), 22(43), *49*
Praglin, J., 216(50), *262*
Prat, H., 193(16), 196(23), 197, 198, 201(16), 205(16), 206(16), 210(16), 214(16), 215(16), 219(16), 220(16), 223(16), 232(16), *261*
Preston, J., 176(156), *189*
Prettre, M., 204(36), *262*, 302(78), *342*
Priestley, M. G., 148(98), *187*
Pritchard, J., 90, *113*
Pshezhetskii, S. Y., 309(105), 312(111), *342*, *343*
Ptushinskii, Y. G., 66(70), *73*
Pulkkinen, E., 41(59), *49*
Pyn'ko, V. G., 130(56), *186*

Q

Quinn, H. W., 22(44), *49*

R

Rachinger, W. A., 141, *187*
Rafter, E. A., 135(69), 141(69), 159, 160(69), *186*
Ranc, R. E., 251(82), 252(82), *263*
Rao, V. V., 309(103), *342*
Raub, E., 172(151), *189*
Read, J. F., 1(3), 3(3), 16(3), *48*
Redhead, P. A., 54(9), *72*
Reed, D., 70(82), *73*
Reinberg, A. R., 272(28), *341*
Reisner, T., 69(79), *73*
Rendulic, K. D., 69(78), *73*
Reusmann, G., 140(75), *187*
Reuveni, A., 276(34), *341*
Reymond, J. P., 227(67), 229(67), 231(67), 233(67), 235(67), 236, 257(67), 258(67), *263*
Reynolds, J. E., 183(164), *189*
Reynolds, P. W., 148, 156, *188*
Rhodin, T. N., 67(74, 75, 76), *73*, 145(94), 146, *187*
Richard, M., 204(35), *262*
Richards, J. L., 142(81), *187*
Richardson, J. T., 302, 324, *342*, *344*
Richman, M. H., 67(73), *73*
Rienäcker, G., 162(134), 168(149), 170(149), *188*, *189*
Rinfret, M., 222(59), *262*
Riter, J. R., Jr., 28(48), *49*
Rizzo, F. E., 118(19), *185*
Roberts, M. W., 138(72), *187*
Rooney, J. J., 2(7), *48*, 301(70), 302(74), *342*
Rose, T., 53(6), *71*
Rossington, D. R., 127(47), 128(47), 135(47), 162(47), 168(47), 169(47), *186*
Rouquerol, J., 200(30), *262*
Rousseau, A., 211(40, 41), 213(40), 216(41), 217(41), 221(40), 222(40), 223(40), 226(65), *262*, *263*
Roux, A., 204(35), *262*
Roux, R., 219(57), 221(57), 222(57), *262*
Rowe, C. A., Jr., 3(15), *48*
Rudman, P. S., 121(31, 32), *185*
Rué, P., 251(82), 252(82), *263*
Runk, R. B., 127(47), 128(47), 135(47), 162(47), 168(47), 169(47), *186*

Russell, D. R., 34(52), *49*
Russell, W. W., 127(48), 128(48), 135(48), 138(48), 149, *186*, *188*
Rylander, P. N., 116(3), *185*

S

Sachtler, W. M. H., 61(41), 70(41), *72*, 116(2), 121(28), 122(33), 126(41), 138(70), 142, 143(28, 41), 150, 151(2, 70, 115), 152(33, 70), 153(121), 154, 155(70), 156(2), 180, 181, *185*, *186*
Sagel, K., 183(163), *189*
St. Quinton, J. M. P., 176(157), *189*
Sancier, K. M., 314(117), 315(122), 321, 322(160), *343*, *344*
Sanders, J. V., 77, 78, 106, *113*
Sarichev, M. E., 312(115), *343*
Sazonov, V. A., 192(7), *261*
Schay, G., 65(52), *73*
Schlosser, E. G., 76, 77, *112*
Schmidt, L. D., 54(11), 66(72), 67(72), *72*, *73*, 148(104), *187*
Schnabel, K. H., 266(15), *340*
Schonland, D., 272(24), *341*
Schoonheydt, R., 315, 324(168), *343*, *344*
Schrage, K., 100, 104, *113*
Schrenk, G. L., 65(85), *73*
Schriesheim, A., 3(15), *48*
Schuit, G. A. S., 321(159), *344*
Schultz, D. G., 225(64), *263*
Schwab, G.-M., 161(132), 163(132), *188*
Scott, J. A. N., 304(85), *342*
Seki, T., 324(167), *344*
Selwood, P. W., 266(16), *340*
Serway, R. A., 272(28), *341*
Seshadri, K. S., 307(95), *342*
Setaka, M., 314(117), *343*
Sewell, P. B., 144(89, 90), *187*
Sharp, M. J., 321(158), *344*
Sheppard, N., 21, 43(64), *49*, *50*
Shimizu, A., 300(63), 303(63), 305(63, 92), 306, *341*, *342*
Shimizu, M., 299(55), *341*
Shunk, F. A., 117(12), *185*
Shvets, V. A., 297, 312, *341*, *343*
Sicard, L., 203(34), *262*

AUTHOR INDEX

Siegel, B. M., 128(50), *186*
Silver, B. L., 276(34), *341*
Simpson, D. M., 43(64), *50*
Sing, K. S. W., 11(36), *49*
Sinicki, G., 202(33), 223(60), 224(33, 60), *262*
Skinner, H. A., 195(20a), *261*
Slot, H. B., 324, *344*
Smith, A. E., 164(141), *188*
Smith, D. R., 319, *343*
Smith, J. V., 118(19), *185*
Smithells, C. J., 127, 129(52), *186*
Sokolskaya, I. L., 70(83), *73*
Sollich-Baumgartner, W. A., 266(12, 13), *340*
Somorjai, G. A., 55(13), 66(71), *72, 73*
Sorokin, Y. A., 309, 312(111), *342, 343*
Speiser, R., 118(20), *185*
Spenadel, L., 87(23), *113*
Spooner, T. A., 155, *188*
Stamires, D., 323, *344*
Stamires, D. N., 316, 317, 323, *343, 344*
Starostina, T. S., 183(165), *189*
Steckelmacher, W., 125(39), *186*
Steiger, R. F., 55(13), *72*
Steiner, H., 1(4), *48*, 321(153), *344*
Steinrücke, E., 42(63), *50*
Stephens, S. J., 127(46), 158(46), 159, *186*
Stetsenko, A. I., 173(153), *189*
Stevenson, D. P., 164(141), *188*
Stoicheff, B. P., 21(42), 22, *49*
Stone, F. G. A., 27, 28, *49*
Stone, F. S., 193(18), 195(21), 238(18), 239(74), 240, 243(78), *261, 263*
Stouthamer, B., 103(42), *113*
Strelko, V. V., 300(60, 61), *341*
Strother, C. O., 4(28), *49*
Suhrmann, R., 140(75), *187*
Sukhanova, R. V., 130(56), *186*
Sundquist, B. E., 120, *185*
Suprenenko, K. A., 300(60, 61), *341*
Sutula, V. D., 65(64), 66(64), *73*
Swanson, L. W., 70(82), *73*
Symons, M. C. R., 266, 292(7), *340*

T

Tachiore, P., 196(22), *261*
Takasu, Y., 146(95b), *187*

Takayasu, O., 154(123), *188*
Takeuchi, T., 154(123), *188*
Tamm, P. W., 54(11), 66(72), 67(72), *72, 73*, 148(104), *187*
Tanaka, H., 321(154), 322(163), 323, *344*
Tanaka, K., 42(63), *50*
Tanei, T., 300, 305(62), *341*
Taylor, E. H., 4(21), 9, *49*, 320, *343*
Taylor, H. A., 4(29), *49*
Taylor, H. S., 4, *49*
Taylor, J. C., 148(99), *187*
Taylor, K. C., 1(3), 3(3), 16(3), *48*
Tebben, J. H., 100(37), *113*
Teichner, S. J., 4, 9, 11, *49*, 192(8), 193(19), 218(19), 219(53, 54, 55), 231(55), 232(55), 233(55), 234(55), 235(55), 237(19) 238(19, 73), 239(19), 240(19, 55, 73), 241(55, 75, 76, 77), 242(8), 244(73), 245(19, 73), 246(19, 53, 73), 247(80), 248(53), 250(8, 54), 251(19, 54, 82), 252(82), 255(53), 256(53), 257(8, 84), *261, 262, 263*
Tench, A. J., 272(22), 274, 275(32), 298(52), 304, 307(90), 311, 312, 313(22), 318(141), 319, *340, 341, 342, 343*
Terenin, A. N., 304(86), *342*
Tezuka, Y., 154(123), *188*
Thom, H. G., 266(14), *340*
Thomas, D. G., 9, *49*
Thomas, D. H., 121(30), 122(30), 126(40), 129(30, 54, 55), 130(55), 131(60), 132(60), 134(40), 135(40), 136(40), 137(60), 139(30, 54, 73), 140(30, 74), 141(30, 40), 142(60), 158(126), 159, 162(30, 40, 126, 135), 165(40, 135), 167(40), 170(54, 126), 171(126), 172(60), 173(60, 73), 174(73), 175(73), 176(73), 182(135), *185, 186, 187, 188*
Thomas, J. M., 192(5), *261*
Thomas, W. J., 192(5), *261*
Thomson, S. J., 192(9), *261*
Thouvenin, Y., 211(40), 213(40), 217(40), 221(40), 222(40), 223(40), *262*
Tien, A., 197, 199, 223, *261*
Tikhomirova, N. N., 324, 325(172), *344*
Tobin, H. H., 266(11), *340*
Toya, T., 65(60), *73*
Toyama, O., 4(31), 9(31), *49*

AUTHOR INDEX

Trapnell, B. M. W., 168(145), *189*, 192(4, 6), *261*
Troiano, P. F., 91(29), 92(29), *113*
Tsitsishvili, G. V., 325(172), *344*
Tsong, T. T., 70(81), *73*
Tupikov, V. I., 312, *343*
Turkevich, J., 299(54), 315(121), 316, 317, 321, 323, *341*, *343*, *344*
Tverdovskii, I. P., 173(153), *189*
Twigg, G. H., 12, *49*, 163, *188*

U

Ulmer, K., 148(101), *187*
Ursu, I., 324(170), *344*
Uytterhoeven, J. B., 324(168), *344*

V

Vainchtock, M. T., 257(84), *263*
Valette, A., 219(57), 221(57), 222(57), *262*
Valouchová, A., 57(35a), 63(35a), *72*
van der Avoird, A., 65(63), *73*
van der Plank, P., 61(41), 70(41), *72*, 116(2), 138(70), 151(2, 70, 115), 152(70), 155(70), 156(2), 181(2), *185*, *186*, *188*
van Hardeveld, R., 76, 77, 78(13), 79, 80(13), 85(13), 87, 89(10), 97, *112*, *113*
van Hooff, J. H. C., 272(23), 297, 314, 321(159), *340*, *341*, *344*
van Leeuwen, P. W. N. M., 42(62), *50*
Vanngard, T., 299(59), *341*
van Reijen, L. L., 321, 322, *344*
van Riet, R., 28(49), *49*
Vasko, N. P., 66(70), *73*
Venkataswarlu, P., 30(50), 32(50), *49*
Verbeek, J. L., 324, *344*
Vergnon, P., 257(84), *263*
Vladinirova, V. I., 301(66), *342*
Voevodskii, V. V., 295(43), 296(43), 316(128), *341*, *343*
Voge, H. H., 163, 164(136, 141), *188*
Vorotyntsev, V. M., 297, 312(49, 116), *341*, *343*
Vuillemin, J. J., 148(98), *187*

W

Wagner, C. N. J., 142(82), *187*
Walker, R. A., 118(21), *185*
Walling, J. C., 165(142), *188*
Walter, D., 42(63), *50*
Wang, K. M., 275(33), 307(99), 311(110), 312, 317, *341*, *342*, *343*
Wannier, G. H., 65(52), *73*
Ward, J. W., 302(80), 309(80), *342*
Wasserman, E., 294(40), 295(40), *341*
Wayman, C. M., 78, *113*, 120, 155(27), *185*
Webb, G., 2(7), *48*, 192(9), *261*
Weber, R. E., 145(92), 146(92, 95), *187*
Webster, D. E., 116(4, 5), 177(5), *185*
Wedler, G., 140(75), *187*, 192(11), *261*
Weeks, R. A., 316(130), *343*
Weinberger, M. A., 307(98), *342*
Weisberg, J., 312(112), *343*
Weller, S. W., 315(124), *343*
Wells, P. B., 1(2), 2(2), *48*
Werner, H., 35(54), *49*
Wertz, J. E., 266, 318, 319(144), *340*, *343*
Weterings, C. A. M., 100(37), *113*
Wethington, J. A., 4(21), 9, *49*
Whalley, L., 129(54), 139(54), 148(105), 158(126), 162(126), 170(54, 126), 171(126), *186*, *187*, *188*
Whiffen, D. H., 272(27), 291(27), 294(27), *341*
White, H. M., 309(101), *342*
Widmer, H., 132(63, 67), *186*
Wilke, G., 42(63), *50*
Wilke, H. G., 140(75), *187*
Willard, J. E., 306, *342*
Willens, R. H., 132(66), *186*
Williamson, W. B., 297(50), *341*
Wilmshurst, T. H., 282, *341*
Wilson, E. B., Jr., 35(57), 36(57), *49*
Wilson, J. N., 164(141), *188*
Wise, H., 321(160), 322(160), *344*
Wolfgang, R., 53(6), *71*
Wong, N. B., 297(51), 298(51), *341*
Wong, P. K., 306, *342*
Wood, D. E., 273(31), 309, *341*
Woodman, J. F., 4, *49*
Wray, E. M., 147(96), 148(96), *187*

Y

Yamashina, T., 146(95b), *187*
Yates, D. J. C., 22(45), *49*, 91, 92, *113*, 316, *343*

Yates, J. T., 91, 92, 111, *113*
Yoshida, K., 116(6, 7), 177(7), *185*
Young, D. M., 192(3), 226(66), *261*, *263*

Z

Zacharov, I. I., 65(64), 66(64), *73*
Zettlemoyer, A. C., 196(24), *261*
Zhabrova, G. M., 301(66), *342*
Zhavoronkova, K. N., 144(84), 154, 157, 158, 171(84), *187*
Zhidomirov, G. M., 299(56), *341*
Zhmud, E. S., 75(3), 111, *112*
Zielenkiewicz, A., 199(27), *261*
Zielenkiewicz, W., 199(27), *261*
Zimmerman, H., 42(63), *50*

Subject Index

A

Acetylenes
  reactions of, on zinc oxide, 46, 47
  spectrum of, 46
Adsorption
  on alloy films, 115–185
  heat of, 191ff, see also Microcalorimetry
    differential, 197
    for gas-solid interactions, 226–237
    magnitude of, 191
Adsorption calorimetry, 57, see also Microcalorimetry
Alcohols, ESR of, 301
Alloy films, see also specific alloys
  Auger electron spectroscopy, 145–147
  characterization of, 134–147
    bulk structure, 139–143
      bulk homogeneity, 139–142
      phase separation, 142, 143
    composition, 134, 135
    surface area, 138, 139
    surface examination, 143–147
      work function measurements, 143
      crystallite size and orientation, 135–138
    electron-probe microanalysis, 144
  evaporated, 115–185
  nucleation and growth of, 131
  preparation of, 117–134
    evaporation
      simultaneous, 125–134
        heated substrates and vacuum effects, 128–132
        vapor quenching, 132–134
      successive, 120–124
    miscibility of components, 117–120
      free energy, 119
Alloys, see also specific alloys
  evaporation, 120–134
    reaction vessel, 128
    sliding cathode phototube, 126
    thermodynamic properties of, 118

Alumina
  ESR of adsorbed hydrogen on, 295, 296
    of alcohols on, 301
    of tetracyanoethylene on, 304
    of transition metal ions on, 320–325
    of trinitrobenzene on, 304, 305
Aluminum, see also Zeolites
  ESR of, 275
Anthracene, ESR of, 301
Auger electron spectroscopy, 55, 70
  of alloy films, 145–147

B

Benzene
  deuteration and exchange of, 100–109
    experimental procedure, 100–102
    on iridium catalysts, 107–109
    on nickel catalysts, 103–106
  ESR of, 300
Bohr magneton, 267
Boltzmann law, 280
Butene
  adsorption on zinc oxide, 42–45
  isomerization of, 45, 46
  reactions of, 41–46
  spectrum of, 43, 44

C

Calorimeters, see Microcalorimeters
Calvet microcalorimeter, 197–201
  electrical calibration of, 235, 236
Carbon dioxide
  adsorption on gallium-doped NiO, 247–251
    on nickel catalysts, 87–96
      dissociative, 93–96
    radical, ESR of, 272, 273, 291
Carbon monoxide
  adsorption on gallium-doped NiO, 246–249
    on Ir, 99, 100
      infrared spectra, 99

on nickel catalysts, 87–96
  absorption bands, 95
  wavenumbers, 91
on Pd, 97–99
  infrared spectra, 98
chemisorption, 63
reduction of NiO by, 240, 241
Catalysis
  on alloy films, 115–185
  heterogeneous, heat-flow microcalorimetry and, *see* Microcalorimetry
Catalysts, *see* specific substances
Catalytic reactions, heterogeneous, 53, 56
Chemisorption
  complexes in catalytic reactions on transition metals, 51–71
  surface interactions, 52, 53
Chromia, as catalyst, 3
Chromium, ESR spectrum of, 293, 295, 320, 321
ClO_2, ESR of, 309
Cobalt, evaporation data, 127
Cobalt–palladium alloy films, *see* Palladium–cobalt alloy films
Copper
  ESR of, 323, 324
  evaporation data, 127
Copper–nickel (Cu–Ni) alloy films, 148–158
  Auger spectra of, 147
  benzene hydrogenation, 151, 152
  butene hydrogenation, 152, 153
  deuterium exchange, 157
  ethylene hydrogenation, 154
  free energy of mixing, 150
  hydrogen chemisorption on, 70, 151
  hydrogenation, 149, 150
  phase diagram, 155
  successive evaporation, 122, 123
  surface area, 138
  thermodynamic properties of, 118
Copper–palladium alloy films, *see* Palladium–copper alloy films
Copper–platinum alloy films, *see* Platinum–copper alloy films
Crystallites, *see also* Alloy films
  electron micrographs of, 135–138
Cyclohexadienyl radicals, ESR of, 300
Cyclopropane, chemisorption, 62, 63

D

Diethyl mercury, spectrum of, 28
Diethyl zinc, spectrum of, 28

E

Electron-probe microanalysis of alloy films, 144
Electron spin resonance (ESR) in catalysis, 265–340
  of adsorbed atoms and molecules, 295–315
    inorganic molecules, 307–315
    monatomic radicals, 295–298
    organic molecules, 298–307
  applications of, 295–325
  coordinate system, 288
  determination of energy levels from spin Hamiltonian, 328–332
  energy absorption, 288–290
    Lorenztion function, 289, 290
  filling of molecular orbitals in, 271
  g tensor, *see* g tensor
  from g and hyperfine tensors to molecular structure, 271–279
  hyperfine tensor, *see* Hyperfine tensor
  polycrystalline spectra, 287–295
  quantitative determination of spin concentration, 286, 287
  quantum mechanics for, 326–328
  relaxation phenomena, 279–282
  saturation in, 280
  surface defects, 315–320
  theoretical basis, 267–282
  of transition metal ions, 320–325
ESR, *see* Electron spin resonance
ESR spectrometer, 282–286
  klystron, 282
Ethane, deuterium distribution in, 13
Ethyl radical, ESR of, 299
Ethylene
  addition of H_2–D_2 mixtures to, 12–13
  adsorption on zinc oxide, 19–22
  chemisorbed, observed bands for, 22
  hydrogenation and isomerization of, 2, 3
    over chromia, 3
    intermediates in, 23–28
    kinetics of, 16–18
    mechanism of, 16–28

over zinc oxide, 3
   at room temperature, 7
reaction with deuterium, 2, 7
spectrum of, in presence of H_2, 23, 24
   of helium, 24
steady state vs. decay rate, 27

G

g tensor, 267, 269, 271–273, 332–336
   values for inorganic radicals, 272
Gas-solid interactions, measurement of differential heat by microcalorimetry, 226–237
Gold, evaporation data, 127
Gold–nickel alloy films, see Nickel–gold alloy films
Gold–silver (Au–Ag) alloy films, 183
   ethylene oxidation, 183
   thermodynamic properties of, 118
Gold–palladium alloy films, see Palladium–gold alloy films
Gold–platinum alloy films, see Platinum–gold alloy films

H

Hall effect, 57, 61, 62
Heat of adsorption, see Adsorption
Heat flux, 207
Heat sink, 195, 196
Hydrogen
   adsorption, 4–8
      heat of, 192
   chemisorption, 6–8
      on transition metals, 57–65
   ESR of, 295
   magnetic fields for, 268
Hydrogenation over zinc oxide, 1–48
Hyperfine coupling constant, 267, 269
Hyperfine interactions, ESR data for, 274
Hyperfine tensor, 267, 273–279, 336–340
   isotropic and anisotropic components of, 277, 278

I

Infrared studies on adsorption of N_2, CO, and CO_2, 86–100
   experimental procedure, 86, 87

Iridium catalysts
   adsorption on, 97, 99–100
      analytical data, 97
   deuteration and exchange of benzene on, 107–109
      isotopic products, 106, 108
      rates of exchange, 107
   preparation of, 111, 112
Iron
   chemisorption on, 63
   evaporation data, 127
Iron–nickel (Fe–Ni) alloy films
   single-crystal, 130
   surface area, 138
Isomerization over zinc oxide, 1–48

J

Joule heating, 209, 220

K

Kirkendall effect, 183
Klystron, 282

L

LEED (low-energy electron diffraction), 55, 56, 70, 145

M

Magnesium ion, ESR of transition metal ions on, 321–325
Magnesium oxide
   adsorption on, 274$f\!f$.
   surface defects, 318–320
Magnetic fields, 268, 269
Manganese, ESR of, 324, 325
Matano interface, 121
Mesitylene, ESR of, 299, 300
Metal catalysts
   crystal properties, 77–79
   particle size, 75–112
Metal crystals, see Surface atoms
Methyl radicals, ESR of, 298, 299
Microcalorimeters for heterogeneous catalysis reactions, 196–205, see also specific types
   adiabatic model, 194, 195

calibration of, 232–237
  electrical, 233–235
cells for, 227–231
heat-flow, 194–196
  heat transfer coefficient, 208
  impulse response, 211–213
  intrinsic sensitivities of, 205
  recording of data, 215–218
  theoretical model, 206
  time constant of, 209, 220
  as transducer, 211
isoperibol, 195
isothermal, 194, 195
preliminary experiments, 232–237
Microcalorimetry, heat-flow
  analysis of data, 214–226
    data correction and determination of thermokinetics, 218–226
      manual, 219–222
      off-line, 224–226
      on-line, 222–224
    methods of recording data, 215–218
  calorimetric cells, 227–231
  differential heats of adsorption, 237–259
    of catalytic reactions, 254–259
    of gas-solid interactions, 226–237
    of interaction between reactant and preadsorbed species, 246–254
  energy spectrum of catalyst surface, 238–246
  heterogeneous catalysis and, 181–260
  modifications of catalyst surface, 254–259
  principles of, 194–196
  theory of, 206–214
    heat transfer equations, 211–214
    Tian equation, 206–211
  volumetric line, 227–231
    expansion coefficient, 229
Molybdenum
  chemisorption complexes on, 61
  ESR of, 321, 322

N

Naphthalene, ESR spectra for, 294, 301
NF_2, ESR of, 309
NH_2, ESR of, 310, 311
  catalysts, 75–112

adsorption on, 87–96
  analytical data, 88
  of CO on, 89–93
deuteration and exchange of benzene on, 103–106
  electron micrographs, 105
  isotopic products, 104
  rate of exchange, 103
preparation and characterization of, 110–112
chemisorption of oxygen and hydrogen on, 57–65
evaporation data, 127
Nickel–copper alloy films, see Copper–nickel alloy films
Nickel–gold (Ni–Au) alloy films, 182, 183
  ethylene oxidation, 183
  phase diagram, 182
  thermodynamic properties of, 118
Nickel–iron alloy films, see Iron–nickel alloy films
Nickel oxide
  adsorption of O_2 on, 238–240
  differential heats of adsorption of O_2 on, 243–245
  doped, 242
    gallium-doped, 246–250, 254–257
    lithium-doped, 250, 251
  reduction by CO, 240, 241
Nickel–palladium alloy films, see Palladium–nickel alloy films
Nickel–platinum alloy films, see Platinum–nickel alloy films
Nitric oxide, ESR of, 271ff.
  on alumina, 307, 308
  on metal oxides, 314, 315
  on zeolites, 308, 309
NO_2 molecule
  coordinate system for, 276
  ESR of, 273, 276–278, 293, 297, 298, 309
Nitrogen
  adsorption
    heat of, 192
    on Ir, 99, 100
    on nickel catalysts, 87–96
    on Pd, 97–99
NMR, see Nuclear magnetic resonance
Nuclear magnetic resonance (NMR), 279
Nuclear magneton, 267

SUBJECT INDEX

O

Olefins, hydrogenation and isomerization of, 1–3
Orbital angular momentum operators, 326, 327
Oxygen
  adsorption
    heat of, 192
      differential, 243–245
      on nickel oxide, 238–240
        gallium-doped, 246–248
  chemisorption on transition metals, 57–65
  ESR of, 271ff., 274, 275, 296–298, 310–314
Oxygen poisoning, 9–12
  activity vs. water adsorption, 10
  rate of water adsorption, 11, 12

P

Palladium
  chemisorption complexes on, 61, 63
  catalysts
    adsorption on, 97–99
      analytical data, 97
      preparation of, 111
  diffusion data, 129
  evaporation data, 127
Palladium–cobalt (Pd–Co) alloy films
  thermodynamic properties of, 118
Palladium–copper (Pd–Cu) alloy films
  thermodynamic properties of, 118
Palladium–gold (Pd–Au) alloy films, 158–161
  electron micrographs of, 137
  formic acid decomposition, 159–161
  oxygen chemisorption, 158, 159
  thermodynamic properties of, 118
Palladium–nickel (Pd–Ni) alloy films
  surface area, 138, 139
  thermodynamic properties of, 118
Palladium–rhodium (Pd–Rh) alloy films, 172–178
  annealed, 178, 179
  CO oxidation, 177
  crystallite orientation in, 137
  ethylene oxidation, 173, 174
    composition range, 175, 176

  miscibility gap, 172
  successive evaporation, 123, 124
  thermodynamic properties of, 118
  X-ray diffraction pattern, 123
Palladium–silver (Pd–Ag) alloy films, 161–172
  CO oxidation, 162, 163
  CO surface potentials, 170–172
  diffusion of atoms, 129
  electron micrographs of, 136
  ethylene oxidation, 163–168
    parameters, 166
  homogeneity of, 130
  parahydrogen conversion, 168–170
    activation energy, 169
  successive evaporation, 121, 122
  thermodynamic properties of, 118
  X-ray diffraction pattern, 121, 122, 167
  uniformity of, 135
Peltier cooling, 209, 224
Perylene, ESR of, 301
Petit–Eyraud microcalorimeter, 203–205
Petit microcalorimeter, 201–203
Platinum
  chemisorption on, 63, 64, 67, 68
  evaporation data, 127
  field ion image, 67, 68
Platinum–copper (Pt–Cu) alloy films
  thermodynamic properties of, 118
Platinum–gold (Pt–Au) alloy films, 178–182
  work function, 180, 181
Platinum–nickel (Pt–Ni) alloy films
  thermodynamic properties of, 118
Platinum–ruthenium (Pt–Ru) alloy films, 182
Propylene
  adsorption on zinc oxide, π-allyl, 29–37
  calculated vs. observed C–C frequencies, 36, 37
  reactions of π-allyls, 37–41
    addition of deuterium, 39–41
    double-bond isomerization, 38, 39
    with zinc oxide, 29–41
  spectra of chemisorbed, 29–33

R

Relaxation phenomena in ESR, 279–282
Rhenium, field ion image, 69

Rhodium
  chemisorption complexes on, 61
  evaporation data, 127
Rhodium–palladium alloy films, see Palladium–rhodium alloy films
Rideal–Eley mechanism, 8, 13
Ruthenium–platinum alloy films, see Platinum–ruthenium alloy films

S

Saturation in ESR, 280
Silica–alumina
  ESR of adsorbed hydrogen, 295
    of polynuclear aromatic compounds on, 301, 307
    of transition metal ions on, 321–325
    of trinitrobenzene on, 304, 305
Silica gel
  ESR of adsorbed hydrogen, 295, 296
    of adsorbed oxygen, 297
    of benzene on, 300
    of Cl atoms, 296
    of methyl and ethyl radicals on, 298, 299
    of polynuclear aromatic compounds on, 306
  surface defects, 315, 316
Silver, evaporation data, 127
Silver–gold alloy films, see Gold–silver alloy films
Silver–palladium alloy films, see Palladium–silver alloy films
SO_2 radical, ESR of, 315
Spin Hamiltonian
  determination of energy levels from, 328–332
  in ESR, 267–271, 291, 295
Spin operators, 267, 326
Spin-lattice relaxation time, 280, 281
Spin-orbit coupling, 281
Spin-spin interactions, 271
Spin-spin relaxation time, 282
Superoxide, see Oxygen
Surface
  energy spectrum of catalyst, 238–246
  modifications of catalyst, 254–259
  "well-defined," 55, 56
Surface atoms, 77–86
  fcc crystals, 79–81
  fcc cubo-octahedron crystals, 382, 383
  statistics of, 79–84
Surface interactions, 52, 53
Surface sites, 77–86
  statistics of, 84–86

T

Taylor series, 281
Tetracyanoethylene, ESR of, 304, 305
Thermograms, see Microcalorimeters
Thermokinetics, 218
Tian equation, 206–211, 220, 221
Titanium oxide, 257, 258
Transition metals
  adsorption calorimetry, 57
  atomic orbitals on surface, 66, 67
  bond energy, 52
  catalytic reactions, 51–71
  contaminants, 55, 56
  electrical resistance and work function, 57–59
  ESR of, 320–325
  fcc lattice, 66–70
  Hall effect, 57
  surface characterization, 65–70
Trinitrobenzene, ESR of, 304, 305
Tungsten, chemisorption, 70

V

Vanadium, ESR of, 322, 323
Vapor quenching, 132–134 327, 328

W

Wave equation, matrix formulation of, 327, 328
Wavefunctions, properties of, 326

Z

Zeeman effects, 267, 283, 284, 291
Zeolites, 272, 274, 276, 302, 303, 307–315
  ESR of adsorbed hydrogen, 295
    of transition metal ions on, 323–325
  NO on, 307–309
  NO_2 on, 309
  oxygen on, 310–314
  surface defects, 316–318

Zinc oxide
    acetylenes on, see Acetylenes
    active sites, 4–16
        adsorption of H_2, D_2, and H_2O on, 14
        model for, 13–16
        reaction of water and hydrogen, 15
    addition of H_2–O_2 mixtures to ethylene, 12, 13
    butene on, see Butene
    ESR of tetracyanoethylene on, 304
    ethylene adsorption on, 19–22
        infrared spectrum, 20, 21
    ethylene hydrogenation, see also Ethylene
        mechanism of, 16–28
    hydrogen adsorption on, 5–8
        vs. time, 5
    hydrogenation and isomerization over, 1–48
    poisoning reactions, 9–12, see also Oxygen poisoning
    reactions with propylene, 29–41, see also Propylene
    spectrum of hydrogen on, 6, 7

Contents of Previous Volumes

Volume 1

The Heterogeneity of Catalyst Surfaces for Chemisorption
  HUGH S. TAYLOR.
Alkylation of Isoparaffins
  V. N. IPATIEFF AND LOUIS SCHMERLING.
Surface Area Measurements. A New Tool for Studying Contact Catalysts
  P. H. EMMETT.
The Geometrical Factor in Catalysis
  R. H. GRIFFITH.
The Fischer-Tropsch and Related Processes for Synthesis of Hydrocarbons by Hydrogenation of Carbon Monoxide
  H. H. STORCH.
The Catalytic Activation of Hydrogen
  D. D. ELEY.
Isomerization of Alkanes
  HERMAN PINES.
The Application of X-Ray Diffraction to the Study of Solid Catalysts
  M. H. JELLINEK AND I. FANKUCHEN.

Volume 2

The Fundamental Principles of Catalytic Activity
  FREDERICK SEITZ.
The Mechanism of the Polymerization of Alkenes
  LOUIS SCHMERLING AND V. N. IPATIEFF.
Early Studies of Multicomponent Catalysts
  ALWIN MITTASCH.
Catalytic Phenomena Related to Photographic Development
  T. H. JAMES.
Catalysis and the Adsorption of Hydrogen on Metal Catalysts
  OTTO BEECK.
Hydrogen Fluoride Catalysis
  J. H. SIMONS.

Entropy of Adsorption
  CHARLES KEMBALL.
About the Mechanism of Contact Catalysis
  GEORGE-MARIA SCHWAB.

Volume 3

Balandin's Contribution to Heterogeneous Catalysis
  B. M. W. TRAPNELL.
Magnetism and the Structure of Catalytically Active Solids
  P. W. SELWOOR.
Catalytic Oxidation of Acetylene in Air for Oxygen Manufacture
  J. HENRY RUSHTON AND K. A. KRIEGER.
The Poisoning of Metallic Catalysts
  E. B. MAXTED.
Catalytic Cracking of Pure Hydrocarbons
  VLADIMIR HAENSEL.
Chemical Characteristics and Structure of Cracking Catalysts
  A. G. OBLAD, T. H. MILLIKEN, Jr., AND G. A. MILLS.
Reaction Rates and Selectivity in Catalyst Pores
  AHLBORN WHEELER.
Nickel Sulfide Catalysts
  WILLIAM J. KIRKPATRICK.

Volume 4

Chemical Concepts of Catalytic Cracking
  R. C. HANSFORD.
Decomposition of Hydrogen Peroxide by Catalysts in Homogeneous Aqueous Solution
  J. H. BAXENDALE.
Structure and Sintering Properties of Cracking Catalysts and Related Materials
  HERMAN E. RIES, Jr.

Acid-Base Catalysis and Molecular Structure
R. P. Bell.
Theory of Physical Adsorption
Terrell L. Hill.
The Role of Surface Heterogeneity in Adsorption
George D. Halsey.
Twenty-Five Years of Synthesis of Gasoline by Catalytic Conversion of Carbon Monoxide and Hydrogen
Helmut Pichler.
The Free Radical Mechanism in the Reactions of Hydrogen Peroxide
Joseph Weiss.
The Specific Reactions of Iron in Some Hemoproteins
Philip George.

Volume 5

Latest Developments in Ammonia Synthesis
Anders Nielsen.
Surface Studies with the Vacuum Microbalance: Instrumentation and Low-Temperature Applications
T. N. Rhodin, Jr.
Surface Studies with the Vacuum Microbalance: High-Temperature Reactions
Earl A. Gulbransen.
The Heterogeneous Oxidation of Carbon Monoxide
Morris Katz.
Contributions of Russian Scientists to Catalysis
J. G. Tolpin, G. S. John, and E. Field.
The Elucidation of Reaction Mechanisms by the Method of Intermediates in Quasi-Stationary Concentrations
J. A. Christiansen.
Iron Nitrides as Fischer-Tropsch Catalysts
Robert B. Anderson.
Hydrogenation of Organic Compounds with Synthesis Gas
Milton Orchin.
The Uses of Raney Nickel
Eugene Lieber and Fred L. Morritz.

Volume 6

Catalysis and Reaction Kinetics at Liquid Interfaces
J. T. Davies.
Some General Aspects of Chemisorption and Catalysis
Takao Kwan.
Noble Metal–Synthetic Polymer Catalysts and Studies on the Mechanism of Their Action
William P. Dunworth and F. F. Nord.
Interpretation of Measurements in Experimental Catalysis
P. B. Weisz and C. D. Prater.
Commercial Isomerization
B. L. Evering.
Acidic and Basic Catalysis
Martin Kilpatrick.
Industrial Catalytic Cracking
Rodney V. Shankland.

Volume 7

The Electronic Factor in Heterogeneous Catalysis
M. McD. Baker and G. I. Jenkins.
Chemisorption and Catalysis on Oxide Semiconductors
G. Parravano and M. Boudart.
The Compensation Effect in Heterogeneous Catalysis
E. Cremer.
Field Emission Microscopy and Some Applications to Catalysis and Chemisorption
Robert Gomer.
Adsorption on Metal Surfaces and Its Bearing on Catalysis
Joseph A. Becker.
The Application of the Theory of Semiconductors to Problems of Heterogeneous Catalysis
K. Hauffe.
Surface Barrier Effects in Adsorption, Illustrated by Zinc Oxide
S. Roy Morrison.
Electronic Interaction between Metallic Catalysts and Chemisorbed Molecules
R. Suhrmann.

Volume 8

Current Problems of Heterogeneous Catalysis
  J. Arvid Hedvall.
Adsorption Phenomena
  J. H. de Boer.
Activation of Molecular Hydrogen by Homogeneous Catalysts
  S. W. Weller and G. A. Mills.
Catalytic Syntheses of Ketones
  V. I. Komarewsky and J. R. Coley.
Polymerization of Olefins from Cracked Gases
  Edwin K. Jones.
Coal-Hydrogenation Vapor-Phase Catalysts
  E. E. Donath.
The Kinetics of the Cracking of Cumene by Silica-Alumina Catalysts
  Charles D. Prater and Rudolph M. Lago.

Volume 9

Proceedings of the International Congress on Catalysis, Philadelphia, Pennsylvania, 1956.

Volume 10

The Infrared Spectra of Adsorbed Molecules
  R. P. Eischens and W. A. Pliskin.
The Influence of Crystal Face in Catalysis
  Allan T. Gwathmey and Robert E. Cunningham.
The Nature of Active Centres and the Kinetics of Catalytic Dehydrogenation
  A. A. Balandin.
The Structure of the Active Surface of Cholinesterases and the Mechanism of Their Catalytic Action in Ester Hydrolysis
  F. Bergmann.
Commercial Alkylation of Paraffins and Aromatics
  Edwin K. Jones.
The Reactivity of Oxide Surfaces
  E. R. S. Winter.
The Structure and Activity of Metal-on-Silica Catalysts
  G. C. A. Schuit and L. L. van Reijen.

Volume 11

The Kinetics of the Stereospecific Polymerization of α-Olefins
  G. Natta and I. Pasquon.
Surface Potentials and Adsorption Process on Metals
  R. V. Culver and F. C. Tompkins.
Gas Reactions of Carbon
  P. L. Walker, Jr., Frank Rusinko, Jr., and L. G. Austin.
The Catalytic Exchange of Hydrocarbons with Deuterium
  C. Kemball.
Immersional Heats and the Nature of Solid Surfaces
  J. J. Chessick and A. C. Zettlemoyer.
The Catalytic Activation of Hydrogen in Homogeneous, Heterogeneous, and Biological Systems
  J. Halpern.

Volume 12

The Wave Mechanics of the Surface Bond in Chemisorption
  T. B. Grimley.
Magnetic Resonance Techniques in Catalytic Research
  D. E. O'Reilly.
Base-Catalyzed Reactions of Hydrocarbons
  Herman Pines and Luke A. Schaap.
The Use of X-Ray K-Absorption Edges in the Study of Catalytically Active Solids
  Robert A. Van Nordstrand.
The Electron Theory of Catalysis on Semiconductors
  Th. Wolkenstein.
Molecular Specificity in Physical Adsorption
  D. J. C. Yates.

Volume 13

Chemisorption and Catalysis on Metallic Oxides
  F. S. Stone.
Radiation Catalysis
  R. Coekelbergs, A. Crucq, and A. Frennet.

Polyfunctional Heterogeneous Catalysis
  PAUL B. WEISZ.
A New Electron Diffraction Technique, Potentially Applicable to Research in Catalysis
  L. H. GERMER.
The Structure and Analysis of Complex Reaction Systems
  JAMES WEI AND CHARLES D. PRATER.
Catalytic Effect in Isocyanate Reactions
  A. FARKAS AND G. A. MILLS.

Volume 14

Quantum Conversion in Chloroplasts
  MELVIN CALVIN.
The Catalytic Decomposition of Formic Acid
  P. MARS, J. J. F. SCHOLLEN, AND P. ZWIETERING.
Application of Spectrophotometry to the Study of Catalytic Systems
  H. P. LEFTIN AND M. C. HOBSON, Jr.
Hydrogenation of Pyridines and Quinolines
  MORRIS FREIFELDER.
Modern Methods in Surface Kinetics: Flash Desorption, Field Emission Microscopy, and Ultrahigh Vacuum Techniques
  GERT EHRLICH.
Catalytic Oxidation of Hydrocarbons
  L. YA. MARGOLIS.

Volume 15

The Atomization of Diatomic Molecules by Metals
  D. BRENNAN.
The Clean Single-Crystal-Surface Approach to Surface Reactions
  N. E. FARNSWORTH.
Adsorption Measurements during Surface Catalysis
  KENZI TAMARU.
The Mechanism of the Hydrogenation of Unsaturated Hydrocarbons on Transition Metal Catalysts
  G. C. BOND AND P. B. WELLS.
Electronic Spectroscopy of Adsorbed Gas Molecules
  A. TERENIN.

The Catalysis of Isotopic Exchange in Molecular Oxygen
  G. K. BORESKOV.

Volume 16

The Homogeneous Catalytic Isomerization of Olefins by Transition Metal Complexes
  MILTON ORCHIN.
The Mechanism of Dehydration of Alcohols over Alumina Catalysts
  HERMAN PINES AND JOOST MANASSEN.
π Complex Adsorption in Hydrogen Exchange on Group VIII Transition Metal Catalysts
  J. L. GARNETT AND W. A. SOLLICH-BAUMGARTNER.
Stereochemistry and the Mechanism of Hydrogenation of Unsaturated Hydrocarbons
  SAMUEL SIEGEL.
Chemical Identification of Surface Groups
  H. P. BOEHM.

Volume 17

On the Theory of Heteroneous Catalysis
  JURO HORUIT AND TAKASHI NAKAMURA.
Linear Correlations of Substrate Reactivity in Heterogeneous Catalytic Reactions
  M. KRAUS.
Application of a Temperature-Programmed Desorption Technique to Catalyst Studies
  R. J. CVETANOVIC AND Y. AMENOMIYA.
Catalytic Oxidation of Olefins
  HERVEY H. VOGE AND CHARLES R. ADAMS.
The Physical-Chemical Properties of Chromia-Alumina Catalysts
  CHARLES P. POOLE, JR. AND D. S. MACIVER.
Catalytic Activity and Acidic Property of Solid Metal Sulfates
  KOZO TANABE AND TSUNEICHI TAKESHITA.
Electrocatalysis
  S. SRINIVASEN, H. WROBLOWA, AND J. O'M. BOCKRIS.

Volume 18

Stereochemistry and Mechanism of Hydrogenation of Naphthalenes on Transition Metal Catalysts and Conformational Analysis of the Products
A. W. Weitkamp.

The Effects of Ionizing Radiation on Solid Catalysts
Ellison H. Taylor.

Organic Catalysis over Crystalline Aluminosilicates
P. B. Venuto and P. S. Landis.

On Transition Metal-Catalyzed Reactions of Norbornadiene and the Concept of π Complex Multicenter Processes
G. N. Schrauzer.

Volume 19

Modern State of the Multiplet Theory of Heterogeneous Catalysis
A. A. Balandin.

The Polymerization of Olefins by Ziegler Catalysts
M. N. Berger, G. Boocock, and R. N. Hawarr.

Dynamic Methods for Characterization of Adsorptive Properties of Solid Catalysts
L. Polinski and L. Naphtali.

Enhanced Reactivity at Dislocations in Solids
J. M. Thomas.

Volume 20

Chemisorptive and Catalytic Behavior of Chromia
Robert L. Burwell, Jr., Gary L. Haller, Kathleen C. Taylor, and John F. Read

Correlation among Methods of Preparation of Solid Catalysts, Their Structures, and Catalytic Activity
Kiyoshi Morikawa, Takayasu Shirasaki, and Masahide Okada

Catalytic Research on Zeolites
J. Turkevich and Y. Ono

Catalysis by Supported Metals
M. Boudart

Carbon Monoxide Oxidation and Related Reactions on a Highly Divided Nickel Oxide
P. C. Gravelle and S. J. Teichner

Acid-Catalyzed Isomerization of Bicyclic Olefins
Jean Eugene Germain and Michel Blanchard

Molecular Orbital Symmetry Conservation in Transition Metal Catalysis
Frank D. Mango

Catalysis by Electron Donor–Acceptor Complexes
Kenzi Tamaru

Catalysis and Inhibition in Solutions of Synthetic Polymers and in Micellar Solutions
H. Morawetz

Catalytic Activities of Thermal Polyanhydro-α-Amino Acids
Duane L. Rohlfing and Sidney W. Fox

Volume 21

Kinetics of Adsorption and Desorption and the Elovich Equation
C. Aharoni and F. C. Tompkins

Carbon Monoxide Adsorption on the Transition Metals
R. R. Ford

Discovery of Surface Phases by Low Energy Electron Diffraction (LEED)
John W. May

Sorption, Diffusion, and Catalytic Reaction in Zeolites
L. Riekert

Adsorbed Atomic Species as Intermediates in Heterogeneous Catalysis
Carl Wagner

QD
501
A33

v.22
1972

JUN 8 1972